024125
ROYAL AGRICULTURAL COLLEGE

D0230535

THE CONTROL OF FAT AND LEAN DEPOSITION

Proceedings of Previous Easter Schools In Agricultural Science, published by Butterworth-Heinemann, Oxford

*SOIL ZOOLOGY Edited by D.K. McL, Kevan (1955)
*THE GROWTH OF LEAVES Edited by F.L. Milthorpe (1956)
*CONTROL OF PLANT ENVIRONMENT Edited by J.P. Hudson (1957)
*NUTRITION OF THE LEGUMES Edited by E.G. Hallsworth (1958)
*THE MEASUREMENT OF GRASSLAND PRODUCTIVITY Edited by J.D. Ivins (1959)
*DIGESTIVE PHYSIOLOGY AND NUTRITION OF THE RUMINANT Edited by D. Lewis (1960)
*NUTRITION OF PIGS AND POULTRY Edited by J.T. Morgan and D. Lewis (1961)
*ANTIBIOTICS IN AGRICULTURE Edited by M. Woodbine (1962)
*THE GROWTH OF THE POTATO Edited by J.D. Ivins and F.L. Milthorpe (1963)
*EXPERIMENTAL PEDOLOGY Edited by E.G. Hallsworth and D.V. Crawford (1964)
*THE GROWTH OF CEREALS AND GRASSES Edited by F.L. Milthorpe and J.D. Ivins (1965)
*REPRODUCTION IN THE FEMALE ANIMAL Edited by G.E. Lamming and E.C. Amoroso (1967)
*GROWTH AND DEVELOPMENT OF MAMMALS Edited by G.A. Lodge and G.E. Lamming (1968)
*ROOT GROWTH Edited by W.J. Whittington (1968)
*PROTEINS AS HUMAN FOOD Edited by R.A. Lawrie (1970)
*LACTATION Edited by I.R.Falconer (1971)
*PIG PRODUCTION Edited by D.J.A. Cole (1972)
*SEED ECOLOGY Edited by W. Heydecker (1973)
HEAT LOSS FROM ANIMALS AND MAN: ASSESSMENT AND CONTROL Edited by J.L. Monteith and L.E. Mount (1974)
*MEAT Edited by D.J.A. Cole and R.A. Lawrie (1975)
*PRINCIPLES OF CATTLE PRODUCTION Edited by Henry Swan and W.H. Broster (1976)
*LIGHT AND PLANT DEVELOPMENT Edited by H. Smith (1976)
PLANT PROTEINS Edited by G. Norton (1977)
ANTIBIOTICS AND ANTIBIOSIS AGRICULTURE Edited by M. Woodbine (1977)
CONTROL OF OVULATION Edited by D.B. Crighton, N.B. Haynes, G.R. Foxcroft and G.E. Lamming (1978)
POLYSACCHARIDES IN FOOD Edited by J.M.V. Blanshard and J.R. Mitchell (1979)
SEED PRODUCTION Edited by P.D. Hebblethwaite (1980)
PROTEIN DEPOSITION IN ANIMALS Edited by P.J. Buttery and D.B. Lindsay (1981)
PHYSIOLOGICAL PROCESSES LIMITING PLANT PRODUCTIVITY Edited by C. Johnson (1981)
ENVIRONMENTAL ASPECTS OF HOUSING FOR ANIMAL PRODUCTION Edited by J.A. Clark (1981)
EFFECTS OF GASEOUS AIR POLLUTION IN AGRICULTURE AND HORTICULTURE Edited by M.H. Unsworth and D.P. Ormrod (1982)
CHEMICAL MANIPULATION OF CROP GROWTH AND DEVELOPMENT Edited by J.S. McLaren (1982)
CONTROL OF PIG REPRODUCTION Edited by D.J.A. Cole and G.R. Foxcroft (1982)
SHEEP PRODUCTION Edited by W. Haresign (1983)
UPGRADING WASTE FOR FEEDS AND FOOD Edited by D.A. Ledward, A.J. Taylor and R.A. Lawrie (1983)
FATS IN ANIMAL NUTRITION Edited by J. Wiseman (1984)
IMMUNOLOGICAL ASPECTS OF REPRODUCTION IN MAMMALS Edited by D.B. Crighton (1984)
ETHYLENE AND PLANT DEVELOPMENT Edited by J.A. Roberts and G.A. Tucker (1985)
THE PEA CROP Edited by P.D. Hebblethwaite, M.C. Heath and T.C.K. Dawkins (1985)
PLANT TISSUE CULTURE AND ITS AGRICULTURAL APPLICATIONS Edited by Lindsey A. Withers and P.G. Alderson (1986)
CONTROL AND MANIPULATION OF ANIMAL GROWTH Edited by P.J. Buttery, N.B. Haynes and D.B. Lindsay (1986)
COMPUTER APPLICATIONS IN AGRICULTURAL ENVIRONMENTS Edited by J.A. Clark, K. Gregson and R.A. Saffell (1986)
MANIPULATION OF FLOWERING Edited by J.G. Atherton (1987)
NUTRITION AND LACTATION IN THE DAIRY COW Edited by P.C. Garnsworthy (1988)
MANIPULATION OF FRUITING Edited by C.J. Wright (1989)
APPLICATIONS OF REMOTE SENSING IN AGRICULTURE Edited by M.D. Steven and J.A. Clark (1990)
GENETIC ENGINEERING OF CROP PLANTS Edited by G.W. Lycett and D. Grierson (1990)
FEEDSTUFF EVALUATION Edited by J. Wiseman and D.J.A. Cole (1990)

** These titles are now out of print but are available in microfiche editions*

The Control of Fat and Lean Deposition

K.N. Boorman, P.J. Buttery and D.B. Lindsay

Butterworth-Heinemann Ltd
Linacre House, Jordan Hill, Oxford OX2 8DP

 PART OF REED INTERNATIONAL BOOKS

OXFORD LONDON BOSTON
MUNICH NEW DELHI SINGAPORE SYDNEY
TOKYO TORONTO WELLINGTON

First published 1992

British Library Cataloguing in Publication Data
A catalogue record for this book is available from the British Library

Library of Congress Cataloguing in Publication Data
A catalogue record for this book is available from the Library of Congress

ISBN 0 7506 0354 2

Printed and bound in Great Britain.

CONTENTS

PREFACE

This volume represents the proceedings of the third Easter School in recent years to consider the control of growth of animals. This Easter School (the 51st in the series) considered the factors controlling protein metabolism in muscle at the cellular level and this was complemented by similar considerations of adipose tissue. Examination of foetal growth and its influence on subsequent postnatal growth and development preceded a detailed discussion of the role of growth hormone and insulin-like growth factors on fat and lean disposition. Particular attention was paid to the influence of diet and the possible role of gut hormones in influencing peripheral tissue metabolism.

The fundamental papers formed the background to a series of reports considering methods of influencing fat and lean deposition in whole animals, for example a) the use of the immune response, the use of exogenously applied materials, transgenesis or b) diet itself. Finally consideration of the consequences of manipulating fat and lean deposition on meat quality served to remind delegates of the importance of ensuring that the major function of production in most animals is to produce human food.

The meeting clearly indicated that there was a role for exploiting recent advances in biotechnology to improve the efficiency of production of meat and in this context it was most pleasing that the OECD were able to incorporate the meeting into their Animal Biotechnology Programme.

ACKNOWLEDGEMENTS

The contributions of those who presented papers at the conference together with their efforts in preparing the written versions for inclusion in the proceedings are gratefully acknowledged. Individual sessions were chaired by Dr G Lobley, Professor P J Reeds, Dr K N Boorman and Professor D Demeyer.

Conferences of this nature are greatly helped by financial support towards the expenses of speakers and reductions in the charges made to junior scientists. The Organisation for Economic Cooperation and Development included as part of their Animal Biotechnology Programme and sponsored the following speakers:

Dr J J Bass
Dr B Brenig
Professor D B Lindsay

and several delegates to the conference. In addition we would also like to acknowledge donations from the following:

BOCM Silcock Limited
Boehringer Ingelheim Vetmedica GmbH
BP Nutrition (UK) Limited
Lilly Research Centre Limited
Pauls Agriculture Limited
Pfizer Limited
Rumenco

Particular thanks go to Mrs Rosemary Reid who acted as conference secretary and who prepared the manuscripts. She was ably assisted by Mrs Anna Caves. Thank you also to members of the University of Nottingham Faculty of Agricultural and Food Sciences who helped behind the scenes.

1

HORMONAL REGULATION OF MUSCLE PROTEIN SYNTHESIS AND DEGRADATION

P.J. REEDS and T.A. DAVIS
USDA/ARS Children's Nutrition Research Center, Department of Pediatrics, Baylor College of Medicine, 1100 Bates Street, Houston, TX 77030, USA

Introduction

Given the importance of skeletal muscle as an "organ" of locomotion, the major protein store of the body, and a major agricultural product, it is not surprising that there is a substantial literature on the regulation of muscle growth (see Campion, Hausman and Martin, 1989 for extensive treatment). It is now well established that protein deposition is a result of the twin processes of protein synthesis and degradation (Waterlow, Garlick and Millward, 1978), and an equally large body of literature (reviewed in Campion *et al*, 1989 and by Sugden and Fuller, 1991) is devoted to the various factors that regulate these two processes in muscle. Some of these factors are listed in Table 1. In considering this table, it is important to distinguish those factors that regulate muscle protein turnover from those that have been associated with changes in muscle protein turnover. In this paper, we will confine our comments to the 3 humoral factors, insulin, thyroid hormone, and glucocorticoids, for which we believe there is sufficient evidence to support a physiological role. We will not comment on the numerous factors that play only a pharmacological or pathophysiological role in muscle protein turnover. We shall discuss the regulation of protein synthesis and degradation in separate sections and also describe a potential scheme for the regulation of muscle growth.

Protein synthesis

Although mRNA levels, and in some cases, specific post-translation processes are important to the ultimate expression of specific functionally active proteins, in this paper we use the term "protein synthesis" to denote the process of translation, *ie* polypeptide synthesis. Peptide-chain translation is the end result of a complex series of interrelated events for which the term "ribosomal cycle" has been coined (Pain, 1986; Figure 1). In essence, translation proceeds through three stages: first, the formation of an "initiation complex" that contains the two ribosomal subunits, the mRNA and the initiator aminoacyl-tRNA; second, the process of peptide chain elongation; and finally, the process of termination in which the 80S

Table 1 FACTORS AFFECTING PROTEIN SYNTHESIS AND DEGRADATION IN SKELETAL MUSCLE

Factors	Synthesis	Degradation
Hormones		
Insulin	Increase	No change
Glucocorticoids	Decrease	Increase
Thyroid hormone	Increase	Increase
IGF	Increase	Decrease
Growth hormone	Increase	No change
Glucagon	Decrease	Increase
Catecholamines	Decrease	Decrease
ß-agonists	Increase	Decrease
Androgens	Increase	No change
Nutrition		
Fasting	Decrease	No change
Refeeding	Increase	No change
Starvation	Decrease	Increase
Low energy diet	Decrease	No change
Low protein diet	Decrease	Decrease
Recovery from malnutrition	Increase	Increase
Low protein quality diet	Decrease	-
Amino acids	Increase	Decrease
Fatty acids	Increase	No change
Ketone bodies	Increase	No change
Physiological states		
Growth	Increase	Increase
Age	Decrease	Decrease
Diabetes	Decrease	No change
Sepsis	Decrease	Increase
Malignancy	Decrease	Increase
Burn	No change	Increase
Heat-shock	Decrease	Increase
Hypoxia	Decrease	Decrease
Work		
Stretch	Increase	Decrease
Mechanical stimulation	Increase	Increase
Compensatory work hypertrophy	Increase	Increase
Exercise training	No change	No change
Acute exercise	Decrease	No change
Unweighting	Decrease	Increase
Denervation	Increase	Increase
Muscular dystrophy	Increase	Increase
Other		
Fiber type - oxidative	Increase	Increase
Calcium	Increase	Increase
Protein kinase C	Increase	No change
Prostaglandins	Increase	Increase
pH	Increase	Decrease
Phosphocreatine	Increase	-

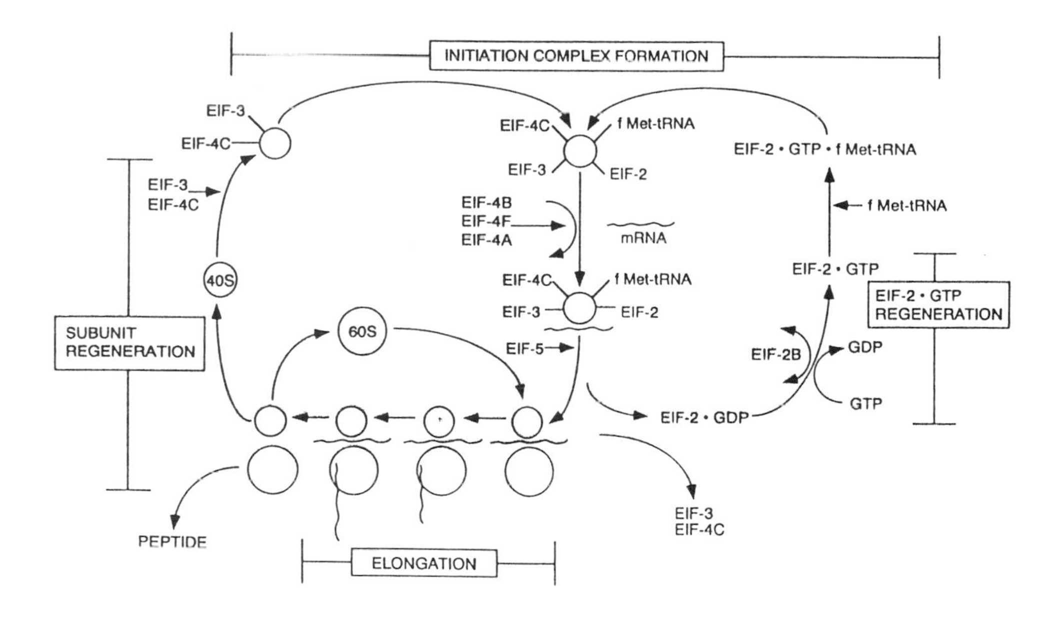

Figure 1 The ribosome cycle. EIF=eukaryotic initiation factors. EIF-3 stimulates the splitting of monosomes into subunits. EIF-2-GTP catalyzes the formation of the initiation complex. EIF-4 promotes incorporation of mRNA into the quaternary initiation complex. EIF-2B catalyzes the GTP-GDP exchange on EIF-2.

monosomes detach from the mRNA and are recycled into a common pool from which they can re-enter the initiation process. In theory, each one or all of these three stages could be regulated. Each stage involves accessory factors, *ie* the aminoacyl-tRNA synthase complex, its product, the aminoacyl-tRNA, and the soluble initiation, elongation, and termination factors (Figure 1). In principle, any one of the factors could be rate-limiting for the overall process.

In some *in vitro* experiments, the concentration of amino acids, and hence that of aminoacyl-tRNA, has been shown to limit the rate of protein synthesis. It is rare, however, to encounter circumstances *in vivo* in which amino acid concentrations fall to levels that might limit translation. Further, in the intact animal it is unlikely that amino acid concentrations alone function as kinetic regulators of translation. Nevertheless, as we shall point out later in this paper, there are important regulatory interactions between amino acid concentrations and hormonal influences upon muscle protein turnover.

Most of the literature on the short-term regulation of protein synthesis by nutrient/hormonal factors suggests that the initiation process is the translational stage that is usually regulated (Jefferson, 1980; Pain, 1986; Kimball and Jefferson, 1988). Furthermore, the majority of evidence from cultured cells and limited evidence from the perfused liver (Kimball, Antonetti, Brawley and Jefferson, 1991) suggest that the phosphorylation state of the initiation factor, EIF-2, is most important in the formation of the initiation complex (see Pain, 1986; Reeds, 1989; Kimball and Jefferson, 1988; Hershey, 1989 for summaries). However, some studies, (*eg* Kelly and Jefferson, 1985; Seedorf, Leberer, Kirschbaum and Pette, 1986) have indicated that the separation of the 40S-60S monomer into the ribosomal subunits could limit initiation, which implies a potential role for EIF-3. In some circumstances, notably those associated with a diminished energy status, a suppression of muscle protein synthesis is associated with proportional reductions in both initiation and elongation (MacLennan and Rennie, 1989). Indeed, MacLennnan and Rennie (1989) have shown that although comparisons of polysome profiles with the translation rate indicate that hormones such as insulin produce disproportionate changes in the rate of initiation, they may also produce changes in the elongation and termination stages.

Table 2 AGE-RELATED CHANGES IN PROTEIN SYNTHETIC CAPACITY AND EFFICIENCY IN SKELETAL MUSCLE OF THE RAT

Age (days)	Capacity (μg protein/mg RNA)	Efficiency (g protein synthesised/ g RNA per day)
1	55	7.7
5	42	7.9
10	34	7.4
16	28	9.6
21	19	12.9
34	11	15.8
48	8	15.6
76	5	15.3
360	4	10.4

Ages 1-21 days were derived from Davis *et al* (1989) and ages 34-360 days were derived from Garlick *et al* (1989)

In considering the long-term relationship between protein synthesis and cellular growth, many studies suggest that there is an upper limit to the rate at which the ribosomal system can catalyze peptide bond formation. It is very rare to find in the literature that the translational efficiency (k_{RNA}) of protein synthesis in any organ has a value greater than 20 g protein synthesized per day per unit of cellular RNA. Once skeletal muscle has completed its metabolic maturation, the translational efficiency of protein synthesis in the muscles of a healthy, well-nourished animal rarely exceeds 15 g protein synthesized per day per unit of cellular RNA (Table 2). Furthermore, the initiation of an increase in protein deposition by a specific stimulation of protein synthesis must involve at least a transient increase in the fractional rate of protein synthesis. Therefore, if it is true that translational efficiency in skeletal muscle cannot exceed 15 g protein per day per g RNA, a regulatory increase in protein synthesis must involve an increase in the concentration of ribosomes (RNA/protein). For example, evidence suggests that a rise in the RNA:protein ratio is the first demonstrable response of the protein synthetic system to an increase in work-load (D.N. McMillan, personal communication; Seedorf *et al*, 1986; Table 3).

Table 3 TIME COURSE OF THE EFFECT OF CHRONIC STIMULATION ON RNA IN SKELETAL MUSCLE

Treatment Duration (days)	Poly (A) RNA	Monosomes	Polysomes
		(% of control)	
0	98	100	86
2	115	190	158
4	105	181	211
6	125	267	247
8	195	250	365
10	205	320	427
14	240	270	500
21	235	364	620
28	205	250	382
52	185	253	233

Derived from Seedorf *et al* (1986)

Thus, in interpreting the literature we must be careful to separate:

(a) the minute-to-minute regulation of the efficiency of translation (k_{RNA}),

(b) the slower changes in protein synthetic capacity (RNA/protein),

(c) an increase in protein synthesis resulting merely from a coordinated and equal increase in mass and ribosome number (see for example Maltin, Hay, Delday, Lobley and Reeds, 1989).

We know that muscle is compartmentalized according to structure; we have recently learned that it is also compartmentalized at the metabolic level. Growing evidence exists that the regulation of protein synthesis, as well as protein degradation, is compartmentalized in skeletal muscle (see Samarel, 1991 for discussion). First, skeletal muscle is an organ consisting of multinucleated cells, and these nuclei can be divided into at least two populations: those located in the motor end-plate region and those distributed more uniformly throughout the fibres. It seems likely that these two populations of nuclei direct the synthesis of different mRNA. Secondly, the subcellular structure of muscle is crucial for its physiological function, and the major protein pools of this tissue are segregated into myofibrillar and sarcoplasmic fractions. Third, within the myofilament, there must be some coordination of the rates of myofibrillar protein synthesis and degradation, because a disturbance in the stoichiometric relationships among these proteins (as might result from discoordinated turnover) would eventually compromise muscle contractile function. Finally, there is increasing evidence that synthesis (Samarel, 1991) and degradation of these two protein compartments is separately regulated. Indeed, immunocytochemical studies (Horne and Hesketh, 1990) have demonstrated the existence of a quantitatively significant pool of ribosomes located within the myofilament, and it is tempting to speculate that these ribosomes are specifically synthesizing myofilament protein.

Hormonal regulation of protein synthesis

Much research into the effects of hormones on protein synthesis has involved either substantial manipulations of endocrine status (such as diabetes, hypophysectomy, thyroidectomy) or the administration of large doses of hormones, and it is difficult to separate any associated disease from the primary and regulatory effects of the hormone under investigation. For example, hypophysectomy leads to multiple changes in hormonal status. Although much of the literature on growth regulation in hypophysectomized animals has examined the growth hormone-IGF axis, some papers suggest that the changes associated with growth hormone treatment of hypophysectomized animals can be ascribed to the effects on other hormones. For example, studies in the pituitary-deficient Snell dwarf mouse (Bates and Holder, 1988) have shown that both growth hormone and thyroxine are effective in restoring growth and stimulating muscle protein synthesis. Diabetic animals also present a complex endocrinological picture, and it has been shown that the poor growth of diabetic rats can be ameliorated by the injection of IGF-I (Scheiwiller, Guler, Merryweather,

Scandella, Mairki, Zapf and Froesch, 1986). As will become apparent, we believe that insulin, the glucocorticoids, and thyroid hormones are the major systemic physiological regulators of muscle growth, and that although evidence is growing in favour of a role for local mediators, the picture is still incomplete.

INSULIN

The implication of insulin in the regulation of muscle growth (Lotspeich, 1949) and protein synthesis (Sinex, MacMullen and Hastings, 1952; Krahl, 1952) has a long history. Insulin exerts a pleiotypic effect: it stimulates amino acid transport, particularly via the A-system, and alters both protein synthesis and degradation. Early research into how insulin regulated protein synthesis depended upon examining the effects of diabetes, and hence is subject to the caveat about disease effects. It is clear, however, that insulin does play a crucial short-term anabolic role in the response to food intake. Garlick, in particular, has carried out a long-term series of investigations that have established the critical role of insulin in response to food intake (Garlick, Fern and Preedy, 1983; Preedy and Garlick, 1986; Garlick and Grant, 1988). These papers have shown that the stimulation of muscle protein synthesis by refeeding is blocked by the co-administration of anti-insulin serum (Preedy and Garlick, 1986), and that the infusion of insulin alone in post-absorptive rats stimulates amino acid incorporation into muscle protein (Garlick *et al*, 1983; Table 4).

Table 4 EFFECT OF INSULIN INFUSION OR REFEEDING IN POSTABSORPTIVE RATS ON MUSCLE PROTEIN SYNTHESIS

Treatment	Plasma Insulin (μUnits/ml)	Protein Synthesis (%/day)
Insulin Infusion	10 ± 1	9.8 ± 0.7
	38 ± 3	10.5 ± 0.4
	70 ± 5	12.8 ± 0.7
	107 ± 5	12.7 ± 1.1
Postabsorptive	13 ± 1	10.2 ± 0.5
Refed	36 ± 3	14.6 ± 0.5

Derived from Garlick *et al* (1983)

Insulin stimulates protein synthesis primarily by stimulating translational efficiency. A number of studies based on polysome profiles (Jefferson, 1980; Pain, 1986) indicate that the main effect of insulin is to stimulate the initiation phase of protein synthesis. By analogy to the response of protein synthesis in cultured cells to some specific inhibitors of translation (Henshaw, Hirsch, Morton and Hiatt, 1971; Kimball *et al*, 1991) and to serum or specific growth factors (*eg* Montine and Henshaw, 1989), it is generally presumed that insulin exerts its actions in muscle by stimulating the dephosphorylation of the α subunit of EIF-2. This proposition is also compatible with studies on the relative proportion of subunits to polysomes in muscle ribosomes isolated from rats in which insulin status has been altered (Harmon, Proud and Pain, 1984). However, recent studies (Karinch and Jefferson, 1991) have suggested that EIF-2α is largely unphosphorylated in skeletal muscle, and that diabetes does not increase EIF-2α phosphorylation. It is possible, therefore, that insulin stimulation of muscle protein synthesis involves a different initiation factor (*eg* EIF-3 or EIF-4).

Despite a general consensus that insulin is an important stimulator of muscle protein deposition, a number of recent studies in adult humans have challenged the notion that the insulin-induction of a positive amino acid balance in skeletal muscle derives predominantly from an effect on protein synthesis (Castellino, Luzi, Simonson, Haymond and DeFronzo, 1987; Gelfand and Barrett, 1987; Tessari, Inchiostro, Biolo, Trevisan, Fantin, Marescotti, Iori, Tiengo and Crepaldi, 1987; Pacy, Nair, Ford and Halliday, 1989; Bennet, Connacher, Smith, Jung and Rennie, 1990; Fryburg, Barrett, Louard and Gelfand, 1990). The vast majority of studies that have measured the direct incorporation of isotopes into muscle protein either *in vitro* (Reeds and Palmer, 1986; Kimball and Jefferson, 1988; Sugden and Fuller, 1991) or *in vivo* (Garlick *et al*, 1983; Sugden and Fuller, 1991) suggest that insulin stimulates muscle protein synthesis, while studies based on amino acid exchange, either in the whole body (Castellino *et al*, 1987; Tessari *et al*, 1987) or in perfused limbs (Gelfand and Barrett, 1987; Pacy *et al*, 1989; Bennet, Connacher, Smith, Jung and Rennie, 1990; Fryburg *et al*, 1990), suggest that insulin only decreases proteolysis.

This dichotomy can be explained in part by differences in methods. First, one must recognize that the results of studies of whole body amino acid turnover reflect a composite of the turnover of amino acids in many body compartments. Evidence exists that physiological levels of insulin suppress hepatic proteolysis (Mortimore and Mondon, 1970; Mortimore and Surmacz, 1983; Lardeaux and Mortimore, 1987); therefore, changes in whole body degradation could merely reflect an effect in the viscera. Second, measurements of protein synthesis from estimations of labelled amino acid exchange across limbs assume that the labelled amino acid is freely and rapidly equilibrated into the intracellular space and thence into aminoacyl-tRNA. Because insulin infusion reduced the circulating amino acid concentrations in many of these studies, real changes in protein synthesis may have been obscured by kinetic artifacts. In support of this proposition, a recent report of studies in adult men (Bennett, Connacher, Scrimgeour, Jung and Rennie, 1990) has shown that when insulin is co-infused with an excess of amino acids, the hormone stimulates muscle protein synthesis. Finally, the differences between the two sets of studies on the stimulation of protein synthesis by insulin may result from an interaction between the age of the subject and the responsiveness of muscle protein synthesis to insulin. The majority of the amino acid exchange studies have used adult subjects, while those based on the direct incorporation of isotope have used young, growing rats. When older, non-growing rats have been studied, there has been little effect of insulin on protein synthesis (Baillie, Maltin and Garlick, 1988).

All the experiments that we have discussed thus far relate to the acute regulation of protein synthesis and degradation. A separate question is whether insulin also mediates the long-term regulation of muscle growth. Although muscle wasting is a specific accompaniment of diabetes and can be reversed by replacement doses of insulin, we do not know whether insulin plays a long-term physiological role in stimulating, as opposed to maintaining, protein synthesis and muscle growth. Studies of diabetes and prolonged starvation demonstrate that changes in ribosome number, measured by the RNA/protein ratio, accompany long-term reductions in insulin-status. Studies in cell culture and *in vivo* suggest that insulin also stimulates the synthesis of ribosomal RNA (Palmer and Bain, 1989) as well as ribosomal protein, and specifically suppresses the degradation of ribosomal proteins (Ashford and Pain, 1986a,b). Even so, these observations indicate a maintenance role for insulin, and although insulin infusion will affect overall nitrogen retention (Fuller, Weekes, Cadenhead and Bruce, 1977), this effect is short-lived and is associated with the development of insulin resistance. Furthermore, although repeated injections of insulin into intact rats lead to increased weight gain, there was little change in whole body or carcass protein in these studies, and the weight gain was almost entirely a reflection of lipid deposition (Woodward and Emery, 1989). Thus, at present, no convincing evidence suggests that insulin levels alone exert positive effects upon growth after they have risen above a level associated with the maintenance of translational efficiency and "normal" ribosome number.

ADRENAL GLUCOCORTICOIDS

There is abundant evidence to suggest that glucocorticoids are capable of inducing muscle wasting (Goldberg, 1969; Odedra and Millward, 1982). Glucocorticoids appear to exert their effect upon protein synthesis in two ways. After a time lag of approximately 4 hours, both dexamethasone and corticosterone reduce protein synthesis *in vivo* (Garlick, Grant and Glennie, 1987; Southorn, Palmer and Garlick, 1990) and *in vitro* (McGrath and Goldspink, 1982; Reeds and Palmer, 1984), and they do so by inhibiting translational efficiency. More prolonged treatments with these glucocorticoids, however, are associated with a reduction in the RNA/protein ratio (Kelly and Goldspink, 1982; Odedra, Bates and Millward, 1983).

The mechanism that produces these effects is not clearly understood. The slight delay in the effect of glucocorticoids on k_{RNA} would be in keeping with other modes of steroid actions, *ie* they must bind to a cytoplasmic receptor which in turn binds to specific promoter sequences on DNA to influence the transcription of mRNA. One of the mRNA codes an endogenous inhibitor of phospholipase A_2 (Hirata, 1981; Palmer, 1987), and it has been argued that the reduction in k_{RNA} is related to an inhibition of prostaglandin production (see Reeds and Palmer, 1986 for discussion). However, the glucocorticoid hormones also appear to influence insulin sensitivity (Odedra, Dalal and Millward, 1982; Southorn *et al*, 1990), although the mechanism of this effect is not understood. The mechanism may be a post-receptor interaction if it is similar to the interaction of glucocorticoids and insulin in stimulating pyruvate dehydrogenase (Begum, Tepperman and Tepperman, 1983) and glycogen synthase (Alvarez, Sanchez-Arias, Guadano, Estevez, Varela, Feliu and Mato, 1991).

It is relatively easy to demonstrate effects of high doses of glucocorticoids on muscle protein deposition and the processes of synthesis and degradation, but it is difficult to determine whether glucocorticoid levels play a physiological role in regulating protein synthesis. Because of the difficulties in carrying out such experiments, large doses of steroids have usually been administered and have induced high plasma levels of glucocorticoids. Furthermore, a very high proportion of circulating glucocorticoids is bound to corticoid-binding globulin. Therefore, unless one measures the free steroid directly, the biologically effective concentration of glucocorticoids cannot be precisely determined. For example, in a recent study (Reeds, Hay, Cadenhead and Fuller, 1991), pigs in which cortisol concentrations were elevated twofold over "normal" levels still exhibited apparently normal rates of protein deposition. Furthermore, adrenalectomy has little effect upon either the rate of protein synthesis or deposition (*eg* Odedra *et al*, 1982; Fletcher and MacKenzie, 1988) despite the fact that adrenalectomy tends to lower feed intake (Fletcher and McKenzie, 1988).

This paper is concerned with the regulation of protein turnover in muscle, but the reader should be aware that the regulation of protein deposition in the body as a whole involves the co-regulation of both protein turnover and amino acid catabolism (see Reeds, Fuller, Cadenhead and Hay, 1987 for discussion). In growing animals, a positive relationship exists between the ratio of nitrogen excretion (or urea synthesis) to protein intake and the circulating concentrations of natural glucocorticoid hormones (*eg* Sharpe, Buttery and Haynes, 1986; Fuller, Reeds, Cadenhead, Seve and Preston, 1987); there is, however, no negative relationship between this ratio and insulin. Indeed, one of the few metabolic processes that does appear to be sensitive to glucocorticoid levels within the physiological range is amino acid catabolism.

However, glucocorticoids do play an important regulatory role in one circumstance. Apart from its obesity, the fa/fa Zucker rat exhibits a specific reduction in skeletal muscle growth (Reeds, Haggarty, Wahle and Fletcher, 1982; Bray, 1989) and also exhibits somewhat elevated glucocorticoid levels. Furthermore, many of the disturbances in protein and lipid metabolism can be reversed by adrenalectomy (Fletcher and Mackenzie, 1988; Bray, 1989). Taken at face value, these observations are difficult to reconcile with the fact that variations in glucocorticoid hormones within the range found in obese Zucker rats do not have any demonstrable effect upon muscle protein turnover and growth in lean animals. Bray (1989), however, has hypothesized that the fundamental defect in the fat Zucker rat is that processes normally insensitive to glucorticoids in the lean genotype have become sensitive to glucocorticoids in the obese genotype. Sharpe *et al* (1986) have shown that adrenalectomy abolishes the ability of the β-adrenergic agonist, clenbuterol, to stimulate muscle growth and that the ability is restored if the animals receive replacement doses of dexamethasone. Perhaps the β-adrenergic agonists act by influencing the sensitivity of protein (and lipid) deposition to circulating glucocorticoids. Such a mechanism would be in keeping with Bray's hypothesis on the interactions between the sympathetic tone and glucocorticoid responses in genetic obesity.

THYROID HORMONES

Both the hypo- and hyperthyroid states are associated with growth disturbances and specific changes in muscle protein synthesis. One of the earliest reports on the use of stable isotopes in the study of endocrine regulation of protein turnover showed that triiodothyronine stimulated whole body protein synthesis (Crispel, Parson, Hollifield and Brent, 1956). Subsequent experiments in the perfused hemicorpus preparation (Flaim, Li and Jefferson, 1978) and *in vivo* (Brown, Bates, Holliday and Millward, 1981; Brown and Millward, 1983) have demonstrated that the injection of thyroid hormones has a direct effect on muscle protein synthesis (Table 5).

Table 5 EFFECT OF THYROID HORMONE ON PROTEIN SYNTHESIS IN SKELETAL MUSCLE

Treatment	Fractional Rate (%/day)	Capacity (μg/g)	Efficiency (g/g per day)
Control	8.5 ± 0.3	5.0 ± 0.9	18.5 ± 2.9
Thyroidectomy	5.0 ± 1.5	3.2 ± 0.5	17.2 ± 5.5
Thyroidectomy + T_3	9.6 ± 1.1	5.2 ± 1.0	19.6 ± 2.0

Derived from Brown *et al* (1981)

The thyroid hormones appear to act on protein synthesis and degradation by means of a mechanism fundamentally different from that of insulin and the glucocorticoids. Alterations in thyroid status require several days to take effect, and they are associated with changes in the RNA/protein ratio in skeletal muscle (resulting largely from changes in cellular RNA concentrations) rather than in k_{RNA} (Brown *et al*, 1981; Table 5). Similar results have been obtained in cardiac muscle (Siehl, Chua, Lautensack-Belser and Morgan, 1985; Sugden and Fuller, 1991) and are in keeping with a transcriptional mode of action. During muscle development, thyroid hormones appear to have an important bearing on the development of the Type II muscle phenotype (for discussion see Pette, 1980).

Relatively little work has been performed on the physiological role of normal variations in thyroid hormone concentrations. In an extensive study of the response of muscle protein synthesis and deposition to changes in protein intake (Jepson, Bates and Millward, 1988), the low rates of muscle protein synthesis which accompanied low intakes of protein were associated with reductions in circulating thyroid hormones. However, this nutritional state was also associated with substantial reductions in insulin concentrations, and the levels of both hormones rose toward what appeared to be an optimum level as dietary protein intake approached a value associated with maximum muscle growth. It is apparent from these studies, therefore, that thyroid hormone concentrations are related to muscle protein synthesis, but only when they are reduced below the range associated with normal muscle growth. Indeed, the impression from the literature is that thyroid hormones play primarily a permissive role in maintaining muscle ribosome number within an optimum range.

Protein degradation

There are many unanswered questions regarding the regulation of protein synthesis, but the mechanism of this process is well established. Our understanding of the mechanisms of protein degradation, however, is less well developed. First, it is extremely difficult to obtain quantitatively accurate measurements of the rate of protein degradation (see Waterlow *et al*, 1978 for extensive discussion), and second, other factors must be considered in an examination of the regulation of proteolysis.

First, several mechanistically distinct pathways of proteolysis exist (Table 6; see Khairallah, Bond and Bird, 1984; Goll, Kleese and Szpacenko, 1989 for extensive treatment). The proteolytic enzymes associated with these pathways appear to be structurally compartmentalized, and separate proteolytic systems have been identified in the lysosomes, nuclei, plasma membrane, cytosol, mitochondria, and endoplasmic reticulum (Beynon and Bond, 1986; Mayer and Doherty, 1986; Gaskell, Heinrich and Mayer, 1987; Stauber, Fritz, Maltin and Dahlmann, 1987). Circumstances are further complicated by the fact that different proteolytic pathways may be of specific importance in different tissues (Wing, Chiang, Goldberg and Dice, 1991).

Table 6 MULTIPLE SYSTEMS OF PROTEIN DEGRADATION

Enzyme systems	Function
Lysosomal cathepsins	Nutrient deprivation
Ubiquitin/ATP-dependent proteases	Short-lived proteins Abnormal proteins
Ca-dependent calpains	Myofibrillar proteins
Multicatalytic/ATP-dependent protease	-
Metalloproteinases	Peptides
Chymotrypsin- and trypsin-like proteinases	Pathological conditions

The lysosomal pathway of protein degradation is one of the best characterized; it utilizes the acid cathepsins. However, this process appears to play a specific autophagic role, and it responds to substantial hormonal or nutrient deprivation and other pathological conditions (Mortimore and Poso, 1984; Furuno and Goldberg, 1986). Although this pathway plays a major role in liver proteolysis, there is little evidence for its involvement in the degradation of myofibrillar proteins in muscle (Lowell, Ruderman and Goodman, 1986a; Wing *et al*, 1991). Another proteolytic pathway requires the provision of ATP and conjugation of ubiquitin to the protein substrate (Fagan, Waxman and Goldberg, 1987; Gehrke and Jennisen, 1987; Rechsteiner, 1987). This pathway appears to target short-lived and/or abnormal proteins, and it has been suggested that it may play an important role in skeletal muscle proteolysis (Wing *et al*, 1991). There is also a multicatalytic, multifunctional protease (otherwise termed the proteosome) that requires ATP but not ubiquitin (Driscoll and Goldberg, 1989). Its function is not clearly understood.

The calcium-dependent proteases (calpains) appear to be important in nonlysosomal proteolysis in skeletal muscle (Zeman, Kameyama, Matsumoto, Berstein and Etlinger, 1985; Furuno and Goldberg, 1986). There is convincing evidence that nonlysosymal proteolysis initiates myofibrillar protein degradation by degrading Z-disks on the periphery of myofibrils and thereby releasing filaments from the surface of the myofibrils (Goll *et al*, 1989). The membrane-associated metalloproteinases, which may have a role in the degradation of small peptides and polypeptides, and the chymotrypsin-like and trypsin-like serine proteinases have also been identified in skeletal muscle (Beynon and Bond, 1986).

The second feature of protein degradation that must be considered is the fact that within a given proteolytic environment, different proteins exhibit markedly different rates of degradation (Schimke and Doyle, 1970; Fagan and Goldberg, 1985; Beynon and Bond, 1986; Vilaro, Llobera, Bengtsson-Olivecrona and Olivecrona, 1988). The factors that regulate the susceptibility of different proteins to proteolysis are important but, at this time, unknown. Current evidence suggests that the susceptibility (or resistance) to proteolysis is affected by the cellular location of the protein (Mayer and Doherty, 1986) and physical properties such as the hydrophobicity of the protein (Schimke and Doyle, 1970; Beynon and Bond, 1986). The presence of specific peptide sequences within the protein can target them for different proteolytic pathways (Rogers, Wells and Rechsteiner, 1986; Dice and Chiang, 1989), which may be selectively activated in some tissues but not in others during circumstances (such as fasting) that are associated with protein loss (Wing *et al*, 1991).

Even though the mechanisms by which protein degradation is regulated are not clearly understood, it is apparently a tightly regulated process. Protein degradation in skeletal muscle responds rapidly and chronically to factors that include nutritional state (Jepson *et al*, 1988; Millward, 1988), workload (Laurent, Sparrow and Millward, 1978), and hormones.

INSULIN

Although it is known that insulin suppresses hepatic protein degradation (Mortimore and Surmacz, 1983), insulin may or may not be a regulator of proteolysis in muscle. It is generally held, however, that insulin suppresses muscle proteolysis. Diabetes is clearly associated with increased muscle proteolysis, but there is little direct evidence that physiological concentrations of insulin have a specific effect on proteolysis. Thus, although a significant number of *in vitro* studies show that protein synthesis responded to insulin concentrations within the physiological range, few studies show a similar sensitivity of muscle protein degradation. In most studies, insulin has been shown to lower protein degradation at concentrations in excess of 1 mU/ml, and this response may reflect insulin binding to IGF-1 receptors (Florini, 1987). Indeed, in the paper by Stirewalt and Low (1983), which is often cited in support of the idea that physiological levels of insulin lower muscle protein degradation, the "reduced" rate of proteolysis in the presence of insulin required two to three hours and achieved statistical significance only because, after prolonged periods of incubation, the rate of proteolysis in the control muscles rose to pathologically high rates.

Nevertheless, recent reports (Gelfand and Barrett, 1987; Fryburg *et al*, 1990; Bennett, Connacher, Scrimgeour, Jung and Rennie, 1990) based upon leucine exchange across the limb tissues of adults are compatible with an acute suppression of proteolysis by elevated insulin levels. On the other hand, based upon phenylalanine kinetic measurements, data on the effect of insulin on protein degradation are conflicting. Perhaps the acute effects of insulin on muscle protein degradation are a specific feature of the adult.

THYROID HORMONES

The evidence that thyroid status affects the rate of muscle protein degradation is much more convincing than the evidence that insulin regulates muscle protein degradation. The hypothyroid state is associated with a reduction in muscle protein degradation, and the administration of thyroid hormone to thyroidectomized animals increases muscle protein degradation (Brown *et al*, 1981; Brown and Millward, 1983; Table 7). A strong positive relationship has been demonstrated between physiological levels of thyroid hormone and insulin and muscle protein degradation in rats fed different amounts of protein (Jepson *et al*, 1988). To some extent this relationship offers further evidence that physiological levels of insulin do not suppress muscle protein degradation.

Table 7 EFFECT OF TRIIODOTHYRONINE REPLACEMENT ON PROTEIN DEGRADATION IN SKELETAL MUSCLE OF THYROIDECTOMIZED RATS

Treatment	T_3 Dose (μ mol)	Degradation (nmol Phe/h per g)
Control	0	6.0
Thyrex	0	3.7
Thyrex	0.30	4.5
Thyrex	0.75	6.5
Thyrex	2.00	6.6
Thyrex	20.00	6.8

Derived from Brown and Millward (1983)

Unfortunately, despite the convincing evidence that thyroid hormones influence muscle proteolysis, the precise protein catabolic pathway that they do influence is unknown. Some evidence indicates that calcium-activated proteases may be involved (Zeman, Bernstein, Ludemann and Etlinger, 1986). It is difficult to determine precisely how thyroid hormones influence proteolysis, because they also influence energy expenditure and balance: the changes in muscle protein turnover associated with manipulations of the thyroid status may include a response to changing energy balance and ATP synthesis. For example, Sugden and Fuller (1991) have pointed out that the increased cardiac growth and protein synthesis associated with the hyperthyroid state may actually result from the increased cardiac work associated with the elevated body energy expenditure.

GLUCOCORTICOID HORMONES

The effects of glucocorticoids on muscle protein degradation are also equivocal, and during the late 70's and early 80's, the existence of such effects was a subject of great controversy (Millward, Garlick, Nnanyelugo and Waterlow, 1976; Odedra and Millward, 1982; Tomas, Munro and Young, 1979). The divergence of opinion may have been a function of the different methods that were used to estimate protein degradation.

We now know that myofibrillar (myosin and actin degradation as measured by methylhistidine release) is regulated quite independently of non-myofibrillar (sarcoplasmic) proteolysis. The data that have led us and others (Goodman, 1987a; Goodman and Gomez, 1987; Kayali, Young and Goodman, 1987; Kadowaki, Harada, Takahashi, Noguchi and Naito, 1989; Hasselgren, James, Benson, Hall-Angeras, Angeras, Hiyama, Li and Fischer, 1989; Hasselgren, Hall-Angeras, Angeras, Benson, James, and Fischer, 1990) to this conclusion are presented in Table 8. First, *in vivo* or *in vitro*, glucocorticoids appear to exert a relatively specific effect on actomyosin turnover (and by implication myofibrillar turnover) and have little effect on sarcoplasmic protein degradation. Thus, the essential difference between the studies of Millward and Tomas was that Millward measured total muscle protein degradation (from the difference between protein synthesis and protein deposition), while Tomas measured actomyosin degradation (from the excretion of 3-methylhistidine). Furthermore, if one compares the effects of starvation, refeeding, and insulin, the evidence strongly favours insulin as a key regulator of protein synthesis in the switch from the fasted to the refed state; the evidence that insulin regulates protein degradation, however, is tenuous at best. Conversely, although the evidence is not strong that glucocorticoids are primary regulators of muscle protein synthesis during a short fast, a comparison of the effects of more prolonged starvation and refeeding with the effects of glucocorticoids on myofibrillar and sarcoplasmic proteolysis suggests that these hormones may be key regulators of protein degradation during acute changes in feed intake. If so, it is an exciting development, even though it poses difficult questions with regards to the sequestration of the proteolytic pathways and the second messenger systems that link hormone-receptor activation to the regulation of protein degradation.

Table 8 EFFECTS OF DIFFERENT AGENTS ON THE PERCENTAGE CHANGE IN MYOFIBRILLAR, TOTAL AND NON-MYOFIBRILLAR PROTEIN DEGRADATION

	Protein Degradation		
Treatment	Myofibrillar	Total	Non-myofibrillar
Fasting [a,b,c]	+220%	+32%	-4%
Starvation [a,d,e]	+485%	+38%	-18%
Refeeding [f]	-64%	-5%	+41%
Insulin [e,g]			
Physiological	+6%	0%	-1%
Pharmacological	+1%	-27%	-40%
Corticosterone [h,i,j]	+45%	+16%	2%
Diabetes [d,k,l]			
Acute	+34%	+10%	-2%
Chronic	-39%	-4%	+8%
Protein-free diet [d]	+18%	-51%	-84%
Stretch [e]	-56%	-24%	-21%
Calcium [c,m]	+3%	+23%	+31%
Sepsis [i,m]	+98%	+46%	+17%

Values in this table are derived from the following references:

[a]Lowell *et al* (1986b)
[b]Lowell *et al* (1986a)
[c]Goodman (1987a)
[d]Kadowaski *et al* (1989)
[e]Hasselgren *et al* (1990)
[f]Goodman and Gomez (1987)
[g]Hasselgren *et al* (1989)
[h]Kayali *et al* (1987)
[i]Kayali *et al* (1990)
[j]Smith *et al* (1990)
[k]Goodman (1987b)

Regulatory interactions

The reductionist nature of mechanistic research leads us to consider the effects of a single regulatory factor (*eg* insulin, glucocorticoids, or thyroid hormones) on a specific metabolic pathway (*eg* protein synthesis or degradation). Ultimately, this is too simplistic a view. Tissues are simultaneously subjected to many potential regulatory factors, and a growing organism must regulate the disposition of organic nutrients to support life and to maintain the appropriate allometry of growth, *ie* an appropriate balance between mass and function. The growth of any organ is regulated by interactions between its potential for growth (set by its genotype and developmental stage), the functional demands placed upon it, and the availability of nutrients. The coordination of these three factors is achieved by an interactive system in which anabolic and anti-anabolic factors are balanced at a number of levels.

First, we would argue not only do the amount and composition of nutrient absorption influence the secretion of the key hormones, but also the circulating concentration of each hormone determines the sensitivity of other endocrine organs to nutrient absorption. Second, we propose that the three hormones we have discussed not only directly affect the maintenance of viable cells (so that they are capable of protein deposition), but also influence the sensitivity of the end-organs of protein storage to other hormones. Finally, the key to protein deposition is the ability to up-regulate and down-regulate the anabolic processes in response to food intake. This issue could readily be the subject of its own paper, and we will therefore confine ourselves to two examples of end-organ sensitivity: one involving a nutrient-hormone interaction and the other involving a hormone-hormone interaction.

We emphasized earlier the strong direct evidence that insulin plays a key role in the short-term utilization of a protein-containing meal. However, while insulin is necessary to activate anabolism, it is not sufficient. Specifically, the rate of protein synthesis in response to maximum insulin stimulation was somewhat less than the rate of protein synthesis after a meal, and the insulin concentration at the maximum protein synthetic rate associated with a meal was substantially lower than the insulin concentration at the maximum protein synthetic rate associated with insulin infusion alone (Garlick *et al*, 1983; Preedy and Garlick, 1986; Table 4). Further work (Garlick and Grant, 1988; Table 9) has offered an explanation for these observations: additional amino acids increase the ability of insulin to increase muscle protein synthesis; and the insulin concentration that achieved a maximal stimulation of protein synthesis was much lower in the presence of amino acids than in their absence. In other words, higher levels of amino acids (the substrate for protein synthesis) do not themselves stimulate protein synthesis; rather, they sensitize the process to the presence of insulin. Thus, amino acids not only increase insulin secretion (Fajans, Floyd, Knopf and Conn, 1967) but enhance their own disposition. Similar amino acid/insulin interactions have now been observed in the limb tissues of adult men (Bennett, Connacher, Scrimgeour, Jung and Rennie, 1990) and may also participate in the regulation of liver proteolysis (Mortimore, Poso, Kadowaki and Wert, 1987). This interaction provides a mechanism whereby tissue responsiveness can be regulated and therefore provides an explanation for the regulation of nutrient channeling (homeorrhesis). Those tissues that are most responsive activate anabolism at low regulator concentrations and hence sequester a higher proportion of a given amino acid intake. The mechanism underlying this interaction is not known, but it is significant that the

Table 9 EFFECT OF AMINO ACIDS ON THE INSULIN SENSITIVITY OF MUSCLE PROTEIN SYNTHESIS

Insulin Infusion (Units/ml)	Plasma Insulin (μU/ml)		Protein Synthesis (%/day)	
	-AA	+AA	-AA	+AA
0	2	5	10	11
30	5	20	11	15
47	41	39	13	15
75	89	89	13	15
105	158	-	14	-

Derived from Garlick and Grant (1988)

large majority of the response to a complete amino acid mixture can be achieved merely by infusing the branched-chain amino acids (Garlick and Grant, 1988) and that leucine is also a specific regulator of insulin secretion.

Recent studies have also provided evidence for a potentially physiologically relevant antagonism interaction between glucocorticoids and insulin. It has been known for many years that these two hormones generally act on protein synthesis and degradation in opposition to one another, and that some of the deleterious effects of diabetes on muscle protein deposition can be avoided by adrenalectomizing the animals. Adrenalectomy, however, is not merely the removal of a suppressive influence, because muscle protein synthesis in aderenalectomized diabetic rats is more sensitive to the stimulatory effects of insulin (Odedra *et al*, 1982; Table 10). Moreover, adrenalectomized non-diabetic rats maintain protein deposition despite a significantly lower food intake, a fact that argues in favour of a more efficient utilization of dietary protein -- an effect that would be consistent with enhanced insulin sensitivity. Recent data on diabetic animals characterized the sensitivity of muscle protein synthesis to insulin within the physiological range (Southorn *et al*, 1990; Table 11). Animals pretreated with corticosterone 12 hours before insulin infusion required significantly higher concentrations of insulin to achieve a maximum rate of protein synthesis.

Table 10 EFFECT OF ADRENALECTOMY ON THE INSULIN SENSITIVITY OF MUSCLE PROTEIN SYNTHESIS IN DIABETIC RATS

	Plasma Insulin (μUnits/ml)		Protein Synthesis (%/day)	
Insulin Status	+Adrenals	-Adrenals	+Adrenals	-Adrenals
Control	35	74	14	12
Diabetic	2	3	6	7
Diabetic + 50 mU/h	23	43	7	11
Diabetic + 75 mU/h	38	65	7	11
Diabetic + 100 mU/h	52	75	8	12

Derived from Odedra *et al* (1982)

Table 11 EFFECT OF CORTICOSTERONE INJECTION ON THE INSULIN SENSITIVITY OF PROTEIN SYNTHESIS IN SKELETAL MUSCLE

Insulin Infusion (mUnits/ml)	Plasma Insulin (μUnits/ml)		Protein Synthesis (%/day)	
	-C	+C	-C	+C
0	8	16	11	8
25	10	18	13	8
50	29	28	14	9
100	67	84	15	11
150	154	122	15	11

Derived from Southorn *et al* (1990)

It is possible, then, that the short-term sensitivity of protein deposition to nutrient absorption is set by regulating the sensitivity and responsiveness of cells to insulin. Long-term increases in glucocorticoid concentration down-regulate insulin sensitivity, and circulating glucocorticoid levels are negatively related to the energy:protein ratio in the diet. On the other hand, long-term elevations in thyroid hormone (which is positively related to protein intake) may up-regulate insulin responsiveness by altering ribosome concentrations and hence the maximum rate of protein synthesis. Finally, amino acid concentrations may be the key to the level of insulin that affects protein synthesis and proteolysis.

Concluding remarks

A notable omission from this paper is a discussion of the role of pituitary growth hormone and the associated role of the insulin-like growth factors. This omission is not intended to deny the importance of these regulators; alterations in growth hormone (*eg* Hammer, Brinster, Rosenfeld, Evans and Mayo, 1985; Boyd and Bauman, 1988; Campbell, Johnson, Taverner and King, 1991; Beermann and Boyd, 1992) and IGF-I status (*eg* Mathews, Hammer, Behringer, Dercole, Bell, Brinster and Palmiter, 1988; Siddiqui, Blair, McCutcheon, Mackenzie, Gluckman and Brier, 1990; Lemmey, Martin, Read, Tomas, Owens and Ballard, 1991) affect protein deposition substantially. Relationships also exist between protein intake and IGF concentrations. Unfortunately, the muscle protein metabolic changes that accompany the changes in growth are incompletely characterized. Studies in adult men suggest that growth hormone acutely stimulates whole body (Horber and Haymond, 1990) and muscle (Fryburg, Gelfand and Barrett, 1991) protein synthesis and at the same time has little effect on protein degradation. Studies in sheep (Pell and Bates, 1987) suggest that the enhancement of growth after prolonged growth hormone treatment is associated with an increase in translational capacity. There is also increasing evidence (*eg* Jennische and Hansson, 1987; DeVol, Rotwein, Sadow, Novakofski and Bechtel, 1990) to suggest that local release of IGF is stimulated during work-induced hypertrophy of skeletal muscle. Because this condition appears to increase primarily muscle ribosome accretion, it could be proposed that local production of IGF-I is important in regulating the capacity of muscle to synthesize protein.

The regulation of critically important processes such as protein synthesis and degradation is necessarily complex. While this review has concentrated on muscle protein turnover, the regulation of this process must be accommodated within the regulation of cellular growth and metabolism in the whole body. Thus long-term growth regulation involves the coordination of the protein metabolic activities of the major organs and also the control of amino acid catabolism and oxidation, processes that are also compartmentalized in different tissues. It is far too simplistic to believe that any single metabolic process or hormonal system is capable of regulating muscle growth. It seems to us then that the next challenge in understanding muscle growth regulation is to understand the physiologically relevant interactions between the different hormones that regulate protein deposition and in particular to accommodate the increasing information on paracrine or local growth regulation within our understanding of the control of muscle protein synthesis and degradation.

References

Alvarez, J.F., Sanchez-Arias, J.A., Guadano, A., Estevez, F., Varela, I., Feliu, J.E. and Mato, J.M. (1991) *Biochemical Journal*, **274**, 369-374

Ashford, A.J. and Pain, V.M. (1986a) *Journal of Biological Chemistry*, **261**, 4059-4065

Ashford, A.J. and Pain, V.M. (1986b) *Journal of Biological Chemistry*, **261**, 4066-4070

Baillie, A.G.S., Maltin, C.A. and Garlick, P.J. (1988) *Proceedings of the Nutrition Society*, **47**, 114A

Bates, P.C. and Holder, A.T. (1988) *Journal of Endocrinology*, **119**, 31-41

Beermann, D.H. and Boyd, R.D. (1992) In *The Control of Fat and Lean Deposition*, (eds P.J. Buttery, K.N. Boorman and D.B. Lindsay), London, Butterworths

Begum, N., Tepperman, H.M. and Tepperman, J. (1982) *Endocrinology*, **114**, 99-107

Bennet, W.M., Connacher, A.A., Scrimgeour, C.M. Jung, R.T. and Rennie, M.J. (1990) *American Journal of Physiology*, **259**, E185-E194

Bennet, W.M., Connacher, A.A., Smith, K., Jung, R.T. and Rennie, M.J. (1990) *Diabetologia*, **33**, 43-51

Benson, D.W., Hasselgren, P., Hiyama, D.T., James, J.H., Li, S., Rigel, D.F. and Fischer, J.E. (1989) *Surgery*, **106**, 87-93

Beynon, R.J. and Bond, J.S. (1986) *American Journal of Physiology*, **251**, C141-C152

Boyd, R.D. and Bauman, D.E. (1988) In *Animal Growth Regulation* (eds D.R. Campion, G.J. Hausman and R.J. Martin), New York, Plenum Press, pp 257-293

Bray, G. (1989) *American Journal of Clinical Nutrition*, **50**, 891-902

Brown, J.G. and Millward, D.J. (1983) *Biochimica Biophysica Acta*, **757**, 182-19

Brown, J.G., Bates, P.C., Holliday, M.A. and Millward, D.J. (1981) *Biochemical Journal*, **194**, 771-782

Campbell, R.G., Johnson, R.J., Taverner, M.R. and King R.H. (1991) *Journal of Animal Science*, **69**, 1522-1531

Campion, D.R., Hausman, G.J. and Martin, R.J. (1989) *Animal Growth Regulation*, New York, Plenum Press

Castellino, P., Luzi, L., Simonson, D.C., Haymond, M. and DeFronzo, R.A. (1987) *Journal of Clinical Investigation*, **80**, 1784-1793

Crispel, K.R., Parson, W., Hollifield, G. and Brent, S. (1956) *Journal of Clinical Investigation*, **35**, 164-168

DeVol, D.L., Rotwein, P., Sadow, J.L., Novakofski, J. and Bechtel, P.J. (1990) *American Journal of Physiology*, **259**, E89-E95
Dice, J.F. and Chiang, H. (1989) *Biochemical Society Symposium*, **55**, 45-55
Driscoll, J. and Goldberg, A.L. (1989) *Proceedings of the National Academy of Sciences USA*, **86**, 787-791
Fagan, J.M. and Goldberg, A.L. (1985) *Proceedings of the National Academy of Sciences USA*, **83**, 2771-2775
Fagan, J.M., Waxman, L. and Goldberg, A.L. (1987) *Biochemical Journal*, **243**, 335-343
Fajans, S.S., Floyd, J.C. Jr, Knopf, R.F. and Conn, J.W. (1967) *Recent Progress in Hormone Research*, **23**, 617-652
Flain, K.E., Li, J.B. and Jefferson, L.S. (1978) *American Journal of Physiology*, **235**, E231-E236
Fletcher, J.M. and McKenzie, N. (1988) *British Journal of Nutrition*, **60**, 563-569
Florini, J.R. (1987) *Muscle Nerve*, **10**, 577-598
Fryburg, D.A., Barrett, E.J., Louard, R.J. and Gelfand R.A. (1990) *American Journal of Physiology*, **259**, E477-E482
Fryburg, D.A., Gelfand, R.A. and Barrett, E.J. (1991) *American Journal of Physiology*, **260**, E499-E504
Fuller, M.F., Weekes, T.E.C., Cadenhead, A. and Bruce, J.B. (1977) *British Journal of Nutrition*, **38**, 489-496
Fuller, M.F., Reeds, P.J., Cadenhead, A., Seve, B. and Preston, T. (1987) *British Journal of Nutrition*, **58**, 287-300
Furuno, K. and Goldberg, A.L. (1986) *Biochemical Journal*, **237**, 859-864
Garlick, P.J. and Grant, I. (1988) *Biochemical Journal*, **254**, 579-585
Garlick, P.J., Fern, M. and Preedy, V.R. (1983) *Biochemical Journal*, **210**, 669-676
Garlick, P.J., Grant, I. and Glennie, R.T. (1987) *Biochemical Journal*, **248**, 439-442
Gaskell, M.J., Heinrich, P.C. and Mayer, R.J. (1987) *Biochemical Journal*, **241**, 817-825
Gehrke, P.P and Jennisen, H.P. (1987) *Biological Chemistry Hoppe-Seyler*, **368**, 691-708
Gelfand, R.A. and Barrett, E.J. (1987) *Journal of Clinical Investigation*, **80**, 1-6
Goldberg, A.L. (1969) *Journal of Biological Chemistry*, **244**, 3223-3229
Goll, D.E., Kleese, W.C. and Szpacenko, A. (1989) In *Animal Growth Regulation*, (eds D.R. Campion, G.J. Hausman and R.J. Martin), New York, Plenum Press, pp 141-182
Goodman, M.N. (1987a) *Biochemical Journal*, **241**, 121-127
Goodman, M.N. (1987b) *Diabetes*, **36**, 100-105
Goodman, M.N. and Gomez, M.D.P (1987) *American Journal of Physiology*, **253**, E52-E58
Hammer, R.E., Brinster, R.L., Rosenfeld, M.G. Evans, R.M. and Mayo, K.E. (1985) *Nature*, **315**, 413-416
Harmon, C.S., Proud, C.G. and Pain, V.M. (1984) *Biochemical Journal*, **223**, 687-696

Hasselgren P., James, J.H., Benson, D.W., Hall-Angeras, M., Angeras, U., Hiyama, D.T., Li, S. and Fischer, J.E. (1989) *Metabolism*, **38**, 634-640
Hasselgren, P., Hall-Angeras, M., Angeras, U., Benson, D., James, J.H. and Fischer, J.E. (1990) *Biochemical Journal*, **267**, 37-44
Henshaw, E.C., Hirsch, C.A., Morton, B.E. and Hiatt, H.H. (1971) *Journal of Biological Chemistry*, **246**, 436-446
Hershey, J.W.B. (1989) *Journal of Biological Chemistry*, **264**, 20823-20826
Hirata, F. (1981) *Journal of Biological Chemistry*, **246**, 436-446
Horber, F.F. and Haymond, M.W. (1990) *Journal of Clinical Investigation*, **86**, 265-272
Horne, Z. and Hesketh, J. (1990) *Biochemical Journal*, **272**, 831-833
Jefferson, L.S. (1980) *Diabetes*, **29**, 1407-1408
Jennische, E. and Hansson, H-A. (1987) *Acta Physiologica Scandinavica*, **130**, 327-332
Jepson, M.M., Bates, P.C. and Millward, D.J. (1988) *British Journal of Nutrition*, **59**, 397-415
Kadowaki, M., Harada, N., Takahashi, S., Noguchi, R. and Naito, H. (1989) *Journal of Nutrition*, **119**, 471-477
Karinch, A.M. and Jefferson, L.S. (1991) *FASEB Journal*, **5**, 1535A
Kayali, A.G., Goodman, M.N., Li, J. and Young, V.R. (1990) *American Journal of Physiology*, **259**, E699-E705
Kayali, A.G., Young, V.R. and Goodman, M.N. (1987) *American Journal of Physiology*, **252**, E621-E626
Kelly, F.J. and Goldspink, D.F. (1982) *Biochemical Journal*, **208**, 147-151
Kelly, F.J. and Jefferson, L.S. (1985) *Journal of Biological Chemistry*, **260**, 6677-6683
Kimball, S.R. and Jefferson, L.S. (1988) *Diabetes/Metabolism Reviews*, **4**, 773-787
Kimball, S.R., Antonetti, D.A., Brawley, R.M. and Jefferson, L.S. (1991) *Journal of Biological Chemistry*, **266**, 1969-1976
Khairallah, E.A., Bond, J.S. and Bird, J.W.C. (1984) *Intracellular Protein Catabolism*, New York, Alan R Liss
Krahl, M.E. (1952) *Science*, **116**, 524-525
Lardeux, B.R. and Mortimore, G.E. (1987) *Journal of Biological Chemistry*, **262**, 14514-14519
Laurent, G.J., Sparrow, M.P and Millward, D.J. (1978) *Biochemical Journal*, **176**, 407-417
Lemmey, A.B., Martin, A.A., Read, L.C., Tomas, F.M., Owens, P.C. and Ballard, F.J. (1991) *American Journal of Physiology*, **260**, E213-E219
Lotspeich, W.D. (1949) *Journal of Biological Chemistry*, **179**, 175-181
Lowell, B.B., Ruderman, N.B. and Goodman, M.N. (1986a) *Biochemical Journal*, **234**, 237-240
Lowell, B.B., Ruderman, N.B. and Goodman, M.N. (1986b) *Metabolism*, **35**, 1121-1127
Mayer, R.J. and Doherty, F. (1986) *FEBS Letters*, **198**, 181-193

MacLennan, P.A. and Rennie, M.J. (1989) *Biochemical Journal*, **260**, 195-200
Maltin, C.A., Hay, S.M., Delday, G.E., Lobley, G.E. and Reeds, P.J. (1989) *Biochemical Journal*, **261**, 965-971
Mathews, L.S., Hammer, R.E., Behringer, R.R., Dercole, A.J., Bell, G.I., Brinster, R.L. and Palmiter, R.D. (1988) *Endocrinology*, **123**, 2827-2833
McGrath, J.A. and Goldspink, D.F. (1982) *Biochemical Journal*, **206**, 641-645
McMillan, D.N. (1988) Personal communication
Millward, D.J. (1988) In *Milk Proteins*, (eds C.A. Barth and E. Schlimme), Darmstadt, Styeinkopf Verlag, pp 49-61
Millward, D.J., Garlick, P.J., Nnanyelugo, D.O. and Waterlow, J.C. (1976) *Biochemical Journal*, **156**, 185-188
Montine, K.S. and Henshaw, E.C. (1989) *Biochimica et Biophysica Acta*, **1014**, 282-288
Mortimore, G.E. and Mondon, C.E. (1970) *Journal of Biological Chemistry*, **245**, 2375- 2381
Mortimore, G.E. and Poso, A.R. (1984) *Federation Proceedings*, **43**, 1289-1294
Mortimore, G.E. and Surmacz, C.A. (1983) In *Protein Metabolism and Nutrition*, (eds M. Arnal, R. Pion and D. Bonin), Paris, INRA, pp 97-115
Mortimore, G.E., Poso, A.R., Kadowaki, M. and Wert, J.J. (1987) *Journal of Biological Chemistry*, **262**, 16322-16327
Odedra, B.R. and Millward, D.J. (1982) *Biochemical Journal*, **204**, 663-672
Odedra, B.R., Bates, P.C. and Millward, D.J. (1983) *Biochemical Journal*, **214**, 617-627
Odedra, B.R., Dalal, S.S. and Millward, D.J. (1982) *Biochemical Journal*, **202**, 363-368
Pacy, P.J., Nair, K.S., Ford, C. and Halliday, D. (1989) *Diabetes*, **38**, 618-624
Pain, V.M. (1986) *Biochemical Journal*, **235**, 625-637
Palmer, R.M. (1987) *The Role of Prostaglandins in the Hormonal Control of Protein Synthesis*, University of Aberdeen, PhD Thesis
Palmer, R.M. and Bain, P.A. (1989) *Prostaglandins*, **37**, 193-203
Pell, J.M. and Bates, P.C. (1987) *Journal of Endocrinology*, **115**, 1-4
Pette, D. (1980) *Plasticity of Muscle*, New York, de Gruyter
Preedy, V.R. and Garlick, P.J. (1986) *Bioscience Reports*, **6**, 177-183
Rechsteiner, M. (1987) *Annual Review of Cell Biology*, **3**, 1-80
Reeds, P.J. (1987) In *Comparative Nutrition*, (eds K.L. Blaxter and I. MacDonald) London, John Libbey, pp 55-72
Reeds, P.J. (1989) In *Animal Growth Regulation*, (eds D.R. Campion, G.J. Hausman and R.J. Martin), New York, Plenum Press, pp 183-210
Reeds, P.J. and Palmer, R.M. (1984) *Biochemical Journal*, **219**, 953-957
Reeds, P.J. and Palmer, R.M. (1986) In *Control and Manipulation of Animal Growth*, (eds P.J. Buttery, D.B. Lindsay and N.B. Haynes), London, Butterworths, pp 161-185
Reeds, P.J., Haggarty, P., Wahle, K.W.J. and Fletcher, J.M. (1982) *Biochemical Journal*, **204**, 393-398

Reeds, P.J., Fuller, M.F., Cadenhead, A. and Hay, S.M. (1987) *British Journal of Nutrition*, **58**, 301-311

Reeds, P.J., Hay, S.M., Cadenhead, A. and Fuller, M.F. (1991) *American Journal of Physiology*, **261** (6)

Rogers, S.W., Wells, R. and Rechsteiner, M. (1986) *Science*, **234**, 364-369

Samarel, A.M. (1991) *FASEB Journal*, **5**, 2020-2028

Scheiwiller, E., Guler, H-P., Merryweather, C., Scandella, C., Mairki, W., Zapf, J. and Froesch, E.R. (1986) *Nature* (London), **323**, 169-171

Schimke, T. and Doyle, D. (1970) *Annual Review of Biochemistry*, **39**, 929-976

Seedorf, V., Leberer, E., Kirschbaum, B.J. and Pette, D. (1986) *Biochemical Journal*, **239**, 115-121

Sharpe, P.M., Buttery, P.J. and Haynes, N.B. (1986) *British Journal of Nutrition*, **56**, 289-304

Siddiqui, R.A., Blair, H.T., McCutcheon, S.N., Mackenzie, D.D.S., Gluckman, P.D. and Brier, B.H. (1990) *Endocrinology*, **124**, 151-158

Sinex, F.M., MacMullen, J. and Hastings, A.B. (1952) *Journal of Biological Chemistry*, **195**, 615-619

Siehl, D., Chua, B.H.L., Lautensack-Belser, N. and Morgan, H.E. (1985) *American Journal of Physiology*, **248**, C309-319

Smith, O.L.K., Wong, C.Y. and Gelfand, R.A. (1989) *Diabetes*, **38**,1117-1122

Smith, O.L.K., Wong, C.Y. and Gelfand, R.A (1990) *Metabolism*, **39**, 641-646

Southorn, B.G., Palmer, R.M. and Garlick, P.J. (1990) *Biochemical Journal*, **272**, 187-191

Stauber, W.T., Fritz, V.K., Maltin, C.A. and Dahlmann, B. (1987) *Histochemical Journal*, **19**, 594-597

Stirewalt, W.S. and Low, R.B. (1983) *Biochemical Journal*, **210**, 323-330

Sugden, P.H. and Fuller, S.J. (1991) *Biochemical Journal*, **273**, 21-37

Tessari, P., Inchiostro, S., Biolo, G., Trevisan, R., Fantin, G., Marescotti, C., Iori, E., Tiengo, A. and Crepaldi, G. (1987) *Journal of Clinical Investigation*, **79**, 1062-1069

Tomas, F.M., Munro, H.N. and Young, V. R. (1979) *Biochemical Journal*, **178**, 139-149

Vilaro, S., Llobera, M., Bengtsson-Olivecrona, G. and Olivecrona, T. (1988) *American Journal of Physiology*, **254**, G711-G722

Waterlow, J.C., Garlick, P.J. and Millward, D.J. (1978) *Protein Turnover in Mammalian Tissues and in the Whole Body*, Amsterdam, North Holland

Wing, S.S., Chiang, H., Goldberg, A.L. and Dice, J.F. (1991) *Biochemical Journal*, **275**, 165-169

Woodward, C.J.H. and Emery, P.W. (1989) *British Journal of Nutrition*, **61**, 437-444

Zeman, R.J., Kameyama, T., Matsumoto, K., Berstein, P. and Etlinger, J.D. (1985) *Journal of Biological Chemistry*, **260**, 13619-13624

Zeman, R.J., Bernstein, P.L., Ludemann, R. and Etlinger, J.D. (1986) *Biochemical Journal*, **240**, 269-272

2
MUSCLE CELL GROWTH

J.M.M. HARPER and P.J. BUTTERY
University of Nottingham School of Agriculture, Sutton Bonington, Loughborough, UK, LE12 5RD

Introduction

There have now been several decades of research into the possibilities of manipulating the growth performance of domestic livestock. The primary aims were initially to improve growth rate and feed conversion efficiency but there has been increasing awareness of the need also to improve the proportion of muscle and decrease that of fat in carcasses, in response to current dietary recommendations. There is currently also a realisation that none of these goals can usefully be achieved without taking into account the consumer acceptability and palatability of the meat produced (see Woods and Warriss, 1992). To these ends, more information is needed about the fundamental processes involved in muscle growth.

Muscle is itself not an homogenous tissue but is composed of many cell types, including fibroblasts, capillary endothelial cells, adipocytes and mononucleate satellite cells, in addition to muscle fibres. In rat muscles these other nuclei are about 35% of the total (Enesco and Puddy, 1964). However muscle fibres account for the major portion of muscle protein. Muscle fibres differ from the classical picture of "an animal cell" in that they are multinucleate and are formed by the fusion of mononucleate muscle precursor cells; this process makes differentiation irreversible. Mature fibres may grow up to several centimetres in length but remain only a few hundred microns in diameter.

The existence of two distinct populations of muscle cells, the fused and the unfused, complicates the analysis of muscle growth and blurs the distinction between cell enlargement (hypertrophy) and cell proliferation (hyperplasia). In contrast to some other tissues there can be increased cell proliferation in parallel with increased accumulation of muscle-specific protein. This review will cover the major processes of muscle growth and how they are affected by the manipulation of whole animal growth.

Patterns of muscle cell development

EMBRYONIC AND FOETAL GROWTH

The factors which control the entry of an embryonic cell into the muscle forming (myogenic) lineage are the subject of much current research. It is likely that one or more genes are activated whose products allow the subsequent expression of muscle cell characteristics (Blau and Baltimore, 1991) but how this "programming" occurs is not yet fully explained.

A family of genes has been identified, which code for helix-loop-helix DNA binding proteins and whose transfection into fibroblasts will convert them into myoblasts (Davis, Weintraub and Lassar, 1987; Braun, Buschhausen-Denker, Bober, Tannich and Arnold, 1989; Wright, Sassoon and Lin, 1989; Braun, Bober, Winter, Rosenthal and Arnold, 1990). They include MyoD and myogenin. These genes are thought to positively autoregulate each other, so stabilising commitment to the myogenic pathway (Thayer, Tapscott, Davis, Wright, Lassar and Weintraub, 1989). Some negative regulators of these genes have also been identified (Gossett, Zhang and Olson, 1988; Kelvin, Simard, Sue-A-Quan and Connolly, 1989; Benezra, Davis, Lockshon, Turner and Weintraub, 1990; Miner and Wold, 1991). Unrelated genes, of the *ski* family also have important roles (Colmenares and Stavnezer, 1989) and transfection of these will also convert fibroblasts to myoblasts (Pearson-White, 1991). Mice carrying the *ski* transgene have hypertrophied skeletal muscles, but interestingly only one type of muscle fibre is selectively enlarged (Sutrave, Kelly and Hughes, 1990). The activity of these regulatory genes is probably controlled by a variety of environmental influences including locally produced growth factors but the exact sequence of events is not yet known.

In the embryo and foetus, myoblasts divide and the volume of the muscle tissue increases. At a gestational stage characteristic of the species concerned, alignment and fusion of these cells begins. As myoblasts become committed to fuse, their patterns of RNA transcription and protein synthesis change (Devlin and Emerson, 1978; Garrels, 1979; Zevin-Sonkin and Yaffé, 1980; Shani, Zevin-Sonkin, Saxel, Carmon, Katcoff, Nudel and Yaffé, 1981). In the myotube muscle-specific proteins accumulate, such as α-actin, myosins, troponin, tropomyosin, desmin, the M isoform of creatine kinase, glycogen phosphorylase and dystrophin (Prives, Silman and Amsterdam, 1976; Whalen, Butler-Browne and Gros, 1976; Allen, Stromer, Goll and Robson, 1978; Devlin and Emerson, 1978; Garrels, 1979; Cossu, Zani, Coletta, Bouchè, Pacifici and Molinaro, 1980; Gard and Lazarides, 1980; Senni, Castrignano, Polana, Cossu, Scarsella and Biagioni, 1987; Miranda, Bonilla, Martucci, Moraes, Hayes and Dimauro, 1988). The synthesis of other proteins, such as ß- and γ-actin is repressed, whilst some such as fibronectin remain unchanged (Gardner and Fambrough, 1983). These changes are observed both *in vivo* and in myoblasts cultured *in vitro*.

The myotubes formed first are known as primary myotubes; secondary myotubes form at later stages round these. In the chick, where detailed studies have been carried out it is found functional innervation is needed for this to occur normally (Crow and Stockdale, 1986). All the multinucleate cells become enclosed in connective tissue sheaths, which are linked together into the framework of the muscle. The collagens of the basement membranes appear to be secreted both by growing myoblasts and by muscle fibroblasts (Beach, Burton, Hendricks and Festoff, 1982; Kuehl, Timpl and Von der Mark, 1982).

As development proceeds, the myotubes become innervated. Initially, several different motor neurones may synapse with a single syncytium but progressively these contacts are eliminated so that in most muscles groups of fibres with the same contraction properties are controlled by single nerves (Gordon and van Essen, 1985). Evidence from *in vitro* co-culture studies suggests that it is at least in part the influence of nerves which converts myotubes into the highly ordered muscle fibres with marginated nuclei characteristic of the post-natal animal. Direct contact between cholinergic neurones and myotubes increases the synthesis and deposition of myosins (Defez and Brachet, 1985; Ecob-Prince, Jenkinson, Butler-Browne and Whalen, 1986) and muscle-specific enzymes (Miranda, Peterson and Masurovsky, 1988). Some changes are perhaps caused by the physical presence of the nerve cell on the structure of the neighbouring muscle, since it has been reported that clustering of acetyl choline receptors can be achieved experimentally with basic polypeptide coated latex beads (Peng, Gao, Xie and Zhao, 1988). Culture medium from nerve cultures can also produce this effect, suggesting that a diffusible controlling factor may be also involved (Thompson and Rapoport, 1988). Denervation studies *in vivo* have shown that nerves regulate the expression of a number of muscle proteins (Leberer and Pette, 1988; Loirat, Lucas-Heron, Ollivier and Leoty, 1988; Schiaffino, Gorza, Pitton, Saggin, Ausoni, Sartore and Lømo, 1988; Leyland, Turner and Beynon, 1990).

POSTNATAL GROWTH

In placental mammals, fibre number is fixed by the time of birth, or within a short time thereafter, depending on the maturity of the species concerned. This is also true of avian species on hatching but in fish the situation is somewhat different (Stickland, 1983; Koumans, Akster, Dulos and Osse, 1990). During postnatal growth, muscle fibres increase in both width and length (Cardasis and Cooper, 1976; Rehfeldt, Fiedler and Wegner, 1987). DNA content of a muscle may increase anything from 2-100 fold (see review by Young, 1985) and the protein:DNA ratio also increases (Winick and Noble, 1965; Cardasis and Cooper, 1976; Trenkle, DeWitt and Topel, 1978). However, the nuclei within muscle fibres have not been observed to divide: radioactive pulse-chase experiments have shown that new nuclei within fibres originate from satellite cells which are a remaining mononucleate myogenic population in the adult (Moss and Leblond, 1971). These cells are characteristically small and spindle-shaped and are enclosed within the basement membranes of fibres (Mauro, 1961): they may be metabolically active or quiescent. It is not clear whether they are a residue of the myoblast population in which fusion has been suppressed or whether they represent an independent lineage in the foetus. They behave similarly to foetal myoblasts when cultured *in vitro* but some differences have been seen within a species in responsiveness to phorbol esters (Cossu, Molinaro and Pacifici, 1983) and insulin-like growth factors (Harper, Soar and Buttery, 1987; Roe, Harper and Buttery, 1989a). Differing forms of acetylcholinesterase are also secreted (Senni *et al*, 1987).

Satellite cells are relatively abundant in the young animal and decrease both as a proportion of total muscle nuclei and in absolute terms with maturation (Allbrook, Han and Helmuth, 1971; Cardasis and Cooper, 1976). However, they remain as a population which can migrate, proliferate and differentiate to form new muscle fibres in response to injury (Reznick, 1969). Proliferation of satellite cells continues only for a finite time as part of normal growth processes; there is a point at which it ceases abruptly, which coincides with the time lipid deposition in muscle starts (Figure 1) (Enesco and Puddy, 1964; Winick and Noble, 1965; Trenkle *et al*, 1978).

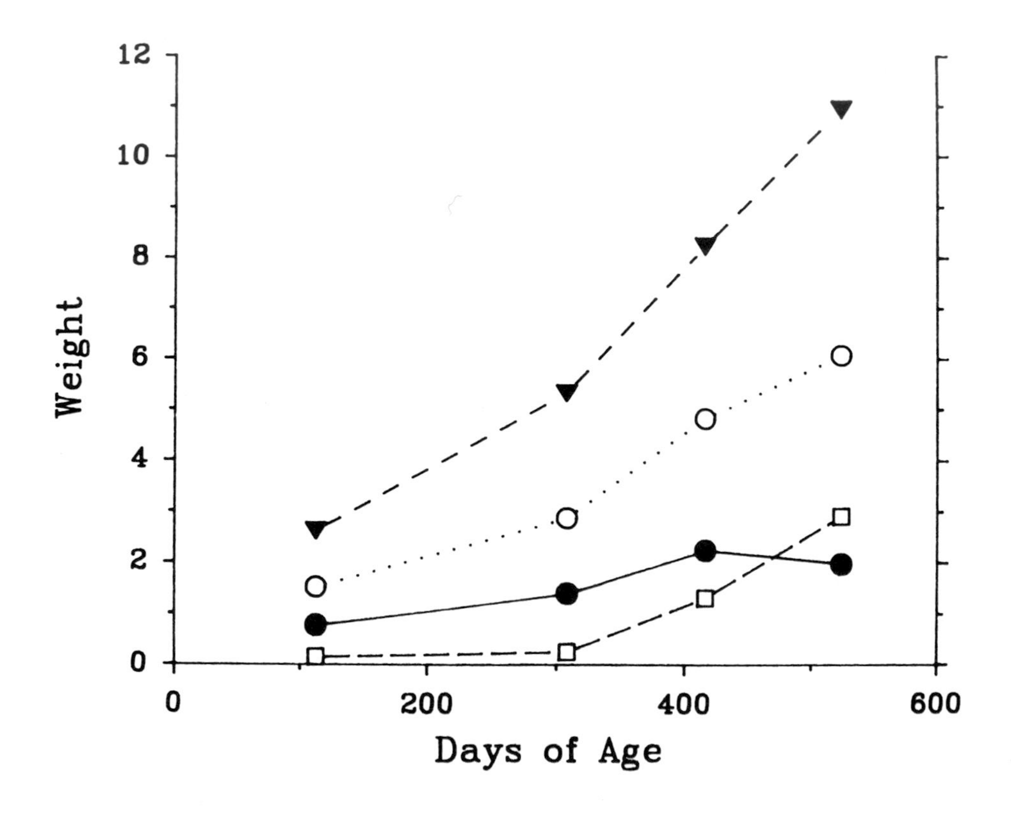

Figure 1 The growth of the longissimus muscle in Angus x Hereford steers (Data from Trenkle *et al*, 1978)

● DNA, g; ▾ Protein g (x 10^{-2}); □ Lipid g (x 10^{-2}); ○ Muscle weight kg

CONTROL OF FIBRE TYPE

In mammals, most skeletal muscle fibres can be divided into three types, according to their contraction speeds and metabolism: slow oxidative, fast oxidative and fast glycolytic (Ashmore, Tompkins and Doerr, 1972). In amphibians and birds other fibre types are also present (Rowlerson and Spurway, 1988). They can be distinguished histochemically by the properties of their myosin ATPase and by the presence or absence of mitochondrial oxidative enzymes. The myosin ATPase reaction is a reflection of the isoforms the fibres contain; the myofibrillar proteins actin and myosin both occur as multigene families. In the case of myosin heavy chain, characteristically foetal and neonatal isoforms have been described, in addition to slow and fast forms in most skeletal muscles, superfast forms in the jaw and tonic forms in the eye socket (Whalen, Schwartz, Bouveret, Sell and Gros, 1979; Rowlerson, Pope, Murray, Whalen and Weeds, 1981; Dix and Eisenberg, 1991). During development a succession of myosin isoforms are expressed which are characteristic of the muscle concerned; many muscles continue to express 'foetal' or 'neonatal' forms in the adult animal (see review by Stockdale and Miller, 1987).

In smaller animals there is a tendency for some muscles to be highly specialised, so that in the rat the soleus is composed predominantly of slow twitch fibres, whilst the *tibialis anterior* and *extensor digitorum longus* (EDL) are typically fast (Ariano, Armstrong and Edgerton, 1973). In the chicken the *anterior latissimus dorsi* (ALD) is slow whilst the *posterior latissimus dorsi* (PLD) is fast. In larger animals, especially ruminants, fibre type tends to be much more mixed, though there may be regional specialisation within muscles (see for example Totland, Kryvi and Slinde, 1988). The first fibres which form in the embryo/foetus initially contain slow myosins, irrespective of the eventual muscle type, secondary myotubes are mixed in their myosin content (Narusawa, Fitzsimons, Izumo, Nadal-Ginard, Rubinstein and Kelly, 1987). Most muscles increase their fast glycolytic (white) fibre content with age (Rehfeldt *et al*, 1987). Results of histological studies suggest that the type of a fibre may be predicted from its intramuscular position, its innervation and its "birth date" (Condon, Silberstein, Blau and Thompson, 1990).

It is a matter of debate whether the contractile characteristics of a muscle are determined primarily by innervation or by regulatory events within the genome of muscle cell lineages. Evidence from chickens and quail suggests that myogenic lineages are important in determining the eventual fibre type of the muscle. Myoblast colonies can be isolated from fast and slow muscles which express only fast or slow myosins on differentiation. These purified cells continued to express a consistent phenotype even though they were grown through many generations *in vitro* (Frémont, Fournier LeRay and Le Douarin, 1983; Miller and Stockdale, 1986). In cats, satellite cells from the jaw muscles continue to express superfast myosins characteristic of the muscle even when they regenerate in the absence of nerves or are transplanted into a fast leg muscle which normally does not express this isoform (Hoh and Hughes, 1991a, b).

In contrast Whalen, Harris, Butler-Browne and Sesodia (1990) found that regeneration of the rat soleus produced the normal fibre type only if functional nerves were present. Chronic electrical stimulation of the nerve supply caused a fast muscle to synthesise slow myosin isoforms (Streter, Elzinga, Mabuchi, Salmons and Luff, 1975). Cross-reinnervation experiments, especially in mammals, have demonstrated that if the nerve from a fast muscle is implanted on a slow muscle and *vice versa* these muscles will change their contractile properties and myosin isoforms over time to correspond with their new innervation (Barany and Close, 1971; Weeds, Trentham, Kean and Buller, 1974). These findings argue that nerves have an overriding influence on fibre type. However there are some indications that this control may be incomplete. Calcium-binding parvalbumin is normally at a high concentration in the fast EDL of the rat but very low in the soleus. On cross-reinnervation changes were seen in the expected directions but the concentration in the EDL was still higher than in the soleus and did not correlate with the histochemical and physiological properties of these muscles (Müntener, Rowlerson, Berchtold and Heizmann, 1987).

It is likely that a complex dual control system exists for the determination of muscle fibre type. It has been shown that, within a muscle fibre, transcription in individual nuclei may respond differently to stimulatory signals (*eg* acetylcholine receptor inducing factor: Harris, Falls and Fischbach, 1989). Myotubes may well contain nuclei from differing muscle lineages and it is an interesting speculation that this might account for the observed differences in transcription (Hall and Ralston, 1989). Might different nuclei be activated by differing patterns of neural stimulation, so leading to predominance of particular myosin isoforms?

The hormonal control of myogenic cell division

MYOGENIC CELL DIVISION

The list of hormones and growth factors which increase muscle precursor cell division is steadily growing: to date no major differences have been shown between foetal myoblasts and satellite cells of the same species in their growth factor responsiveness. However, some variation between species has been noted and there appears to be a wide gap between the behaviour of primary cell cultures and some continuous muscle cell lines.

Muscle cell mitogens have all been identified as a result of *in vitro* studies, in which the factor was added to cells in conjunction with a low concentration of serum. Blood serum itself contains a mixture of hormones and peptides which collectively have a potent mitogenic effect on most cell types in culture, including myogenic ones. Its inclusion in test media allows the detection of factors which may promote passage through a particular part of the cell cycle but which by themselves cannot cause complete cell division. For each factor it is important to assess the physiological importance of the response. Does it have bearing on normal muscle growth or is it part of an "emergency package" mobilised only in times of physical stress or injury?

Mitogenesis by fibroblast growth factor (FGF) was initially described in bovine foetal myoblasts (Gospodarowicz, Wesemen, Moran and Lindstrom, 1976). Since then FGF has been shown to have similar effects in rat and bovine satellite cell cultures (Allen and Boxhorn, 1989; Greene and Allen, 1991); similar molecules are active on chick myoblasts (Kardami, Spector and Strohman, 1985). FGF is also a mitogen for some continuous muscle cell lines (MM-14: Linkhart, Clegg and Hauschka, 1981; BC_3H-1: Kelvin, Simard and Connolly, 1989) but not L6 cells (Florini, Ewton, Falen and Van Wyk, 1986). Both acidic and basic forms are active at low concentrations but basic FGF is the more potent (Olwin and Hauschka, 1986). The physiological relevance of the response is unclear since free FGF is not present in the tissue fluids of mammals but sequestered either inside cells or in extracellular matrix, where it binds heparin-like molecules (see review by Klagsbrun, 1989). However the release of active FGF from extracellular matrix by a heparanase from white blood cells has recently been described (Ishai-Michaeli, Eldor and Vlodavsky, 1990). Heparin has also been shown to enhance the mitogenic activity of acidic FGF in some cell types (Ulrich, Lagente, Lenfant and Courtois, 1986).

As with many other cell types, it is also found that insulin-like growth factors (IGFs) promote cell division. This has been demonstrated in satellite cells and myoblasts from rat, human, bovine and avian species (Schmid, Steiner and Froesch, 1983; Hill, Crace, Strain and Milner, 1986; Allen and Boxhorn, 1989; Duclos, Wilkie and Goddard, 1991; Greene and Allen, 1991). Both IGF-I and IGF-II are active at nanomolar concentrations. It seems that the effects of both are mediated through the Type I IGF receptor in L6 cells (Ewton, Falen and Florini, 1987; Kiess, Haskell, Lee, Greenstein, Miller, Aarons, Rechler and Nissley, 1987) and in chick myoblasts (Duclos *et al*, 1991). IGFs are secreted in culture by foetal myoblasts from rats and humans (Hill, Crace and Milner, 1985). During regeneration of damaged muscle in postnatal life, polysomal IGF peptide is seen in activated satellite cells (Jennische and Hansson, 1987). In foetal development, growing muscle contains IGF peptides though most IGF mRNA synthesis appears to occur in the surrounding connective tissue cells (Han, D'Ercole and Lund, 1987; Han, Hill, Strain, Towle, Lauder, Underwood and D'Ercole, 1987). There is therefore a plausible paracrine or autocrine role for IGFs in muscle development and regeneration *in vivo*, in addition to any effect which may be ascribed to circulating IGFs. Growth hormone was implicated as a mitogen in some early studies *in vitro* (de la Haba, Cooper and Elting, 1968) but it has subsequently been shown that pure preparations lack direct effects on muscle cell division (Gospodarowicz *et al*, 1976).

Insulin also promotes cell division in myogenic precursor cells including bovine myoblasts (Gospodarowicz *et al*, 1976), perinatal rat myoblasts (Crace, Hill and Milner, 1985), rat satellite cells (Dodson, Allen and Hossner, 1985), chick satellite cells (Duclos *et al*, 1991) and L6 cells (Florini and Roberts, 1979); however it is often active *in vitro* only at concentrations very much higher than those typically found in blood plasma. It seems plausible that this effect is mediated by cross-reaction with the Type I IGF receptor (Florini and Ewton, 1981; Duclos *et al*, 1991). In human foetal myoblasts, insulin has no direct effect on DNA synthesis (Hill *et al*, 1986).

Variations between cells of different species are also evident in the case of transforming growth factor beta (TGF-beta). This factor, which has major effects on cell differentiation (see section below), has been reported to inhibit proliferation in the L6 rat cell line, rat and bovine satellite cells and rat and porcine myoblasts (Allen and Boxhorn, 1987, 1989; Pampusch, Hembree, Hathaway and Dayton, 1990; Greene and Allen, 1991) but not in the C2 or BC_3H-1 cell lines or ovine satellite cells (Olson, Sternberg, Hu, Spizz and Wilcox, 1986; Hathaway, Hembree, Pampusch and Dayton, 1991). In combination with FGF, TGF-ß is a mitogen for satellite cells from cattle but not those from rats (Allen and Boxhorn, 1989; Greene and Allen, 1991).

Allen and co-workers, in the studies mentioned above, have also examined the ability of FGF and IGFs to promote cell division in primary muscle cultures in the absence of serum. They concluded that FGF could cause cell division in a defined medium containing dexamethasone, low concentrations of insulin and transferrin; however its effects were greatly enhanced by the addition of IGF-I. IGF-I in serum-free medium was weakly mitogenic for rat satellite cells but not those from cattle. This is in contrast to the L6 rat continuous cell line, where IGFs appear to be the principal mitogen and response to FGF is not seen (Florini and Roberts, 1979). It is possible that in L6 cells coupling of cell surface receptors is altered: they contain an unusual receptor for IGF-I (Burant, Treutelaar, Allen, Sens and Buse, 1987). Rozengurt and co-workers have shown that in a fibroblast cell line, any combination of factors which activates all the necessary second messenger pathways will cause DNA synthesis (reviewed by Rozengurt and Mendoza, 1985). Results from the BC_3H-1 line have led to a model in which FGF is responsible chiefly for the initiation of the cell cycle (progression into G1, figure 2) whilst other factors act at later stages (Lathrop, Thomas and Glaser, 1985; Kelvin, Simard and Connolly, 1989); it should however be remembered that BC_3H-1 cells are not completely normal myoblasts.

Recent studies in our laboratories have shown that epidermal growth factor (EGF) is a mitogen for clonally purifed foetal sheep myoblasts and have given some indications that this is also true for ovine satellite cells (Heywood, Haji Baba, Harper and Buttery, 1991). This is in conflict with earlier studies in bovine and rat myoblasts (Gospodarowicz and Mescher, 1977; Allen, Dodson, Luiten and Boxhorn, 1985) and L6 cells (Florini, Ewton, Falen and Van Wyk, 1986) but agrees with findings in the BC_3H-1 cell line (Kelvin, Simard and Connolly, 1989; Kelvin, Simard, Sue-A-Quan and Connolly, 1989). EGF is present, albeit at low levels, in muscle (Frati, Cenci, Sbaraglia, Venza Teti and Covelli, 1976) and in blood plasma and is released from blood platelets during clotting (Oka and Orth, 1983), so a physiological role is possible.

Recent studies by other workers show that ACTH (adrenocorticotropin) and α- ß- and γ- MSH can increase cell division in mouse myoblast cultures (Cossu, Cusella-DeAngelis, Senni, DeAngelis, Vivarelli, Vella, Bouchè, Boitani and

Molinaro, 1989). This finding was particularly interesting because whereas FGF, IGFs and EGF also cause proliferation of muscle fibroblasts, the authors report that these hormones acted only on myoblasts. The first 24 amino acids of ACTH have independently been found to stimulate 2-deoxyglucose uptake in L6 cells (Levin, Bistritzer, Hanukoglu, Max and Roeder, 1990). Leukaemia inhibitory factor and interleukin-6 have also been implicated as mitogens in mouse satellite cells (Austin and Burgess, 1991). Transferrin promotes division of muscle cells *in vitro* but since the apo-protein is inactive, whereas haem derivatives do promote myoblast growth, it is likely that this reflects a nutritional requirement for iron (Verger, Sassa and Kappas, 1983; Popiela, Taylor, Ellis, Beach and Festoff, 1984; Matsuda, Spector and Strohman, 1984; Byatt, Schmuke, Comens, Johnson and Collier, 1990). In contrast, the structurally related lactoferrin has been shown to have mitogenic activity in L6 cells even in its apoprotein form (Byatt *et al*, 1990). In chick satellite cells ractopamines and a beta-adrenergic agonist (isoproterenol) have also been reported as mitogens (Grant, Helferich, Merkel and Bergen, 1990) but other workers failed to find an effect of beta agonists on chick myoblasts (Young, Moriarty, McGee, Farrer and Richter, 1990).

The role of glucocorticoids in muscle cell proliferation is an intriguing one. Whilst they have little effect on their own they are essential for the optimum growth of both muscle cell lines and primary cells in serum-free culture (Florini and Roberts, 1979; Florini and Ewton, 1981; Allen *et al*, 1985; McFarland, Pesall, Norberg and Dvoracek, 1991). Evidence in two continuous muscle cell lines shows they up-regulate the type I IGF receptor (Whitson, Stuart, Huls, Sams and Cintron, 1989).

Apart from soluble growth factors and hormones, components of the extracellular matrix could also be regulatory for growth; laminin has already been found to stimulate migration and division of MM-14 cells (Öcalan, Goodman, Kühl, Hauschka and von der Mark, 1988), whilst fibronectin promotes muscle cell attachment (Chiquet, Puri and Turner, 1979).

Undoubtedly, more factors also remain to be identified. For example, serum from fasted pigs contains a protein which inhibits the proliferation of the L6 cell line *in vitro* (White, Kretchmar, Allen and Dayton, 1989). Another significant instance comes from culture studies from double-muscled cattle (Quinn, Ong and Roeder, 1990). It was found that muscle cells from foetuses of the double-muscled genotype proliferated faster and for longer *in vitro* than controls; this effect could be attributed to a stimulatory factor secreted by muscle fibroblasts, which was produced in larger than normal amounts by the fibroblasts from double-muscled animals.

It is possible that this factor could be intimately associated with the genetic determination of growth potential. A study by Penney, Prentis, Marshall and Goldspink (1983) examined two strains of mice, selected for large and small body size. It showed that the large animals had a greater number of fibres in all muscles examined than the small ones, which was reflected in a calculated greater number of muscle nuclei. This was attributable to a greater rate of myoblast proliferation in foetuses of the large genotype. Studies comparing broiler and layer strains of chicken also found a greater rate of myoblast proliferation *in vitro* in the broiler strain (Ridpath, Huiatt, Trenkle, Robson and Bechtel, 1984). Conversely, the poor muscle fibre growth seen in runting is a result of diminished myoblast proliferation *in utero* (Aberle, 1984, and see Bell, 1992). Myoblast proliferation before birth is a critical determinant of muscle growth in later life. Stimulation of satellite cell division may be an important component in the mode of action of some anabolic treatments (see below).

MUSCLE CELL DIFFERENTIATION

The control of muscle precursor cell differentiation is vital to the growth of an animal since it affects the balance between cell proliferation and the production and enlargement of mature muscle fibres. It also ensures the maintenance of a reserve of satellite cells in the older animal. For a long time, there was a debate as to whether differentiation was a process requiring a final round of DNA synthesis, to commit irrevocably a cell to this pathway (Holtzer, Sanger, Ishikawa and Strahs, 1972). Current evidence suggests this is not the case but rather the length of time a cell spends in G0/G1 of the cell cycle (Figure 2), before DNA synthesis, is critical in determining whether it will re-enter the cycle or differentiate (Nadal-Ginard, 1978). Changes in transcription begin to occur before cell fusion and can be reversed if the hormonal environment of the cell is altered so as to promote proliferation (Emerson and Beckner, 1975; Zevin-Sonkin and Yaffé, 1980; Nguyen, Medford and Nadal-Ginard, 1983).

FGF in particular has been found to regulate differentiation *in vitro*. It represses the expression of muscle-specific proteins, even under conditions non-permissive for DNA synthesis (Lathrop *et al*, 1985; Spizz, Roman, Strauss and Olson, 1986; Clegg, Linkhart, Olwin and Hauschka, 1987). In BC_3H-1 cells it inhibits the expression of the regulatory protein myogenin (Brunetti and Goldfine, 1990).

IGFs are also muscle cell mitogens but they have the reverse effect to FGF. In L6 cells fusion is increased by concentrations of IGF-I lower than those needed for cell division (Florini, Ewton, Falen and Van Wyk, 1986). This occurs irrespective of the cell density, by a mechanism independent of proliferation (Turo and Florini, 1982). In chick myoblasts too it is found that IGF-I and IGF-II increase the synthesis of acetylcholinesterase, a muscle-specific enzyme (Schmid *et al*, 1983). In bovine satellite cells, increases in desmin have similarly been reported (Greene and Allen, 1991). Synthesis of polyamines appears essential to IGF-induced differentiation, since the process can be blocked by inhibiting the enzyme ornithine decarboxylase (Ewton, Erwin, Pegg and Florini, 1982). IGFs increase transcription of the regulatory protein myogenin; antisense RNA to its message has been found to block IGF-induced differentiation in a strain of L6 cells (Florini, Ewton and Roof, 1991).

Insulin at high concentrations has similar effects to IGFs in chick and rat cells but not in ovine satellite cells (Dodson, Mathison, Brannon, Martin, Wheeler and McFarland, 1988). ACTH appears similar to IGFs in promoting both proliferation

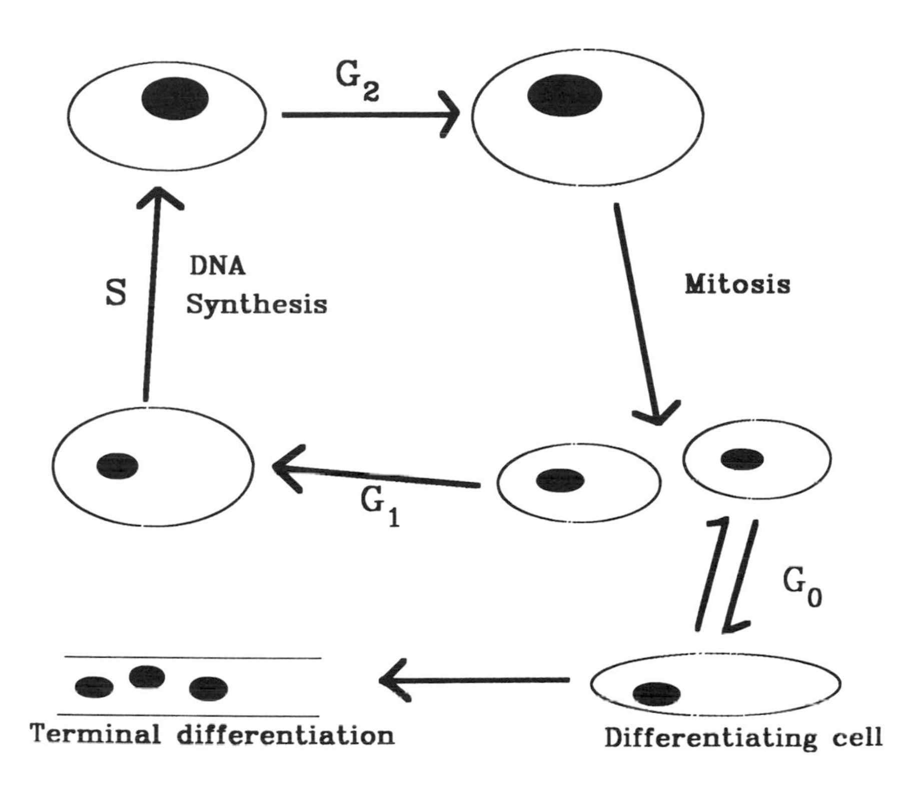

Figure 2 The cell cycle in myogenic cells

and differentiation (Cossu *et al*, 1989) but its mode of action is not yet known.

The transforming growth factor beta (TGF-ß) family of growth factors provides a third type of control on the differentiation process. TGF-ß has variously been reported to inhibit, to stimulate, or to have no effect on muscle cell division (see previous section) but in all cases studied it is a potent inhibitor of differentiation and fusion, even in the absence of mitogens. It over-rides the IGF mediated induction of differentiation (Florini, Roberts, Ewton, Falen, Flanders and Sporn, 1986; Massagué, Cheifetz, Endo and Nadal-Ginard, 1986; Olson *et al*, 1986; Allen and Boxhorn, 1987; Hathaway *et al*, 1991). It acts at an early stage in the differentiation pathway so that, for example, the transcription of creatine kinase mRNA is blocked (Spizz, Hu and Olson, 1987). Under the influence of TGF-ß the transcription, synthesis and secretion of extracellular matrix proteins, especially collagens and fibronectin, are increased (Ignotz, Endo and Massagué, 1987). Like

Table 1 CHANGES IN HORMONE RECEPTOR NUMBER WITH DIFFERENTIATION

Receptor	Cell type	Change	Reference
IGF Type I	BC3H-1	↓	Rosenthal, Brunetti, Brown, Mamula & Goldfine (1991)
	C2	↑ (transient)	Tollefsen, Sadow & Rotwein (1989)
	L6	↓	Beguinot, Kahn, Moses & Smith (1985)
		→	Ewton, Spizz, Olson & Florini (1988)
	turkey satellite	↓	Minshall, McFarland & Doumit (1990)
IGF Type II	C2	↑	Tollefsen, Sadow & Rotwein (1989)
	L6	↓	Beguinot *et al* (1985)
Insulin	BC3H-1*	↑	Standaert, Schimmel & Pollet (1984)
		↑	DeVroede, Romanus, Standaert, Pollet, Nissley & Rechler (1984)
	L6	↑	Klip, Li & Walker (1983)
		↑	Beguinot *et al* (1985)
	L6-A1	↓ (slight)	Ewton *et al* (1988)
	chick myoblast	↑	Sandra & Przybylski (1979)
TGF-beta	BC3H-1*	↓ (slight)	Ewton *et al* (1988)
	C2	↓	Ewton *et al* (1988)
		↓	Hu, Spizz & Olson (1987)
	DD-C2*	→	Hu *et al* (1987)
	L6-A1	↓	Ewton *et al* (1988)
	L6-E9	→	Massagué *et al* (1986)
	L8	→	Massagué *et al* (1986)
FGF	BC3H-1*	→	Olwin & Hauschka (1988)
	C2	↓	Hu *et al* (1987)
	C2C12	↓	Olwin & Hauschka (1988)
	MM14	↓	Lim & Hauschka (1984)
		↓	Olwin & Hauschka (1988)
	DD-MM14*	→	Olwin & Hauschka (1988)
EGF	C2	↓	Hu *et al* (1987)
	DD-C2*	→	Hu *et al* (1987)
	MM14	↓	Lim & Hauschka (1984)
		↓	Olwin & Hauschka (1988)
Beta-adrenergic	chick myoblast	↑	Parent, Tallman Hennebury & Fishman (1980)

* non-fusing cell lines

FGF, TGF-ß is rarely found free in the body but tends to complex with matrix proteins, especially decorin, which acts as a negative regulator (Yamaguchi, Mann and Ruoslahti, 1990). It has been speculated that matrix-bound FGF and TGF-ß could act together to preserve satellite cells in their quiescent state.

The role of glucocorticoids in muscle cell differentiation is unclear. One study in L6 cells reported that cortisol acted synergistically with insulin to promote fusion and creatine kinase accumulation (Ball and Sanwal, 1980). A second showed that dexamethasone inhibited fusion in this cell line (Schonberg, Smith, Krichevsky and Bilezikian, 1981). The key difference between the studies probably lies in the culture conditions used.

Several groups of workers have made studies of changes in hormone receptor numbers in continuous myoblast cell lines during differentiation. As can be seen from Table 1, results have sometimes been contradictory between cell types and even between strains of L6 cells. The difficulties involved in measuring receptor number should not be under-estimated in a system where substantial changes in cell geometry occur on fusion. There is a consistent decrease in the numbers of EGF and FGF receptors in fusing cells, whereas in non-fusing cell types (marked *) this does not occur: this suggests that changes in RNA transcription as such are not responsible. However, increases in insulin receptor number are seen in both fusing and non-fusing cells. Studies on IGF receptor populations have yielded the most inconsistent results suggesting that, if they do occur, they may not be essential to the differentiation process.

THE CONTROL OF PROTEIN DEPOSITION *IN VITRO*

Whilst many studies have examined the relationship of hormone concentrations to muscle protein deposition in the whole animal, relatively few have looked at the range of factors to which the individual muscle cell is responsive. Confluent, differentiated cultures of muscle cells can be used as a model to study this, bearing in mind of course that these are aneural and therefore immature in relation to muscle fibres *in vivo*.

Classically insulin is regarded as the major hormone with anabolic effects on muscle (see Reeds and Davis, 1992). Studies in the isolated epitrochlearis muscle from rats showed increase in protein synthesis at concentrations within the physiological range (1nM), after an initial delay (Stirewalt and Low, 1983). Effects at the same insulin concentration were seen in isolated forelimb muscles from rabbits (Reeds and Palmer, 1983). In mice the soleus responded but not the EDL (Ballard, Nield and Tomas, 1983). However, effects at physiological concentrations of insulin have been seen in remarkably few of the cell culture systems studied. A notable exception is the BC3H-1 cell line (Standaert *et al*, 1984) but this is non-fusing and in many respects intermediate between smooth and striated muscle cells in its properties. In L6 cells (Ballard and Francis, 1983) or in sheep satellite cells (Roe *et al*, 1989a) concentrations of 5-10nM (approx 1mU/ml) or above are needed to see significant increases in protein synthesis or decreases in breakdown. In human muscle myotubes, some effects on nutrient transport has been noted with insulin concentrations in the nanomolar range (Shimizu, Webster, Morgan, Blau and Roth, 1986) and insulin-responsive (Glut-4) glucose transporters have been detected in muscle membranes (Douen, Burdett, Ramlal, Rastogi, Vranic and Klip, 1991). It is nevertheless interesting to speculate whether some of the effect seen on muscle *in vivo* might be achieved indirectly by alterations in the permeability barrier presented by the capillary endothelial cells; endothelial cells possess numerous insulin receptors (Jialal, Crettaz, Hachiya, Kahn, Moses, Buzney and King, 1985; Vilaro, Palacin, Pilch, Testar and Zorzano, 1989).

In contrast, effects of IGF-I are seen at concentrations which are in good agreement with the sensitivity of other biological systems - with half-maximal effects occurring in the low nM range (see Czech, 1989). How this relates to biologically available concentrations in muscles is not known because of the almost complete binding of IGFs to their carrier proteins. Up to five different carrier proteins have now been identified in human plasma (Liu, Powell and Hintz, 1990) and each has a characteristic distribution between blood and tissue fluids. Low molecular weight IGF binding proteins are also secreted by differentiating primary myoblasts and myogenic cell lines (Hill, Crace, Fowler, Holder and Milner, 1984; Hill, Crace, Nissley, Morrell, Holder and Milner, 1985; McCusker and Clemmons, 1988; Tollefsen, Lajara, McCusker, Clemmons and Rotwein, 1989). There is species variation in the exact molecular weight of the binding proteins, so that cross-identification is difficult but it is clear from work in fibroblasts that at least one of the smaller ones can act to enhance IGF action *in vitro* (Elgin, Busby and Clemmons, 1987) whilst others may be inhibitory (DeMellow and Baxter, 1988; Mohan, Bautista, Wergedal and Baylink, 1989). To date, there are no published studies of the effects of carrier proteins on IGF action in cultured skeletal muscle cells. In cultures derived from ovine satellite cells, IGF-I increases protein synthesis by about 20% and decreases breakdown; amino acid and glucose uptake are also enhanced (Roe *et al*, 1989a; Roe, Harper and Buttery, 1989b). In foetal myoblast cultures, results for protein synthesis are similar but those for breakdown are more variable: small decreases or increases in breakdown may be seen but more usually there is no effect (Harper, Roe, Heywood and Buttery, 1987; J. Harper, unpublished observations). This indicates that the two processes are probably regulated by independent intracellular pathways. Protein metabolism in L6 cells is also IGF responsive (Roeder, Blann, Bauer and Hossner, 1986; Harper, Soar and Buttery, 1987). Both IGF-I and IGF-II cause enhancement of protein synthesis and nutrient transport via the type I IGF receptor (Ewton *et al*, 1987; Kiess *et al*, 1987).

Somatostatin has been reported to inhibit the insulin-stimulated component of amino acid uptake and protein synthesis in rat muscle cells (Spencer, Hill, Garssen and Williams, 1987). However we have found this result hard to repeat (Haji Baba, Hawkey, Buttery and Fletcher, 1991). As with cell division studies, growth hormone does not have any direct effects on protein metabolism in cultured muscle cells (Harper, Soar and Buttery, 1987; Roe *et al*, 1989a). Acute insulin-like effects on nutrient transport and protein synthesis have been reported in isolated rat diaphragm (Cameron, Kostyo, Adamafio, Brostedt, Roos, Skottner, Forsman, Fryklund and Skoog, 1988) but the possibility cannot be excluded that its effects were mediated by locally produced IGFs. Recently, specific GH receptors detected a low abundance in the myoblast-like BC_3H-1 cell line. Insulin-like effects were seen on glucose transport, but only at a high concentration (0.1 μM; Adamafio, Towns and Kostyo, 1991).

Epidermal growth factor (EGF) has proved to be a potent anabolic agent for both foetal and postnatal muscle cultures from the sheep, with marked increases (16%)in protein synthesis and decreases in breakdown (Harper, Soar and Buttery, 1987; Roe *et al*, 1989a). Only low concentrations of the factor are needed: half-maximal effects are seen with 0.1 nM. The effects of saturating concentrations of EGF and IGF-I are additive, showing that they act through independent second messenger systems. EGF increases the creatine kinase content of cultures of both foetal and postnatal origin, showing that it is active on the differentiated muscle cells in the cultures (Roe *et al*, 1989a; J. Harper and C. Heywood, unpublished observations). Similar results have been reported in abstract form for the addition of EGF to growing cultures of human muscle cells (Askanas, Gallez-Hawkins and King Engel, 1984). These results are an example of *in vitro* findings which would not easily be predicted from existing knowledge *in vivo*. Protein metabolism in L6 cells does not respond to EGF (Harper, Soar and Buttery, 1987) and we have found they lack EGF receptors (Heywood, Harper and Buttery, 1989).

The catabolic effect of glucocorticoids on muscle is well known: they enhance degradation of myofibrillar proteins especially in fast fibres (Kayali, Young and Goodman, 1987; Seene, Umnova, Alev and Pehme, 1988). In cultured muscle cells they increase protein breakdown (Ballard and Francis, 1983; Roeder, Thorpe, Byers, Schelling and Gunn, 1986; Harper, Roe, Heywood and Buttery, 1987). However they can actually increase protein synthesis in the presence of growth factors (Roeder, Thorpe, Byers, Schelling and Gunn, 1986; Palmer, Bain and Southorn, 1990) and increase glutamine synthetase in L6 cells (Feng, Hilt and Max, 1990). It can be demonstrated *in vitro* that the action of dexamethasone, a synthetic glucocorticoid, is antagonistic to that of insulin: it decreases the sensitivity of protein synthesis in L6 cells to the stimulatory effects of insulin (Palmer *et al*, 1990) and insulin decreases the sensitivity of their protein breakdown to dexamethasone (H. Greathead, J. Harper and P.J. Buttery, unpublished results). Unlike peptide hormones, whose effect is immediate, steroids are found to take some hours to show their effects (Ballard and Francis, 1983; Roeder, Thorpe, Byers, Schelling and Gunn, 1986).

The actions of known anabolic agents have been tested on muscle cells in culture. Roeder, Thorpe, Byers, Schelling and Gunn (1986) reported that protein synthesis in L6 myoblasts was increased with progesterone, and in myotubes with oestradiol. However the high concentrations used (1 μM) make it unlikely that the effect was physiologically relevant. We have not seen any effect on protein synthesis using 10 nM oestradiol in foetal ovine muscle cultures. Ballard and Francis (1983) did not see any effects of oestrogens or androgens on protein degradation in L6 cells.

Varying results have been reported with a number of beta-adrenergic agonists used as repartitioning agents. *In vivo* treatment with these agents usually leads to muscle growth through a decrease in protein breakdown, but a short-term increase in synthesis is also sometimes observed. All muscles do not respond equally (see Lindsay, Hunter and Sillence, 1992). Some authors report an absence of effects in muscle culture (McElligott, Chaung and Barreto, 1989). However others, using cimaterol in chick cells, have shown a shift towards myofibrillar protein synthesis and a decreased degradation rate after long-term (6 days) exposure to the drug in the presence of serum (Young *et al*, 1990). In two rodent muscle cell lines, L6 and G8-1, protein synthesis was increased during short-term co-incubation with cimaterol but protein breakdown was unaltered over an 18 hour exposure period (Harper, Mackinson and Buttery, 1990). The same effects were seen with ovine satellite cell-derived cultures but not foetal myoblasts (Symonds, Roe, Heywood, Harper and Buttery, 1990). Similarly, in foetal bovine or embryonic chick myotubes, cimaterol had no effect on protein synthesis, though here protein degradation was accelerated (Béchet, Listrat, Deval, Ferrara and Quirke, 1990). Thus, in the case of ß-agonists, use of *in vitro* protein metabolism studies has revealed the likelihood that there are both direct and indirect components in the response of muscle.

With single peptide hormones and growth factors in culture an increase in protein synthesis is often accompanied by a decrease in protein breakdown, whereas in the whole animal muscle protein synthesis and breakdown more usually rise and fall together, albeit to slightly differing extents. One of the tasks facing cell biologists in the future will be to find out how the responses to individual factors integrate to give the total response seen *in vivo* and to what extent, for example, the slower acting steroid hormones are involved.

Mechanical factors and muscle growth

It is well known from the field of human experience that physical activity produces enlargement of muscles: this is a local effect on the muscles involved and therefore cannot be attributed primarily to endocrine changes. However, acute exercise does increase insulin binding to muscle. This change is probably mediated by increased circulating catecholamine concentrations (Webster, Vigna and Paquette, 1986). Experiments with rats have shown that when a denervated limb is immobilised, muscles which are so stretched enlarge, whilst those which are relaxed atrophy. This is reflected in measured rates of protein synthesis (Goldspink, 1978). Even in the mature animal, muscle fibres adjust their length and myofibrillar protein content in proportion to the load they usually bear. Recent studies in the suspended rat hindlimb show that the soleus, a slow 'postural' muscle becomes faster as judged by its myosin ATPase when inactive (Diffee, Caiozzo, Herrick and Baldwin, 1991).

Passive stretch also increases protein synthesis and deposition in cultured myotubes (Vandenburgh, 1983). In many primary culture systems, especially chick myoblasts, spontaneous contraction of myotubes is seen. If this is blocked by tetrodotoxin treatment then accumulation of a number of myofibrillar proteins is decreased as a result of increased degradation (Crisona and Strohman, 1983). A number of studies have been carried out to enhance spontaneous contractions by electrical stimulation of cultures. This is a technically difficult procedure and results have been variable but at least in one case an increase in myofibrillar protein deposition was seen (Brevett, Pinto, Peacock and Stockdale, 1976) and in others synthesis of acetylcholinesterase was decreased (Walker and Wilson, 1975; Linden and Fambrough, 1979).

It therefore appears that the tension in an individual muscle fibre can by some means be relayed internally to control rates of protein synthesis and degradation in the cell. The details of how this is achieved are not yet known, though activation of a number of membrane-based second messenger pathways appears to be involved. It has been shown that electrical stimulation of muscle leads to the formation of diacylglycerol and the activation of protein kinase C (Cleland, Appleby, Rattigan and Clark, 1989). Phorbol esters, which activate protein kinase C cause increased rates of protein synthesis in isolated mouse hindlimb muscles (MacLennan, McArdle and Edwards, 1991). In isolated rabbit forelimb muscles intermittent stretching increases protein synthesis, this effect can be blocked by inhibitors of the enzyme cyclooxygenase, suggesting that the arachidonic acid pathway is also involved. Prostaglandins A_1 and $F_{2\alpha}$ directly increase protein synthesis (Smith, Palmer and Reeds, 1983). In overloaded rat hindlimb muscles local IGF mRNA transcription is induced, so autocrine mechanisms are also possible (DeVol, Rotwein, Sadow, Novakofski and Bechtel, 1990).

In animals trained for endurance exercise there is an increase in the number of fibres histologically staining as oxidative (Faulkner, Maxwell, Brook and Lieberman, 1971; Rehfeldt and Bunger, 1990) and in man too training for long-distance running increases the activity of the mitochondrial enzyme succinic dehydrogenase (Costill, Fink and Pollock, 1976). An increase in mitochondrial number and size is responsible (Holloszy, 1967; Gollnick and King, 1969). It should be emphasised, however, that significant changes of this kind are not seen with mild to moderate exercise.

This is not the whole story of muscle activity, as shown by some interesting results in guinea pigs (Faulkner, Maxwell and Lieberman, 1972). One group was trained from a young age on a treadmill and compared with a second group which led a sedentary life. After seven weeks muscle fibre diameter was increased and exercised guinea pigs had more fibres in their plantaris muscle than controls. However this was apparently caused by a prevention of natural fibre atrophy and loss, rather than the growth of new fibres. After the active animals ceased training, their muscle characteristics rapidly changed and became indistinguishable from controls.

Other studies have followed the hypertrophy of the ALD in chickens following its passive stretch by weighting the wing. It is clear that in this case, activation of satellite cells occurs, as well as increased protein deposition (Laurent, Sparrow and Millward, 1978). Some workers have shown the formation of new fibres in response to overload but this is a matter of debate (Sola, Christensen and Martin, 1973; Barnett, Holly and Ashmore, 1980; Kennedy, Eisenberg, Reid, Sweeney and Zak, 1988; Winchester, Davis, Alway and Gonyea, 1991): it may depend on the exact experimental treatment used and how much damage is inflicted on the muscle. A recent study (Winchester *et al*, 1991) followed the timecourse of the response to overload stretch and showed that satellite cell activation was a rapid but a transitory phase, preceding the increase in muscle protein. New fibre formation was not seen in this study. How the physiological state of the muscle is relayed to satellite cells is unclear but local IGF or prostaglandin production may be involved (Smith *et al*, 1983; DeVol *et al*, 1990). Mechanisms are easier to hypothesise in cases where vigorous exercise causes muscle damage and focal formation of new fibres does occur, since damaged muscle fibres contain satellite cell mitogens (Bischoff, 1990). Innervation is not needed for stretch-induced hypertrophy (Sola *et al*, 1973) and nerves normally seem to have an inhibitory rather than a stimulatory effect on satellite cell proliferation (Ontell, 1973).

Genetic selection and muscle growth

Studies in mice selected for high and low body weight have shown that larger animals have heavier muscles. This increase in mass is usually brought about by a combination of increases in fibre number, fibre diameter and fibre length, whilst sarcomere length is unaltered (Byrne, Hooper and McCarthy, 1973; Hooper, 1976; Hooper and Hurley, 1983; Penney *et al*, 1983). Similar results have been reported in chickens (Smith, 1963). Myoblasts of rapidly growing broiler strains proliferate more rapidly *in vitro* (Ridpath *et al*, 1984), differentiate less readily and accumulate more protein in differentiating cultures (Orcutt and Young, 1982). In Japanese quail, selection for body weight resulted in a proportionate increase in DNA, RNA and protein in the *pectoralis major* and *supracoracoides* muscles, also indicating an increase in satellite cell proliferation during growth (Fowler, Campion, Marks and Reagan, 1980). Fibre number was not counted in this study but in the *semimembranosus*, which was studied histologically, some evidence of an increase in fibre diameter was seen.

There are indications that the exact result obtained when selecting for growth rate may depend on the selection criteria employed. Mice were selectively bred from a common stock for either increased protein content, increased body weight or a combined index of bodyweight and endurance fitness on a treadmill. After forty generations, these three groups had increased their mean body weight by 14-21% and showed significant increases in the cross-sectional area of the EDL. Where the selection was for protein content, the main change was in fibre diameter. The nucleus:cytoplasm ratio was also decreased. The other two selection systems caused an increase in fibre number in the EDL, with a relatively minor increase in diameter. The *rectus femoris* behaved differently. The major effect was on fibre diameter in all three treatment groups (Rehfeldt and Bunger, 1990).

No changes in the proportions of fibre types was seen in this study. In chickens, comparison of a broiler and a layer strain showed that the proportion of fast fibres in the sartorius muscle which were white (glycolytic) was increased in the fast-growing broiler birds (Aberle, Addis and Shoffner, 1979). In quail selected for body size, the increase in fibre number in the semimembranosus was also mainly in fast types whilst slow fibres increased in diameter (Fowler *et al*, 1980). How these changes might affect meat quality is not yet clear, but the possibility that selection for rapid growth may have adverse effects must be considered.

Attempts have been made to relate growth rate in domestic species to IGF receptor number in differentiated muscle cells but the situation remains unclear. A study with membranes prepared from embryonic chicken muscle showed a decline in affinity and possibly also number of receptors in fast-growing broiler strains, when compared with layers (Trouten-Radford, Zhao and McBride, 1991). In contrast, a trend towards increased binding capacity was shown with increased efficiency of gain in rams: here membranes from cultured satellite cell myotubes were used (Mathison, Mathison, McNamara and Dodson, 1989).

Cellular events during manipulation of animal growth

The literature on the manipulation of livestock growth by the use of diet and growth promoters is extensive and has recently been reviewed elsewhere (see Buttery, Dawson, Beever and Bardsley, 1991; Buttery, Dawson and Harper, 1990; Pell and Bates, 1990; Beermann and Boyd, 1992; Lindsay *et al*, 1992). This section does not aim to be comprehensive, but rather to concentrate on those studies which have highlighted the cellular processes involved in muscle growth.

One of the most simple ways of altering the growth of an animal is by diet. If growing rats are fed deficient diets, the accumulation of DNA and protein in their muscles is reduced (Winick and Noble, 1966; Howarth and Baldwin, 1971). Glore and Layman (1983) used a graded range of dietary insufficiencies and showed that fibre diameter was reduced. However in all but the most severe nutritional deprivation the ratio of DNA:protein remained constant. This means that satellite cell division must have been reduced in proportion to the available nutrients. In pigs too, inadequate dietary protein reduces muscle fibre diameter (Staun, 1972). Prolonged malnutrition, especially in early life, leads to stunting, characterised by reduced RNA, DNA and protein content in muscles of the adult rat (Winick and Noble, 1966). "Catch-up growth", an unusually rapid proliferation of satellite cells, can occur after a short period of malnutrition if a sufficient food supply is restored. DNA synthesis responds more rapidly than protein synthesis (Howarth and Baldwin, 1971). It is also worth noting that a high plane of nutrition does not always lead to increased muscle growth and increased fibre diameter. Indeed when pigs were fed at an energy level in excess of their requirement fibre diameter at slaughter weight was reduced (Staun, 1972). This was a result of the end weight being attained at an earlier age, due to increased deposition of fat.

The anabolic agents trenbolone acetate and zeranol enhance protein deposition in skeletal muscle by decreasing protein breakdown to a greater extent than protein synthesis. They have very similar effects at the level of muscle composition. In lambs, four weeks' treatment with either of these compounds resulted in a reduced DNA:protein ratio in the *longissimus dorsi* compared to control animals of the same age (Sinnett-Smith, Dumelow and Buttery, 1983). This suggests the increased muscle size was achieved without additional satellite cell proliferation. In female rats treated with trenbolone acetate an increase in fibre diameter and a trend towards more Type I (slow oxidative) fibres were observed (Table 2).

Table 2 EFFECT OF TREATMENT WITH TRENBOLONE ACETATE ON FIBRE TYPE IN THE *EXTENSOR DIGITORUM LONGUS* OF FEMALE RATS

	% Fibre Type	
	Corn Oil Placebo	Trenbolone Acetate (1.0 mg/kg day)
Slow oxidative	26.8 ± 1.47	34.5 ± 1.57
Fast glycolytic	32.6 ± 2.29	26.8 ± 2.40
Fast oxidative	39.6 ± 2.71	37.5 ± 1.68

from Pearson (1982)

A decrease in the DNA:protein ratio was also found for the semitendinosus muscle when lambs were treated for seven weeks with the repartitioning beta-agonist cimaterol (Beerman, Butler, Hogue, Fishell, Dalrymple, Ricks and Scanes, 1987). However, by twelve weeks, although the muscle was still larger than in control animals, its DNA:protein ratio was back to control values. This implies that additional satellite proliferation must have occurred. The delay in the appearance of this effect argues that it is an indirect result of beta-agonist treatment. In this experiment, muscle fibre diameter was increased and the proportion of type I (slow oxidative) fibres was decreased in the muscles studied. A study in rats showed similarly an increase in fibre diameter and an increase in fast fibres with beta-agonist treatment (Maltin, Delday and Reeds, 1986). In another experiment in lambs (Kim, Lee and Dalrymple, 1987), the results differed. There was no change in the ratio of fibre types but growth of fibre diameter was chiefly in fast fibres. Beta-agonist treatment has also been shown to change the calpain enzyme system in muscle, a small increase in calpain II together with a very marked increase in the natural inhibitor of the calpains, calpastatin, being noted (Higgins, Lasslett, Bardsley and Buttery, 1988; Wang and Beermann, 1988). Recent results suggest that beta-agonist treatment increases the concentration of calpastatin mRNA in muscle (Parr, Bardsley, Gilmour and Buttery, 1991), but whether this is directly or indirectly mediated is unknown.

Growth hormone treatment may have immediate effects on satellite cell proliferation. In a histological study, the circulating growth hormone level in rats was raised by implanting them with growth hormone-secreting tumour cells (McCusker and Campion, 1986). Two ages of rat were studied and both grew much faster than the respective non-implanted controls. Muscle fibre number in the soleus was unchanged but the fibres became much thicker. Especially in younger animals, there was evidence of more satellite cells per cross-sectional area. A study in pigs showed no changes in fibre composition with GH treatment (Beerman, Fishell, Roneker, Boyd, Armbruster and Souza, 1990).

Future prospects

Studies on the cellular growth of muscle in mammals and birds have shown clearly that precursor cell proliferation is a key factor, both before and after birth. Failure of this process leads to reduction in the overall growth potential of the animal concerned. Conversely, there is evidence linking genetically determined large body size to an increased rate of myoblast division in early life. A large number of hormones and growth factors have now been recognised as muscle cell mitogens, but work needs to be carried out to clarify which of them are of major importance in the whole animal.

Treatments which increase protein deposition in muscle without activating satellite cells are likely to produce only limited (though economically important) effects because of the constraints of cellular physiology. These anabolic effects are also more likely to be reversed during a treatment withdrawal period prior to slaughter, since there are no permanent changes in muscle structure. It is therefore important as part of studies on potential anabolic agents to ascertain whether they act on satellite cells.

Recent interest has been intensely focussed on the manipulation of the growth hormone/IGF axis as a means of improving animal growth, and it is encouraging to note that enhancement of GH and hence IGF concentrations in body fluids does promote satellite cell division *in vivo*, as would be predicted from studies *in vitro*. However, little attention has been focussed on ACTH and its derivative peptides. It has been generally believed that their influence would be negative, since glucocorticoids, whose release ACTH stimulates, are viewed as 'catabolic' though there are some suggestions that ACTH may contribute to the anabolic effects of growth hormone (Sillence and Etherton, 1989). Locally produced and matrix-bound growth factors are much harder to manipulate *in vivo*, but their role too needs definition. Also, little is known of how for example adipose tissue, muscle and connective tissue may interact in the whole animal, and whether "cross-talk" between them may have a role in growth and nutrient partitioning. Better understanding of these processes may help in the development of new humane and consumer acceptable methods for manipulating farm animal growth.

The value of fundamental study of the genetics and cell biology of muscle growth has been demonstrated by the recent success in expressing a recombinant muscle regulatory gene in mice, and the resulting selective enhancement of muscle mass. However, application of techniques such as genetic manipulation to commercial animal production still seems a long way in the future, and it is clear any such development must include a thorough consideration of the health implications for the animals concerned (see review by Clark, Archibald, McClenaghan, Simons, Whitelaw and Wilmut, 1990).

References

Aberle, E.D. (1984) *Journal of Animal Science*, **59**, 1651-1656

Aberle, E.D., Addis, P.B. and Shoffner, R.N. (1979) *Poultry Science*, **58**, 1210-1212

Adamafio, N.A., Towns, R.J. and Kostyo, J.L. (1991) *Growth Regulation*, **1**, 17-22

Allbrook, D.B., Han, M.F. and Helmuth, A.E. (1971) *Pathology*, **3**, 233-243

Allen, R.E. and Boxhorn, L.K. (1987) *Journal of Cellular Physiology*, **133**, 567-572

Allen, R.E. and Boxhorn, L.K. (1989) *Journal of Cellular Physiology*, **138**, 311-315

Allen, R.E., Stromer, M.H., Goll, D.E. and Robson, R.M. (1978) *The Journal of Cell Biology*, **76**, 98-104

Allen, R.E., Dodson,M.V., Luiten, L.S. and Boxhorn, L.K. (1985) *In Vitro Cellular and Developmental Biology*, **21**, 636-640

Ariano, M.A., Armstrong, R.B. and Edgerton, V.R. (1973) *Journal of Histochemistry and Cytochemistry*, **21**, 51-55

Ashmore, C.R., Tompkins, G. and Doerr, L. (1972) *Journal of Animal Science*, **34**, 37-41

Askanas, V., Gallez-Hawkins, G. and King Engel, W. (1984) *Annals of Neurology*, **16**, 143

Austin, L. and Burgess, A.W. (1991) *Journal of the Neurological Sciences*, **101**, 193-197

Ball, E.H. and Sanwal, B.D. (1980) *Journal of Cellular Physiology*, **102**, 27-36

Ballard, F.J. and Francis, G.L. (1983) *Biochemical Journal*, **210**, 243-249

Ballard, F.J., Nield, M.K. and Tomas, F.M. (1983) *Muscle & Nerve*, **6**, 520-523

Barany, M. and Close, R.I. (1971) *Journal of Physiology*, **213**, 455-474

Barnett, J.G., Holly, R.G. and Ashmore, C.R. (1980) *American Journal of Physiology*, **239**, C39-C45

Beach, R.L., Burton, W.V., Hendricks, W.J. and Festoff, B.W. (1982) *Journal of Biological Chemistry*, **257**, 11437-11442

Béchet, D.M., Listrat, A., Deval, C., Ferrara, M. and Quirke, J.F. (1990) *American Journal of Physiology*, **259**, E822-E827

Beermann, D.H. and Boyd, R.D. (1992) In *The Control of Fat and Lean Deposition*, (eds P. J. Buttery, K. N. Boorman and D. B. Lindsay), London, Butterworths

Beermann, D.H., Butler, W.R., Hogue, D.E., Fishell, V.K., Dalrymple, R.H., Ricks, C.A. and Scanes, C.G. (1987) *Journal of Animal Science*, **65**, 1514-1524

Beermann, D.H., Fishell, V.K., Roneker, K., Boyd, R.D., Armbruster, G. and Souza, L. (1990) *Journal of Animal Science*, **68**, 2690-2697

Beguinot, F., Kahn, C.R., Moses, A.C. and Smith, R.J. (1985) *Journal of Biological Chemistry*, **260**, 15892-15898

Bell, A.W. (1992) In *The Control of Fat and Lean Deposition*, (eds P. J. Buttery, K. N. Boorman and D. B. Lindsay), London, Butterworths

Benezra, R., Davis, R.L., Lockshon, D., Turner, D.L. and Weintraub, H. (1990) *Cell*, **61**, 49-59

Bischoff, R. (1990) *Journal of Cell Biology*, **111**, 201-207

Blau, H.M. and Baltimore, D. (1991) *Journal of Cell Biology*, **112**, 781-783

Braun, H.M., Bober, E., Winter, B., Rosenthal, N. and Arnold, H.H. (1990) *EMBO Journal*, **9**, 821-831

Braun, H.M., Buschhausen-Denker, G., Bober, E., Tannich, E. and Arnold, H.H. (1989) *EMBO Journal*, **8**, 701-709

Brevett, A., Pinto, E., Peacock, J. and Stockdale, F. (1976) *Science*, **193**, 1152-1154

Brunetti, A. and Goldfine, I.D. (1990) *Journal of Biological Chemistry*, **265**, 5960-5963

Burant, C.F., Treutelaar, M.K., Allen, K.D., Sens, D.A. and Buse, M.G. (1987) *Biochemical and Biophysical Research Communications*, **147**, 100-107

Buttery, P.J., Dawson, J.M. and Harper, J.M.M. (1990) *Proceedings of the New Zealand Society for Animal Production*, **50**, 59-72

Buttery, P.J., Dawson, J.M., Beever, D.E. and Bardsley, R.G. (1991) *Proceedings of the 6th International Symposium on Protein Metabolism and Nutrition, Herning, Denmark 9-14 June 1991*, **Vol 1**, pp 88-102

Byatt, J.C., Schmuke, J.J., Comens, P.G., Johnson, D.A. and Collier, R.J. (1990) *Biochemical and Biophysical Research Communications*, **173**, 548-553

Byrne, I., Hooper, J.C. and McCarthy, J.C. (1973) *Animal Production*, **17**, 187-196

Cameron, C.M., Kostyo, J.L., Adamafio, N.A., Brostedt, P., Roos, P., Skottner, A., Forsman, A., Fryklund, L. and Skoog, B. (1988) *Endocrinology*, **122**, 471-474

Cardasis, C.A. and Cooper, G.W. (1976) *Journal of Experimental Zoology* **191**, 347-358

Chiquet, M., Puri, E.C. and Turner, D.C. (1979) *Journal of Biological Chemistry*, **254**, 5475-5482

Clark, A.J., Archibald, A.L., McClenaghan, M., Simons, J.P., Whitelaw, C.B.A. and Wilmut, I. (1990) *Proceedings of the New Zealand Society for Animal Production*, **50**, 167-179

Clegg, C.H., Linkhart, T.A., Olwin, B.B. and Hauschka, S.D. (1987) *Journal of Cell Biology*, **105**, 949-956

Cleland, P.J.F., Appleby, G.J., Rattigan, S. and Clark, M.G. (1989) *Journal of Biological Chemistry*, **264**, 17704-17711

Colmenares, C. and Stavnezer, E. (1989) *Cell*, **59**, 293-303

Condon, K., Silberstein, L., Blau, H.M. and Thompson, W.J. (1990) *Developmental Biology*, **138**, 256-274

Cossu, G., Zani, B., Coletta, M., Bouchè, M., Pacifici, M. and Molinaro, M. (1980) *Cell Differentiation*, **9**, 357-368

Cossu, G., Molinaro, M. and Pacifici, M. (1983) *Developmental Biology*, **98**, 520-524

Cossu, G., Cusella-DeAngelis, M.G., Senni, M.I., DeAngelis, L., Vivarelli, E., Vella, S., Bouchè, M., Boitani, C. and Molinaro, M. (1989) *Developmental Biology*, **131**, 331-336

Costill, D.L., Fink, W.J. and Pollock, M.L. (1976) *Medicine and Science in Sports*, **8**, 96-100

Crace, C.J., Hill, D.J. and Milner, R.D.G. (1985) *Journal of Endocrinology*, **104**, 63-68

Crisona, N.J. and Strohman, R.C. (1983) *The Journal of Cell Biology*, **96**, 684-692

Crow, M.T. and Stockdale, F.E. (1986) *Developmental Biology*, **113**, 238-254

Czech, M.P. (1989) *Cell*, **59**, 235-238

Davis, R.L., Weintraub, H. and Lassar, A.B. (1987) *Cell*, **51**, 987-1000

De la Haba, G., Cooper, C.W. and Elting, V. (1968) *Journal of Cellular Physiology*, **72**, 21-28

Defez, R. and Brachet, P. (1985) *International Journal of Developmental Neuroscience*, **4**, 161-168

DeMellow, J.S.M. and Baxter, R.C. (1988) *Biochemical and Biophysical Research Communications*, **156**, 199-204

DeVol, D., Rotwein, P., Sadow, J.L., Novakofski, J. and Bechtel, P.J. (1990) *American Journal of Physiology*, **259**, E89-E95

DeVroede, M.A., Romanus, J.A., Standaert, M.L., Pollett, R.J., Nissley, S.P. and Rechler, M.M. (1984) *Endocrinology*, **114**, 1917-1919

Devlin, R.B. and Emerson, C.P. Jr. (1978) *Cell*, **13**, 599-611

Diffee, G.M., Caiozzo, V.J., Herrick, R.E. and Baldwin, K.M. (1991) *American Journal of Physiology*, **260**, C528-C533

Dix, D.J. and Eisenberg, B.R. (1991) *Anatomical Record*, **230**, 52-56

Dodson, M.V., Allen, R.E. and Hossner, K.L. (1985) *Endocrinology*, **117**, 2357-2363

Dodson, M.V., Mathison, B.A., Brannon, M.A., Martin, E.L. Wheeler, B.A.and McFarland, D.C. (1988) *Tissue & Cell*, **20**, 909-918

Douen, A.G., Burdett, E., Ramlal, T., Rastogi, S., Vranic, M. and Klip, A. (1991) *Endocrinology*, **128**, 611-616
Duclos, M.J., Wilkie, R.S. and Goddard, C. (1991) *Journal of Endocrinology*, **128**, 35-42
Ecob-Prince, M.S., Jenkinson, M., Butler-Browne, G.S. and Whalen, R.G. (1986) *Journal of Cell Biology*, **103**, 995-1005
Elgin, R.G., Busby W.H. Jr. and Clemmons, D.R. (1987) *Proceedings of the National Academy of Sciences USA*, **84**, 3254-3258
Emerson, C.P. and Beckner, S.K. (1975) *Journal of Molecular Biology*, **93**, 431-447
Enesco, M. and Puddy, D. (1964) *American Journal of Anatomy*, **114**, 235-244
Ewton, D.Z., Erwin, B.G., Pegg, A.E. and Florini, J.R. (1982) *Journal of Cell Biology*, **95**, 366a
Ewton, D.Z., Falen, S.L. and Florini, J.R. (1987) *Endocrinology*, **120**, 115-123
Ewton, D.Z., Spizz, G., Olson, E.N. and Florini, J.R. (1988) *Journal of Biological Chemistry*, **263**, 4029-4032
Faulkner, J.A., Maxwell, L.C., Brook, D.A. and Lieberman, D.A. (1971) *American Journal of Physiology*, **221**, 291-297
Faulkner, J.A., Maxwell, L.C. and Lieberman, D.A. (1972) *American Journal of Physiology*, **222**, 836-840
Feng, B., Hilt, D. C., Max, S.R. (1990) *Journal of Biological Chemistry*, **265**, 18702-18705
Florini, J.R. and Ewton, D.Z., (1981) *In Vitro*, **17**, 763-768
Florini, J.R. and Roberts, S.B. (1979) *In Vitro*, **15**, 983-992
Florini, J.R., Ewton, D.Z., Falen, S.L. and Van Wyk, J.J. (1986) *Journal of American Physiology*, **250**, C771-C770
Florini, J.R., Roberts, A.B., Ewton, D.Z., Falen, S.L., Flanders, K.C. and Sporn, M.B. (1986) *Journal of Biological Chemistry*, **261**, 16509-16513
Florini, J.R., Ewton, D.Z., and Roof, S.L. (1991) *Molecular Endocrinology*, **5**, 718-724
Fowler, S.P., Campion, D.R., Marks, H.L. and Reagan, J.O. (1980) *Growth*, **44**, 235-252
Frati, L., Cenci, A., Sbaraglia, G., Venza Teti, D. and Covelli, I. (1976) *Life Sciences*, **18**, 905-912
Frémont, P.H., Fournier LeRay, C. and Le Douarin, G.H. (1983) *Cell Differentiation*, **13**, 325-339
Gard, D.L. and Lazarides, E. (1980) *Cell*, **19**, 263-275
Gardner, J.M. and Fambrough, D.M. (1983) *Journal of Cell Biology*, **96**, 474-485
Garrels, J.I. (1979) *Developmental Biology*, **73**, 134-152
Glore, S.R. and Layman, D.K. (1983) *Growth*, **47**, 403-410
Goldspink, D.F. (1978) *Biochemical Journal*, **174**, 595-602
Gollnick, P.D. and King, D.W. (1969) *American Journal of Physiology*, **216**, 1502-1509

Gordon, H. and van Essen, D.C. (1985) *Developmental Biology*, **111**, 42-50
Gospodarowicz, D. and Mescher, A.L. (1977) *Journal of Cellular Physiology*, **93**, 117-128
Gospodarowicz, D., Weseman, J., Moran, J.S. and Lindstrom, J. (1976) *Journal of Cell Biology*, **70**, 395-405
Gossett, L.A., Zhang, W. and Olson, E.N. (1988) *Journal of Cell Biology*, **106**, 2127-2137
Grant, A.L., Helferich, W.G., Merkel, R.A. and Bergen, W.G. (1990) *Journal of Animal Science*, **68**, 652-658
Greene, E.A. and Allen, R.E. (1991) *Journal of Animal Science*, **69**, 146-152
Haji Baba, A.S., Hawkey, R.K., Buttery, P.J. and Fletcher, J.M. (1991) *Proceedings of the 6th International Symposium on Protein Metabolism and Nutrition*, Herning, Denmark, 9-14 June 1991, **Vol 2**, pp 207-209
Hall, Z.W. and Ralston, E. (1989) *Cell*, **59**, 771-772
Han, V.K.M., D'Ercole, A.J. and Lund, P.K. (1987) *Science*, **236**, 193-197
Han, V.K.M., Hill, D.J., Strain, A.J., Towle, A.C., Lauder, J.M., Underwood, L.E. and D'Ercole, A.J. (1987) *Pediatric Research*, **22**, 245-249
Harper, J.M.M., Roe, J.A., Heywood, C.M. and Buttery, P.J. (1987) *Proceedings of the 4th International Symposium on Protein Metabolism and Nutrition, Rostock, GDR*, S39-40
Harper, J.M.M., Soar, J.B. and Buttery, P.J. (1987) *Journal of Endocrinology*, **112**, 87-96
Harper, J.M.M., Mackinson, I. and Buttery, P.J. (1990) *Domestic Animal Endocrinology*, **7**, 477-484
Harris, D.A., Falls, D.L. and Fischbach, G.D. (1989) *Nature*, **337**, 173-175
Hathaway, M.R., Hembree, J.R., Pampusch, M.S. and Dayton, W.R. (1991) *Journal of Cellular Physiology*, **146**, 435-441
Heywood, C.M., Harper, J.M.M. and Buttery, P.J. (1989) *Journal of Endocrinology*, **123 Suppl**, Abstr 110
Heywood, C.M., Haji Baba, A.S., Harper, J.M.M. and Buttery, P.J. (1991) *Proceedings of the 6th International Symposium on Protein Metabolism and Nutrition*, Herning Denmark, 9-14 June 1991, **Vol 2**, pp 201-203
Higgins, J.A., Lasslett, Y.V., Bardsley, R.G. and Buttery, P.J. (1988) *British Journal of Nutrition*, **60**, 645-652
Hill, D.J., Crace, C.J., Fowler, L., Holder, A.T. and Milner, R.D.G. (1984) *Journal of Cellular Physiology*, **119**, 349-35
Hill, D.J., Crace, C.J. and Milner, R.D.G. (1985) *Journal of Cellular Physiology*, **125**, 337-344
Hill, D.J., Crace, C.J., Nissley, S.P., Morrell, D., Holder, A.T. and Milner, D.G. (1985) *Endocrinology*, **117**, 2061-2072
Hill, D.J., Crace, Strain, A.J. and Milner, R.D.G. (1986) *Journal of Clinical Endocrinology and Metabolism*, **62**, 753-760
Hoh, J.F.Y. and Hughes, S. (1991a) *Muscle & Nerve*, **14**, 316-325
Hoh, J.F.Y. and Hughes, S. (1991b) *Muscle & Nerve*, **14**, 398-406
Holloszy, J.O. (1967) *Journal of Biological Chemistry*, **242**, 2278-2282

Holtzer, H., Sangar, J.W., Ishikawa, H. and Strahs, K. (1972) *Cold Spring Harbor Symposia on Quantitative Biology*, **37**, 549-566
Hooper, A.C.B. (1976) *Growth*, **40**, 33-39
Hooper, A.C.B. and Hurley, M.P. (1983) *Animal Production*, **36**, 223-227
Howarth, R.E. and Baldwin, R.L. (1971) *Journal of Nutrition*, **101**, 477-484
Hu, J.S., Spizz and Olson, E.N. (1987) *Journal of Cell Biology*, **105**, 274a
Ignotz, R.A., Endo, T. and Massagué, J. (1987) *Journal of Biological Chemistry*, **262**, 6443-6446
Ishai-Michaeli, R., Eldor, A. and Vlodavsky, I. (1990) *Cell Regulation*, **1**, 833-842
Jennische, E. and Hansson, H.-A. (1987) *Acta Physiologica Scandinavica*, **130**, 327-332
Jialal, I., Crettaz, M., Hachiya, H.L., Kahn, C.R., Moses, A.C., Buzney, S.M. and King, G.L. (1985) *Endocrinology*, **117**, 1222-1229
Kardami, E., Spector, D. and Strohman, R.C. (1985) *Proceedings of the National Academy of Sciences USA*, **82**, 8044-8047
Kayali, A.G., Young, V.R. and Goodman, M.N. (1987) *American Journal of Physiology*, **252**, E621-E626
Kelvin, D.J., Simard, G. and Connolly, J.A. (1989) *Journal of Cellular Physiology*, **138**, 267-272
Kelvin, D.J., Simard, G., Sue-A-Quan, A. and Connolly, J.A. (1989) *Journal of Cell Biology*, **108**, 169-176
Kennedy, J.M., Eisenberg, B.R., Reid, S.K., Sweeney, L.J. and Zak, R. (1988) *American Journal of Anatomy*, **181**, 203-215
Kiess, W., Haskell, J.F., Lee, L., Greenstein, L.A., Miller, B.E., Aarons, A.L., Rechler, M.M. and Nissley, S.P. (1987) *Journal of Biological Chemistry*, **262**, 122745-12751
Kim, Y.S., Lee, Y.B. and Dalrymple, R.H. (1987) *Journal of Animal Science*, **65**, 1392-1399
Klagsbrun, M. (1989) *Progress in Growth Factor Research*, **1**, 207-235
Klip, A., Li, G. and Walker, D. (1983) *Canadian Journal of Biochemistry and Cell Biology*, **61**, 644-649
Koumans, J.T.M., Akster, H.A., Dulos, G.J. and Osse, J.W.M. (1990) *Cell and Tissue Research*, **261**, 173-181
Kuehl, U., Timpl, R. and Von der Mark, H. (1982) *Developmental Biology*, **93**, 344-354
Lathrop, B., Thomas, K. and Glaser, L. (1985) *Journal of Cell Biology*, **101**, 2194-2198
Laurent, G.J., Sparrow, M.P. and Millward, D.J. (1978) *Biochemical Journal*, **176**, 407-417
Leberer, E. and Pette, D. (1988) *Biochemical Journal*, **235**, 67-73
Levin, P.A., Bistritzer, T., Hanukoglu, L., Max, S.R. and Roeder, L.M. (1990) *Hormone and Metabolic Research*, **22**, 608-611
Leyland, D.M., Turner, P.C. and Beynon, (1990) *Biochemical Journal*, **272**, 231-237

Lim, R.W. and Hauschka, S.D. (1984) *Developmental Biology*, **105**, 48-58
Linden, D.C. and Fambrough, D.M. (1979) *Neuroscience*, **4**, 527-538
Lindsay, D.B., Hunter, R.A. and Sillence, M.N. (1992) In *The Control of Fat and Lean Deposition*, (eds P. J. Buttery, K. N. Boorman and D. B. Lindsay), London, Butterworths
Linkhart, T.A., Clegg, C.H. and Hauschka, S.D. (1981) *Developmental Biology*, **86**, 19-30
Liu, F., Powell, D.R. and Hintz, R.L. (1990) *Journal of Clinical Endocrinology and Metabolism*, **70**, 620-628
Loirat, M-J., Lucas-Heron, B., Ollivier, B. and Leoty, C. (1988) *Bioscience Reports*, **8**, 369-378
MacLennan, P.A., McArdle, A. and Edwards, R.H.T. (1991) *Biochemical Journal*, **275**, 477-483
Maltin, C.A., Delday, M.I. and Reeds, P.J. (1986) *Bioscience Reports*, **6**, 293-299
Massagué, J., Cheifetz, S., Endo T. and Nadal-Ginard, B. (1986) *Proceedings of the National Academy of Sciences USA*, **83**, 8206-8210
Mathison, B.D., Mathison, B.A., McNamara, J.P. and Dodson, M.V. (1989) *Domestic Animal Endocrinology*, **6**, 191-201
Matsuda, R., Spector, D. and Strohman, R.C. (1984) *Developmental Biology*, **103**, 267-275
Mauro, A (1961) *Journal of Biophysical and Biochemical Cytology*, **9**, 493-495
McCusker, R.H. and Campion, D.R. (1986) *Journal of Endocrinology*, **111**, 279-285
McCusker, R.H. and Clemmons, D.R. (1988) *Journal of Cellular Physiology*, **137**, 505-512
McElligott, M.A., Chaung, L-Y and Barreto, Jr, A (1989) *Biochemical Pharmacology*, **34**, 2199-2205
McFarland, D.C., Pesall, J.E., Norberg, J.M. and Dvoracek, M.A. (1991) *Comparative Biochemical Physiology*, **99A**, 163-167
Miller, J.B. and Stockdale, F.E. (1986) *Journal of Cell Biology*, **103**, 2197-2208
Miner, J.H. and Wold, B.J. (1991) *Molecular and Cellular Biology*, **11**, 2482-2851
Minshall, R.D., McFarland, D.C. and Doumit, M.E. (1990) *Domestic Animal Endocrinology*, **7**, 413-424
Miranda, A.F., Bonilla, E., Martucci, G., Moraes, C.T., Hays, A.P. and Dimauro, S. (1988) *American Journal of Pathology*, **132**, 410-416
Miranda, A.F., Peterson, E.R. and Masurovsky, E.B. (1988) *Tissue & Cell*, **20**, 179-191
Mohan, S., Bautista, C.M., Wergedal, J. and Baylink, D.J. (1989) *Biochemistry*, **86**, 8338-8342
Moss, F.P. and Leblond, C.P. (1971) *Anatomical Record*, **170**, 421-436
Müntener, M., Rowlerson, A.M., Berchtold, M.W. and Heizmann, C.W. (1987) *Journal of Biological Chemistry*, **262**, 465-469
Nadal-Ginard, B. (1978) *Cell*, **15**, 855-864

Narusawa, M., Fitzsimons, R.B., Izumo, S., Nadal-Ginard, B., Rubinstein, N.A. and Kelly, A.M. (1987) *Journal of Cell Biology*, **104**, 447-459
Nguyen, H.T., Medford, R.M. and Nadal-Ginard, B. (1983) *Cell*, **34**, 281-293
Öcalan, M., Goodman, S.L., Kühl, U., Hauschka, S.D. and von der Mark, K. (1988) *Developmental Biology*, **125**, 158-167
Oka, Y. and Orth, D.N. (1983) *Journal of Clinical Investigation*, **72**, 249-259
Olson, E.N., Sternberg, E., Hu, J.S., Spizz, G. and Wilcox, C. (1986) *Journal of Cell Biology*, **103**, 1799-1805
Olwin, B.B. and Hauschka, S.D. (1986) *Biochemistry*, **25**, 3487-3492
Olwin, B.B. and Hauschka, S.D. (1988) *Journal of Cell Biology*, **107**, 761-769
Ontell, M. (1973) *Anatomical Record*, **178**, 211-228
Orcutt, M.W. and Young, R.B. (1982) *Journal of Animal Science*, **54**, 769-776
Palmer, R.M., Bain, P.A. and Southorn, B.G. (1990) *Comparative Biochemistry and Physiology*, **97C**, 369-372
Pampusch, M.S., Hembree, J.R., Hathaway, M.R. and Dayton, W.R. (1990) *Journal of Cellular Physiology*, **143**, 524-528
Parent, J.B., Tallman, J.F., Henneberry, R.C. and Fishman, P.H. (1980) *Journal of Biological Chemistry*, **255**, 7782-7786
Parr, T., Bardsley, R.G., Gilmour, R.S. and Buttery, P.J. (1991) *Journal of Muscle Research and Cell Motility*, **12**, 82
Pearson, J.T. (1982) PhD thesis, University of Nottingham
Pearson-White, S.H. (1991) *Journal of Cellular Biochemistry*, **15C suppl**, 42
Pell, J.M. and Bates, P.C. (1990) *Nutrition Research Reviews*, **3**, 163-192
Peng, H.B., Gao, K.X., Xie, M.Z. and Zhao, D.Y. (1988) *Developmental Biology*, **127**, 452-455
Penney, R.K., Prentis, P.F., Marshall, P.A. and Goldspink, G. (1983) *Cell and Tissue Research*, **228**, 375-388
Popiela, H., Taylor, D., Ellis, S., Beach, R. and Festoff, B. (1984) *Journal of Cellular Physiology*, **119**, 234-240
Prives, J., Silman, I. and Amsterdam, A. (1976) *Cell*, **7**, 543-550
Quinn, L.S., Ong, L.D. and Roeder, R.A. (1990) *Developmental Biology*, **140**, 8-19
Reeds, P.J. and Davis, T.A. (1992) In *The Control of Fat and Lean Deposition*, (eds P. J. Buttery, K. N. Boorman and D. B. Lindsay), London, Butterworths
Reeds, P.J. and Palmer, R.M. (1983) *Biochemical and Biophysical Research Communications*, **116**, 1084-1090
Rehfeldt, C. and Bunger, L. (1990) *Archiv für Tierzucht*, **33**, 507-516
Rehfeldt, C., Fiedler, I. and Wegner, J. (1987) *Zeitschrift für Microskopisch-anatomische Forschung*, **101**, 669-680
Reznick, M. (1969) *Journal of Cell Biology*, **40**, 568-571
Ridpath, J.F., Huiatt,T.W., Trenkle, A.H., Robson, R.M. and Bechtel, P.J. (1984) *Differentiation*, **26**, 121-126
Roe, J.A., Harper, J.M.M. and Buttery, P.J. (1989a) *Journal of Endocrinology*, **122**, 565-571

Roe, J.A., Harper, J.M.M. and Buttery, P.J. (1989b) *Journal of Endocrinology*, **123**, Suppl, Abstract 109
Roeder, R.A., Blann, D.L., Bauer, C.A. and Hossner, K.L. (1986) *Federation Proceedings*, **45**, 233
Roeder, R.A., Thorpe, S.D., Byers, F.M., Schelling, G.T. and Gunn, J.M. (1986) *Growth*, **50**, 485-495
Rosenthal, S.M., Brunetti, A., Brown, E.J., Mamula, P.W. and Goldfine, I.D. (1991) *Journal of Clinical Investigation*, **87**, 1212-1219
Rowlerson, A., Pope, B., Murray, J., Whalen, R.B. and Weeds, A.G. (1981) *Journal of Muscle Research and Cell Motility*, **2**, 415-438
Rowlerson, A.M., and Spurway, N.C. (1988) *Histochemical Journal*, **21**, 461-474
Rozengurt, E. and Mendoza, S.A. (1985) *Journal of Cell Science*, **Suppl 3**, 229-242
Sandra, A. and Przybylski, R.J. (1979) *Developmental Biology*, **68**, 546-556
Schiaffino, S., Gorza, L., Pitton, G., Saggin, L., Ausoni, S., Sartore, S. and Lømo, T. (1988) *Developmental Biology*, **127**, 1-11
Schmid, C., Steiner, T. and Froesch, E.R. (1983) *FEBS Letters*, **161**, 117-121
Schonberg, M., Smith, T.J., Krichevsky, A. and Bilezikian, J.P. (1981) *Cell Differentiation*, **10**, 101-107
Seene, T., Umnova, M., Alev, K. and Pehme, A. (1988) *Journal of Steroid Biochemistry*, **29**, 313-317
Senni, M.I., Castrignano, F., Polana, G., Cossu, G., Scarsella, G. and Biagioni, S. (1987) *Differentiation*, **36**, 194-198
Shani, M., Zevin-Sonkin, D., Saxel, O., Carmon, Y., Katcoff, D., Nudel, U. and Yaffé, D. (1981) *Developmental Biology*, **86**, 483-492
Shimizu, M., Webster, C., Morgan, D.O., Blau, H.M. and Roth, R.A. (1986) *American Journal of Physiology*, **251**, E611-E615
Sillence, M.N. and Etherton, T.D. (1989) *Journal of Endocrinology*, **123**, 113-119
Sinnett-Smith, P.A., Dumelow, N.W. and Buttery, P.J. (1983) *British Journal of Nutrition*, **50**, 225-234
Smith, J.H. (1963) *Poultry Science*, **42**, 283-290
Smith, R.H., Palmer,R.M. and Reeds, P.J. (1983) *Biochemical Journal*, **214**, 153-161
Sola, O.M., Christensen, D.L. and Martin, A.W. (1973) *Experimental Neurology*, **41**, 76-100
Spencer, G.S.G., Hill, D.J., Garssen, G.J. and Williams, J.P.G. (1987) *Acta Endocrinologica (Copenh)*, **114**, 470-474
Spizz, G., Hu, J-S. and Olson, E.N. (1987) *Developmental Biology*, **123**, 500-507
Spizz, G., Roman, D., Strauss, A. and Olson, E.N. (1986) *Journal of Biological Chemistry*, **261**, 9483-9488
Standaert, M.L., Schimmel, S.D. and Pollet, R.J. (1984) *Journal of Biological Chemistry*, **259**, 2337-2345
Staun, H. (1972) *World Review of Animal Production*, **VIII**, 18-27
Stickland, N.C. (1983) *Journal of Anatomy*, **137**, 323-333

Stirewalt, W.S. and Low, R.B. (1983) *Biochemical Journal*, **210**, 323-330
Stockdale, F.E. and Miller, J.B. (1987) *Developmental Biology* **123**, 1-9
Streter, F.A., Elzinga, M., Mabuchi, K., Salmons, S. and Luff, A.R. (1975) *FEBS Letters*, **57**, 107-111
Sutrave, P., Kelly, A.M. and Hughes, S.H. (1990) *Genes and Development*, **4**, 1462-1472
Symonds, M.E., Roe, J.A., Heywood, C.M., Harper, J.M.M. and Buttery, P.J. (1990) *Biochemical Pharmacology*, **40**, 2271-2276
Thayer, M.J., Tapscott, S.J., Davis, R.L., Wright, W.E., Lassar, A.B. and Weintraub, H. (1989) *Cell*, **58**, 241-248
Thompson, J.M. and Rapoport, S.I. (1988) *Synapse*, **2**, 7-10
Tollefsen, S.E., Lajara, R., McCusker, R.H., Clemmons, D.R. and Rotwein, P. (1989) *Journal of Biological Chemistry*, **264**, 13810-13817
Tollefsen, S.E., Sadow, J.L. and Rotwein, P. (1989) *Proceedings of the National Academy of Sciences USA*, **86**, 1543-1547
Totland, G.K., Kryvi, H. and Slinde, E. (1988) *Meat Science*, **23**, 303-315
Trenkle, A., DeWitt, D.L. and Topel, D.G. (1978) *Journal of Animal Science*, **46**, 1597-1603
Trouten-Radford, L., Zhao, X. and McBride, B.W. (1991) *Domestic Animal Endocrinology*, **8**, 129-137
Turo, K.A. and Florini, J.R. (1982) *American Journal of Physiology*, **243**, C278-C283
Ulrich, S., Lagente, O., Lenfant, M. and Courtois, Y. (1986) *Biochemical and Biophysical Research Communications*, **137**, 1205-1213
Vandenburgh, H.H. (1983) *Journal of Cellular Physiology* **116**, 363-371
Verger, C., Sassa, S. and Kappas, A. (1983) *Journal of Cellular Physiology*, **116**, 135-141
Vilaro, S., Palacin, M., Pilch, P.F., Testar, X. and Zorzano, A. (1989) *Nature*, **342**, 798-800
Walker, C.R. and Wilson, B.W. (1975) *Nature*, **256**, 215-216
Wang, S.Y. and Beermann, D.H. (1988) *Journal of Animal Science*, **66**, 2545-2550
Webster, B.A., Vigna, S.R. and Paquette, T. (1986) *American Journal of Physiology*, **250**, E186-E204
Weeds, A.G., Trentham, D.R., Kean, C.J.C. and Buller, A.J. (1974) *Nature*, **247**, 135-139
Whalen, R.G., Butler-Browne, G.S. and Gros, F. (1976) *Proceedings of the National Academy of Sciences USA*, **73**, 2018-2022
Whalen, R.G., Schwartz, K., Bouveret. P., Sell, S.M. and Gros, F. (1979) *Proceedings of the National Academy of Sciences USA*, **76**, 5197-5201
Whalen, R.G., Harris, J.B., Butler-Browne, G.S. and Sesodia, S. (1990) *Developmental Biology*, **141**, 24-40
White, M.E., Kretchmar, D.H., Allen, C.E., and Dayton, W.R. (1989) *Journal of Animal Science*, **67**, 3144-3154

Whitson, P.A., Stuart, C.A., Huls, M.H., Sams, C.F. and Cintron, N.M. (1989) *Journal of Cellular Physiology*, **140**, 8-17
Winchester, P.K., Davis, M.E., Alway, S.E.and Gonyea, W.J. (1991) *American Journal of Physiology*, **260**, C206-C212
Winick, M. and Noble, A. (1965) *Developmental Biology*, **12**, 451-466
Winick, M. and Noble, A. (1966) *Journal of Nutrition*, **89**, 300-306
Woods, J.D. and Warriss, P.D. (1992) In *The Control of Fat and Lean Deposition*, (eds P. J. Buttery, K. N. Boorman and D. B. Lindsay), London, Butterworths
Wright, W.E., Sassoon, D.A. and Lin, V.K. (1989) *Cell*, **56**, 607-617
Yamaguchi, Y., Mann, D.M.and Ruoslahti, E. (1990) *Nature*, **346**, 281-284
Young, R.B., Moriarty, D.M., McGee, C.E., Farrar, W.R. and Richter, W.R. (1990) *Journal of Animal Science*, **68**, 1158-1169
Young, V.R. (1985) *Journal of Animal Science*, **61, suppl 2**, 39-56
Zevin-Sonkin, D. and Yaffe, D. (1980) *Developmental Biology*, **74**, 326-334

3
CONTROL OF LIPOGENESIS AND LIPOLYSIS

R.G. VERNON
Hannah Research Institute, Ayr, KA6 5HL, UK

Introduction

As the rate of muscle development slackens in the growing animal, the rate of fat deposition in adipose tissue accelerates. If unrestricted fat can accumulate until it comprises more than fifty percent of body weight (even more in some grotesquely obese humans). Adipose tissue is not deposited randomly throughout the body but develops at a number of distinct sites, some in the abdominal cavity (*eg* perirenal, omental), some under the skin and some between or within muscle beds. The pattern of adipose tissue depot distribution appears to be common to essentially all mammals and is thought to have developed at an early stage of mammalian evolution and been retained thereafter (Pond and Mattacks, 1985). That said, in fat animals some depots can become so large that they become contiguous with other depots; pigs and some northern species (*eg* bears, reindeer) at the beginning of the winter have a continuous layer of subcutaneous adipose tissue.

While there is considerable interest in decreasing adiposity in meat animals, it is worth remembering that the tissue has a variety of functions in addition to being a store of energy. For example, subcutaneous adipose tissue can act as insulation; some depots, *eg* those of the eye-socket may have a protective function; cardiac and some inter- and intra-muscular depots may have a specialised role in providing fatty acids as a fuel for adjacent muscle fibres; adipose tissue is essential for the growth of the mammary gland probably because adipose tissue produces and secretes a variety of hormones and other factors (*eg* IGF-I, oestrogens, prostaglandins, adenosine, adipsin); adipose tissue may also be important for the immune system for it has been shown to produce several components of the alternative pathway of complement formation (see Flint and Vernon, 1992).

Adipose tissue metabolism

The principal pathways involved in the synthesis and hydrolysis of triacylglycerol (fat) are described in Figure 1. Synthesis of triacylglycerol requires a source of fatty acid and glycerol 3-phosphate. The latter is derived from glucose as adipose tissue has little or no glycerol kinase activity and so cannot re-use glycerol released by lipolysis as a precursor (Vernon, 1980). Fatty acids may be derived from plasma triacylglycerols of chylomicrons and very low density lipoproteins and may be produced within the fat cell from a variety of precursors. Plasma triacylglycerols have to be hydrolysed prior to uptake and this is catalysed by the enzyme lipoprotein lipase. This enzyme is secreted by adipocytes and is transported to the endothelial cells lining the capillaries where it is attached to the outer surface of the plasma membrane facing the lumen of the capillary. Fatty acids and monoacylglycerols released are transported to the fat cells where there is thought to be a fatty acid translocase located in the plasma membrane which transports them into the cell (Scow and Blanchette-Mackie, 1985). Monoacylglycerols are also taken up by the cell and then hydrolysed to fatty acid and glycerol.

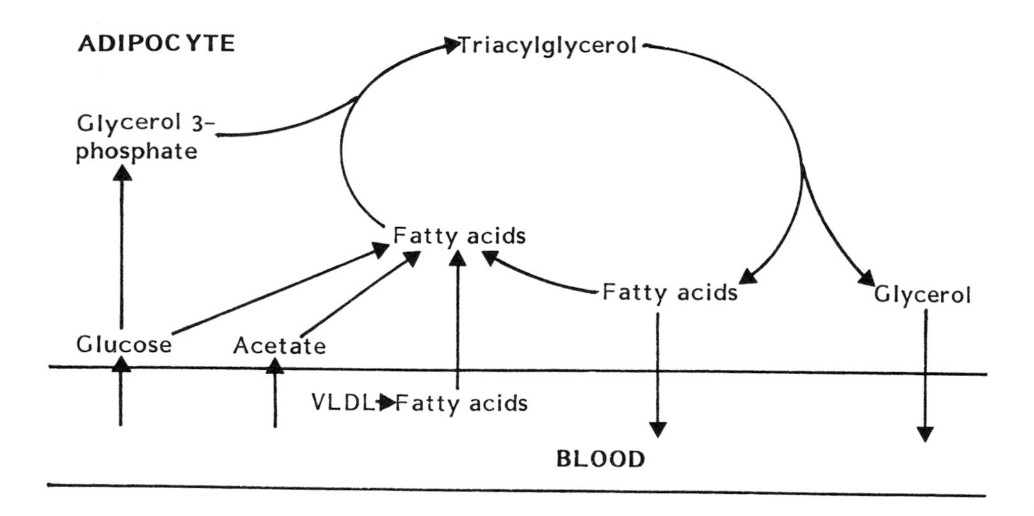

Figure 1 Pathways for synthesis and hydrolysis of triacylglycerol

Fatty acid synthesis (lipogenesis) requires a source of cytosolic acetyl CoA and NADPH. The former can be produced from acetate, glucose, lactate, and a number of amino acids, but only acetate and glucose, and possibly lactate are important precursors, and for these significance varies from species to species and with age within a species (see Vernon, 1980; 1986). Thus for adult ruminants acetate is the most important precursor whereas glucose and lactate may make a more substantial contribution in the foetal animal (Vernon, 1986). It is also apparent that adipose tissue is not a significant competitor of muscle for amino acids (Vernon, 1986). Glucose is an important precursor in pigs and is the major precursor in birds. Pigs and ruminants are similar in that adipose tissue is the major site of lipogenesis in the non-lactating animal whereas in birds the liver is the site and there is little or no lipogenesis in adipose tissue (see Vernon, 1980; Butterwith, 1988). For birds then, essentially all fatty acids for deposit in adipose tissue are derived from plasma lipid through the action of lipoprotein lipase (Butterwith, 1988). The relative importance of lipogenesis and lipoprotein lipase as sources of fatty acids in pigs and ruminants will vary with diet (the greater the amount of fat in the diet the greater the importance of lipoprotein lipase), but even in ruminants which normally consume relatively low amounts of fat, lipoprotein lipase activity may supply in excess of fifty percent of fatty acids deposited (Vernon, 1980), although another estimate suggests a rather lower contribution (Vezinhet, Nougues and Teyssier, 1983).

Triacylglycerols are hydrolysed by the action of hormone-sensitive lipase; this cleaves two molecules of fatty acids, the resulting monoacylglycerol being hydrolysed to glycerol and fatty acid by monoacylglycerol lipase. The rate-limiting step is thought to be cleavage of the first fatty acid moiety by hormone sensitive lipase and hydrolysis is normally thought to go to completion; the monoacylglycerol lipase is very active and monoacylglycerol is normally present in very low amounts in the tissue. Essentially all the glycerol and some fatty acids produced are released from the cell whereas some fatty acid is re-esterified. Esterification and lipolysis occur simultaneously and continuously so there is a constant turnover of triacylglycerol, the relative rates of the two processes determine if there is net loss or accumulation of lipid.

The regulation of lipogenesis

Lipogenesis is subject to several possible limitations: the supply of cytosolic acetyl CoA and NADPH, and the regulation of synthesis from these precursors. The complexity of control of acetyl CoA supply depends on the precursor. For acetate, a single enzyme, acetyl CoA synthase is involved and this enzyme is not thought to be under any form of sophisticated control. Its relatively low affinity for acetate does mean that activity will depend on acetate concentration over much of the normal physiological range (see Vernon, 1980). For glucose, on the other hand, control may be exerted at numerous steps, the most important being glucose transport, phosphofructokinase, pyruvate dehydrogenase and, in some species, ATP-citrate lyase. In ruminants the very low activity of ATP-citrate lyase must restrict the translocation of acetyl CoA from the mitochondria to the cytosol as it is transported mainly as citrate (Vernon, 1980). This limits lipogenesis from a variety of precursors including amino acids and lactate as well as glucose, all of which are metabolised to mitochondrial acetyl CoA. For glucose, however, the major restrictions in ruminants probably occur prior to ATP-citrate lyase for there is thought to be little oxidation of glucose carbon in the mitochondria (Yang, White and Muir, 1982); glucose transport (Sasaki, 1990), phosphofructokinase (Smith, 1984) and pyruvate dehydrogenase (Robertson, Faulkner and Vernon, 1982) are all probably involved. Lactate is something of a paradox: studies have shown that lactate is released from ruminant adipose tissue *in vivo* (Khachadurian, Adrouni and Yacoubian, 1966) and *in vitro* (Yang and Baldwin, 1973; Robertson *et al*, 1982). On the other hand, a number of studies have shown that lactate (at high concentrations) is a much better precursor than glucose for lipogenesis, particularly in the bovine (Prior, 1978; Whitehurst, Beitz, Pothoven, Ellison and Crump, 1978; Smith and Prior, 1986). Studies with the autoperfused inguinal fat pad in the dog may well provide the solution to this paradox for these showed that net uptake or release of lactate depended on the plasma lactate concentration with output below about 2 mM and uptake above (Fredholm, 1971). The NADPH required for lipogenesis can be derived from three systems, glucose-6-phosphate and 6-phosphogluconate dehydrogenases; malic enzyme; NADP-isocitrate dehydrogenase. Their relative importance varies with species, isocitrate dehydrogenase being of special importance in ruminants (Vernon, 1980). For the rat at least, it is thought that the rate of NADPH production is determined by the rate of lipogenesis rather than the reverse. Cytosolic acetyl CoA and NADPH may be used for syntheses other than lipogenesis, so the true control of the process must lie subsequent to their production, hence acetyl CoA carboxylase which catalyses the conversion of acetyl CoA to malonyl CoA, the first committed step on route to fatty acid production, is now thought to be the major control point.

Acetyl CoA carboxylase is a large (approximately 260 kDA) protein which is subject to both chronic and acute controls. The amount of the enzyme is under hormonal control although its half-life (about 40 h) is surprisingly long for a key regulatory enzyme (Volpe and Vagelos, 1976). However, the enzyme exists in active and inactive states regulated by phosphorylation status (phosphorylation of key serine residues results in inactivation) and is also subject to acute, allosteric control being activated by citrate and inhibited by long-chain fatty acyl CoAs (Hardie, 1989). Activation by citrate also results in polymerisation of the enzyme, at least *in vitro*. The role of fatty acyl CoAs as allosteric modulators has been questioned as these are mostly associated with binding proteins within cells. However, incubation of adipose tissue *in vitro* with physiological concentrations of fatty acids causes a readily reversible inhibition of lipogenesis in both rats and sheep, most probably due to inhibition of acetyl CoA carboxylase (Vernon, 1980).

Much recent attention has focused on control by changes in phosphorylation status. Early studies showed that incubation of rat adipocytes with catecholamines caused phosphorylation of the enzyme with a decline in the activation status of the enzyme (Lee and Kim, 1979). Further studies have shown that in cell-free systems acetyl CoA carboxylase can be phosphorylated by a number of protein kinases including cyclic AMP-dependent kinase (A-kinase) (which is activated by catecholamines) and a recently discovered AMP-stimulated kinase (Hardie, 1989). The various kinases phosphorylate different serine residues of acetyl CoA carboxylase and this allowed assessment of which kinases are likely to be acting on the enzyme *in vivo*. Surprisingly, it appears that only serines phosphorylated by AMP-stimulated kinase are phosphorylated in intact cells including adipocytes (Hardie, 1989). The mechanism whereby hormones, such as catecholamines, which activate A-kinase exert their effect on acetyl CoA carboxylase is thus unclear for they cause increased phosphorylation on those serine residues which are phosphorylated by AMP-stimulated kinase. As there is no obvious effect of these hormones on the activity of AMP-stimulated kinase itself, it has been suggested that they may exert their effects through inhibition of a specific phosphatase (phosphatase-1 in adipocytes) which dephosphorylates acetyl CoA carboxylase (Hardie, 1989); this seems a curious mechanism and has not been proven.

While catecholamines inhibit lipogenesis and decrease the activity of acetyl CoA carboxylase, insulin activates the enzyme and increases lipogenic flux. The mechanism, however, is still obscure. Various possibilities have been suggested: insulin causes dephosphorylation of the enzyme at the AMP-stimulated kinase sites, perhaps by activating phosphatase-1; insulin has been reported to cause the covalent binding of a low-molecular weight activator to the enzyme; intriguingly, insulin also causes phosphorylation of the enzyme on at least two distinct serine residues (see Hardie, 1989). It is possible that all three suggestions may be involved for they are not mutually exclusive. Thus although it is now clear that changes in phosphorylation status play a major role in the acute control of acetyl CoA carboxylase, the detailed mechanisms of how insulin and indeed catecholamines elicit such changes remain to be elucidated.

The above detail has been elucidated for laboratory species but is probably of equal relevance to ruminants and other farm animals. The enzyme has been cloned and sequenced for the rat (Lopez-Casillas, Bai, Luo, Kong, Hermodson and Kim, 1988) and chicken (Takai, Yokoyama, Wada and Tanabe, 1988) and over 95% homology prevails for the amino acid sequence. We have recently cloned and sequenced part of the goat acetyl CoA carboxylase gene and again a very high degree of homology between this sequence and corresponding sequences of the rat and chicken is apparent (Barber, Travers and Vernon, unpublished observations). Furthermore the enzyme has been purified from bovine adipose tissue and its properties were very similar to those of the enzyme from rat tissues (Moss, Yamagishi, Kleinschmidt and Lane, 1972).

Lipogenesis is inhibited by catacholamines in ruminant (Vernon, 1980) and pig (Liu, Boyer and Mills, 1989) adipose tissue but it has not yet been checked if this is due to a decrease in activation status of acetyl CoA carboxylase. The role of insulin as an acute regulator of lipogenesis on the other hand has been controversial. Most studies found little or no effect of insulin on the rate of lipogenesis in adipose tissue from various ruminants when incubated *in vitro* for 2 or 3 h (see Vernon, 1980); there have been a few notable exceptions (*eg* Yang and Baldwin, 1973). However, this apparent insensitivity to insulin is probably an artefact, for prolonged incubation with insulin (*eg* for 24 h) results in a highly reproducible increase in the rate of lipogenesis in adipose tissue for sheep and cattle (see Vernon and Sasaki, 1991) and pigs (Walton and Etherton, 1986). Importantly, half-maximum effects in these studies were seen with physiological concentrations of insulin. Time-course studies have shown that 4 to 5 h of incubation is required for an effect of insulin to be manifest (Vernon and Finley, unpublished observation). The effects of insulin, at least over the first 24 h to 48 h of incubation are due to an increase in the activation status of acetyl CoA carboxylase (Figure 2). The reason for the lack of response to insulin in short-term incubations is unknown. One possibility is that adipose tissue from ruminants is kept at about 37°C during transport, preparation etc, this coupled with the fact that there is usually a relatively long time period between slaughter and the arrival of tissue in the laboratory, may result in substantial anoxia in the

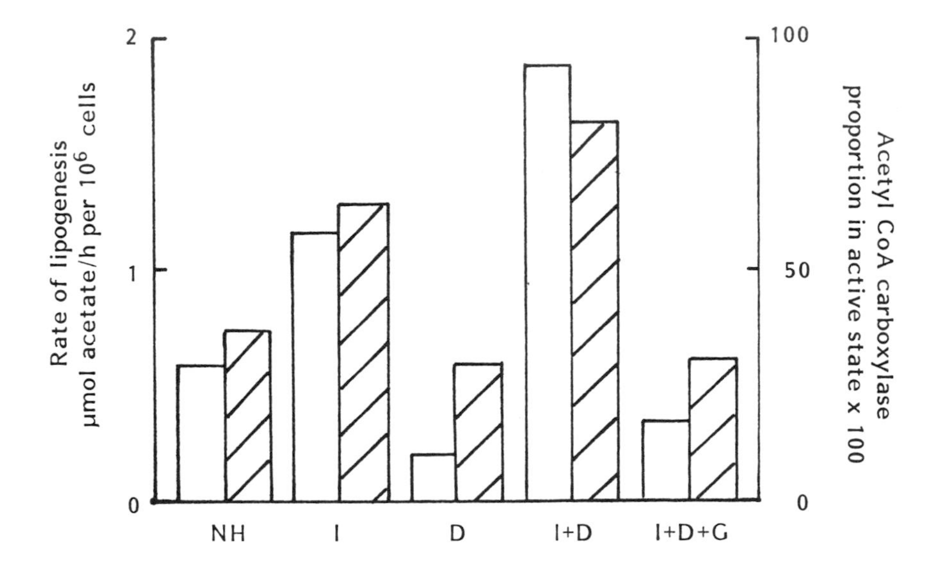

Figure 2 Effects of incubation for 48 h in the absence (NH) or presence of insulin (I), dexamethasone (a glucocorticoid analogue) (D) and growth hormone (G) on the rate of lipogenesis (open columns) and acetyl CoA carboxylase activation status (hatched columns) in sheep adipose tissue. Results from Vernon and Finley (1988) and Vernon, Barber and Finley (1991).

tissue. This is likely to increase AMP-stimulated kinase activity (the enzyme is present in sheep adipose tissue, Vernon and Lindsay, unpublished observation) with a concomitant phosphorylation and inactivation of acetyl CoA carboxylase. A period of preincubation may thus be necessary to restore ATP and AMP levels to normal and to decrease the inhibitory activity of AMP-stimulated kinase.

That insulin should have an effect on ruminant adipose tissue is hardly surprising; the tissue is well supplied with insulin receptors which have properties very similar to those of other species (see Vernon and Sasaki, 1991). The effects of insulin on adipose tissue metabolism are modulated by other hormones. The glucocorticoid analogue, dexamethasone, has a complex effect, in the absence of insulin and in the presence of low concentrations of insulin, dexamethasone inhibits lipogenesis but at high concentrations of insulin it can act synergistically with insulin to increase the rate of lipogenesis (Walton, Etherton and Evock, 1986; Vernon and Finley, 1988). The synergistic effect is peculiar to lipogenesis, however, for dexamethasone inhibits insulin-stimulation of glucose utilisation at all concentrations of insulin (Vernon and Taylor, 1988). The mechanism of the synergistic effect of dexamethasone is not known, but it is expressed as an enhancement of the activation of acetyl CoA carboxylase caused by insulin (Vernon, Barber and Finley, 1991). Growth hormone inhibits lipogenesis in adipose tissue from a variety of species including ruminants, pigs and rats (see Vernon and Flint, 1989). The hormone is a powerful insulin antagonist and inhibits the activation of acetyl CoA carboxylase by insulin, possibly by inhibiting the synthesis of a protein required for mediation of the insulin induced activation of the enzyme (Vernon, Barber and Finley, 1991). Growth hormone can also inhibit lipogenesis in the absence of insulin; in sheep adipose tissue this effect is blocked by actinomycin D, an inhibitor of gene transcription (Snoswell, Finley and Vernon, 1990) and is exacerbated by dexamethasone. Exposure of sheep adipose tissue to growth hormone increases the activity of S-adenosyl methionine decarboxylase an enzyme of polyamine metabolism (Snoswell *et al*, 1990). Polyamines have been implicated in some of the effects of growth hormone on lipolysis in chicken adipose tissue (Campbell and Scanes, 1988) but it is not known if they are involved in the mechanism of growth hormone action in sheep. Several mechanisms can be excluded. Growth hormone had no effect on insulin binding in pig (Magri, Adamo, Leroith and Etherton, 1990) or sheep adipocytes (Wastie, Buttery and Vernon, unpublished observation) or on insulin receptor tyrosine kinase activity of pig adipocytes (Magri *et al*, 1990); growth hormone had no discernable effect on the activity of AMP-dependent kinase in sheep adipose tissue (Vernon and Lindsay, unpublished observation). Growth hormone effects on adipocytes are most probably direct rather than via IGF-I, for growth hormone receptors have been demonstrated on adipocytes in a number of species (Vernon and Flint, 1989). IGF-I at low concentrations had no effect on lipogenesis in bovine (Etherton and Evock, 1986) or ovine (Vernon and Finley, 1988) adipose tissue but at high concentrations it had an insulin-like effect on lipogenesis suggesting mediation by insulin receptors. IGF-I receptors appear to be absent from rat and human adipocytes (see Vernon and Flint, 1989) and the above suggests that they are missing from ruminant adipocytes also. Despite this, adipocytes do appear to synthesise IGF-I (Doglio, Dani, Fredrikson, Grimaldi and Ailhaud, 1987) and growth hormone stimulates IGF-I production by rat (Yang and Novakofski, 1990) and sheep (Beattie and Vernon, unpublished observation) adipose tissue; the role of this locally produced IGF-I is uncertain but it may be involved in adipogenesis.

The above discussion pertains to the relatively short-term control of lipogenesis via changes in the activation status of acetyl CoA carboxylase. In addition, total activity (*ie* that measured after activation of any inactive enzyme), which is probably a reflection of enzyme concentration, is also under chronic endocrine control. Thus maintenance of sheep adipose tissue in culture for six days in the presence of insulin and dexamethasone markedly increased the total activity of acetyl CoA carboxylase; this was prevented by the presence of an inhibitor of gene transcription suggesting the synthesis of enzyme protein was occurring (Vernon, Barber and Finley, 1991). The increase in activity was also prevented by growth hormone, suggesting that this key hormone antagonises the ability of insulin to both activate and increase synthesis of acetyl CoA carboxylase (Vernon, Barber and Finley, 1991).

Thus insulin, catecholamines and growth hormone all have major roles in regulating the lipogenic flux in adipocytes and exert a major effect by controlling the activation status of acetyl CoA carboxylase (Figure 2). In the longer term insulin and growth hormone also modulate the total activity and hence amount of the enzyme. Dexamethasone also has a role modulating the effect of both insulin and growth hormone. In addition a variety of other factors may also be involved. Adenosine, an autocrine/paracrine factor in adipose tissue, which antagonises the effects of catecholamines (see below) stimulates lipogenesis in sheep adipose tissue (Plested, Brindley and Vernon, unpublished observation). The gut hormone, gastric inhibitory peptide, stimulates lipogenesis in sheep adipose tissue whereas epidermal growth factor was inhibitory (Haji Baba and Buttery, 1991). However, the physiological importance of these factors remains to be determined.

Lipolysis: acute control

The hydrolysis of triacylglycerol by hormone-sensitive lipase is also under complex acute and chronic endocrine control. In ruminants (see Vernon, 1980) and pigs (Mersmann, Phinney and Brown, 1975; 1976) catecholamines are potent stimulators of lipolysis whereas glucagon has only a slight effect; in poultry the converse prevails with glucagon being the major lipolytic hormone (Butterwith, 1988). Both glucagon and catecholamines exert their effects through essentially the same pathway (Figure 3) both interacting with a specific receptor in the plasma membrane (glucagon and ß-adrenergic respectively); this causes dissociation and activation of a GTP-binding protein (G_s) which interacts with and activates adenylate cyclase. Adenylate cyclase catalyses the synthesis of cyclic AMP which activates cyclic AMP-dependent kinase (A-kinase), which phosphorylates hormone-sensitive lipase on specific serine residues with concomitant activation (Figure 3). Recent studies suggest that this chain of events also causes translocation of hormone-sensitive lipase from the cytosol to the surface of the fat droplet (Londos, Egan, Greenberg and Wek, 1991). Perilipin, a protein, located on the surface of the fat droplet is phosphorylated by A-kinase and may be a 'docking protein' for hormone-sensitive lipase (Londos *et al*, 1991).

The ß-adrenergic/glucagon signal transduction system is subject to acute modulation by a number of factors. Adenosine and prostaglandin E_2 (both produced in adipose tissue) acting through their own receptors, and also catecholamines acting via α_2-adrenergic receptors, activate another GTP-binding protein, G_i, which inhibits adenylate cyclase activity (Figure 3) and so reduces the cyclic AMP concentration. In addition, insulin, by an as yet unidentified mechanism, activates cyclic AMP-phosphodiesterase activity, again decreasing cyclic AMP concentration. Thus the cyclic AMP concentration within the cell and hence the activation status of A-kinase will depend on the strengths of the various signals channelling through G_s, G_i and phosphodiesterase.

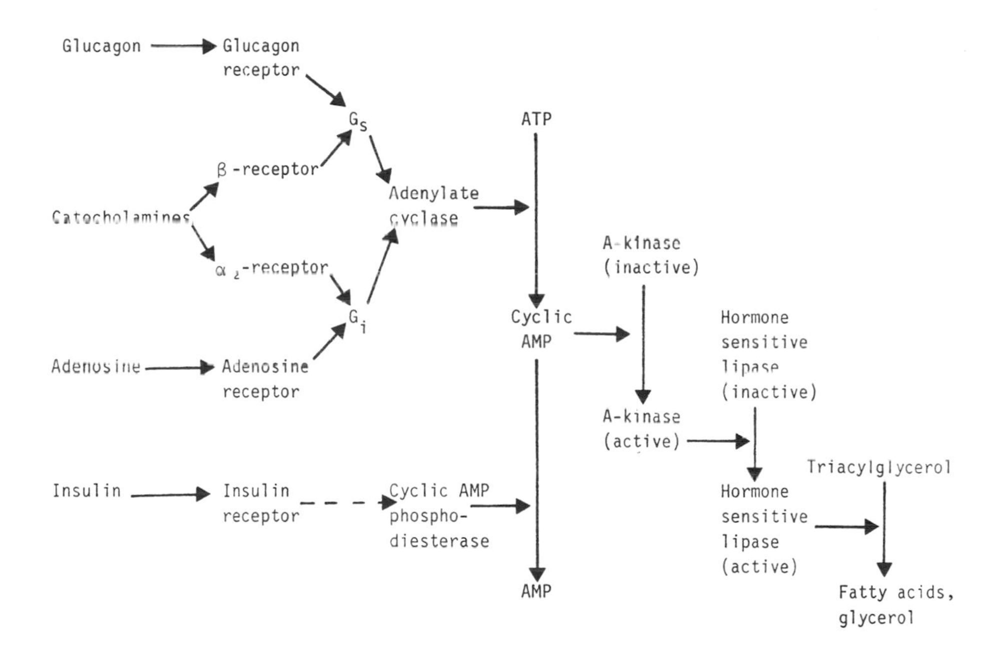

Figure 3 Lipolytic signal transduction cascade

As if the above is not complex enough, insulin also appears to decrease lipolysis by a cyclic AMP-independent mechanism, probably involving activation of a phosphatase and dephosphorylation of hormone sensitive lipase (Stralfors and Honnor, 1989). Furthermore there may be a role for AMP-stimulated protein kinase, for this enzyme phosphorylates a serine which is next but one amino acid to a key serine phosphorylated by A-kinase (Garton, Campbell, Carling, Hardie, Colbran and Yeaman, 1989). Phosphorylation by AMP-stimulated kinase has no apparent effect on hormone-sensitive lipase activity, but it does impede phosphorylation by A-kinase and so may have an antilipolytic role *in vivo* (Garton *et al*, 1989). This may seem odd, for whereas A-kinase has a pivotal role, integrating a variety of signals into one, and promoting lipolysis and inhibiting lipogenesis (albeit indirectly) when activated, AMP-stimulated kinase appears to suppress both lipolysis and lipogenesis. One possible explanation can be derived from the observation that high concentration (about 1 mM or more) of fatty acids inhibit lipolysis (see Vernon and Clegg, 1985). As noted above fatty acids also inhibit lipogenesis but this is achieved with lower concentrations. Fatty acyl CoA esters activate AMP-stimulated protein kinase, possibly by activating a kinase kinase (Hardie, 1989). A putative sequence of events following exposure of a cell to catecholamines would be for an initial increase in A-kinase activity resulting in activation of hormone-sensitive lipase and a rise in fatty acid concentration. This would increase AMP-stimulated kinase activity with a concomitant decrease in lipogenesis. If fatty acid concentrations rise to very high levels, then AMP-stimulated kinase may increase to the point where it can compete with A-kinase for phosphorylation of the neighbouring serine residues, thus blunting the effects of A-kinase and curtailing the rate of lipolysis. In this context AMP-stimulated kinase could be seen to have a protective function analagous to its presumed role in anoxia.

In contrast to insulin, IGF-I and IGF-II have been claimed to have acute lipolytic effects in sheep adipose tissue (Lewis, Molan, Bass and Gluckman, 1988) but we have been unable to reproduce these surprising findings (Vernon and Finley, unpublished observations).

While most of the mechanisms described above have been elucidated in laboratory species, the same mechanisms most probably operate in ruminants and pigs. Adipocytes from ruminants possess adenosine and α_2-adrenergic receptors (Watt, Finley, Cork, Clegg and Vernon, 1991) and adenosine (Vernon, Finley and Watt, 1991) and α_2-adrenergic agonists (Fain, Mohell, Wallace and Mills, 1984; Chilliard and Flechet, 1988; Watt *et al*, 1991) are antilipolytic. The role of prostaglandin E_2 is less certain; the substance was not antilipolytic *in vitro* in bovine adipose tissue (DiMarco, Whitehurst and Beitz, 1986) but microdialysis studies with sheep adipose tissue *in vivo* have shown that infusion of noradrenaline results in a rise in prostaglandin E_2 and this is associated with a decrease in the rate of lipolysis (Thompson and Vernon, unpublished observation). Hormone-sensitive lipase has been purified from porcine (Lee, Yeaman, Fredrikson, Stralfors and Belfrage, 1985) and bovine (Cordle, Colbran and Yeaman, 1986) adipose tissue; properties of both are very similar to those of the rat, but differ from those of the chicken (Berglund, Khoo, Jenson and Steinberg, 1980) (molecular weight of mammalian enzyme is about 84 kDalton compared with 45 kDalton for the chicken) but both mammalian and avian enzymes are phosphorylated by A-kinase. Studies on phosphorylation sites of AMP-stimulated kinase and A-kinase were performed on hormone-sensitive lipase from bovine adipose tissue (Garton *et al*, 1989). Thus control of lipolysis in adipocytes from different mammals is likely to be very similar whereas there may be some differences in chickens.

Lipolysis: chronic control

The rate of lipolysis depends not only on the concentrations of the various stimulators and inhibitors outwith the adipocyte; it also depends on the ability of the various components of the signal transduction systems to transmit signal. Response and sensitivity to catecholamines decreased with age in pig adipocytes (Mersmann *et al*, 1975). Seasonal changes in response to adrenaline have been observed in reindeer adipocytes (greatest response in summer and autumn, lowest in winter) (Larsen, Nilsson and Blix, 1985). Seasonal changes in response to α_2-agonists and adenosine appear to occur in sheep adipose tissue, again with a minimum response in the winter (Vernon and Finley, unpublished observations). Changes in response and sensitivity to catecholamines and adenosine occur during the pregnancy lactation cycle in sheep adipose tissue (see Vernon, Finley and Watt, 1991). More detailed studies with growing and fattening animals may thus reveal further changes in response or sensitivity to the various signals. The factors and mechanisms responsible for changes in responsiveness of the signal transduction systems have not been fully elucidated in farm animals, but growth hormone, thyroid hormones, glucocorticoids, sex steroids and even insulin may be involved.

Studies with cattle and pigs showed that administration of growth hormone for a number of days increased the lipolytic response to catecholamines *in vivo* (see Vernon and Flint, 1989; Sechen, Dunshea and Bauman, 1990). Maintenance of sheep adipose tissue in culture for 48 h in the presence of growth hormone also increases the response and sensitivity to the ß-agonist, isoproterenol, and furthermore increased ligand binding to the ß-adrenergic receptor (Watt *et al*, 1991). Varying serum growth hormone concentration *in vivo* in rats also alters ligand binding to the ß-receptor of adipocytes (Watt, Madon, Flint and Vernon, 1990). These various studies suggest that growth hormone enhances transduction of the ß-adrenergic signal by increasing the number of ß-adrenergic receptors. Further studies have failed to reveal an effect of growth hormone on a number of other components (α_2- and adenosine receptors, G_i activity, adenylate cyclase activity; total A-kinase activity in sheep adipose tissue *in vitro* and rat adipose tissue *in vivo* (Vernon, Watt and Finley, unpublished observations), suggesting that growth hormone's effect may be specific for just one component, the ß-receptor.

Studies with rats have shown that glucocorticoids increase ß-receptor number but decrease maximum adenylate cyclase activity (Giudicelli, Lacasa, de Mazancourt, Pasquier and Pecquery, 1989; Ros, Northup and Malbon, 1989). In agreement with these findings, prolonged incubation of sheep adipose tissue with the glucocorticoid analogue, dexamethasone, increased response and sensitivity to isoproterenol and decreased maximum adenylate cyclase activity (Finley, Lindsay and Vernon, 1990; Vernon, Watt and Finley, unpublished observations). Prolonged exposure to dexamethasone also increases response to α_2-adrenergic agents and to adenosine in sheep adipose tissue (Vernon *et al*, 1991). Conversely studies with rat adipose tissue have shown that adrenalectomy increases response to adenosine (Saggerson, 1980; DeMazancourt, Lacasa, Giot and Giudicelli, 1989).

Castration in rats and hamsters decreased response to both ß- and α_2-agonists but had no effect on response to adenosine (Giudicelli *et al*, 1989; Xu, De Pergola and Bjorntorp, 1991); the effects on response to ß-agonists are at least partly due to a decrease in ß-receptor number (Xu *et al*, 1991). Treatment of castrated rats with testosterone restored response to ß-agonists (Xu *et al*, 1991) but in hamsters there was only a partial response seemingly due to greater effect on response to α_2-agonists (Giudicelli *et al*, 1989), that is testosterone can alter the relative response to α_2 and ß-agonists. Differences between rats and hamsters is probably due to the virtual absence of α_2-receptors in rat adipocytes. Ovariectomy also decreases response to ß-agonists in rat and hamster adipocytes but this does not appear to involve a loss of ß-receptors, rather there appears to be a fall in maximum adenylate cyclase activity which is restored by oestradiol treatment (Giudicelli *et al*, 1989; Lacasa, Agli, Pecquery and Giudicelli, 1991). In contrast to testosterone, oestradiol does not appear to have any effect on response to α_2-agonists (Giudicelli *et al*, 1989). The role of testosterone and oestradiol in the control of lipolysis in farm animals has not been studied, but is likely to be important; castration for example promotes fattening while treatment with sex steroids decreases fat deposition (see Vernon, 1980; Roche and Quirke, 1986).

Hypothyroidism in rats results in a decreased response to ß-agonists and an increased response to adenosine and prostaglandin E_2 (see Malbon, Rapiejko and Watkins, 1988; Vernon *et al*, 1991). These effects are not due to changes at the receptor level, the effects on adenosine and prostaglandin E_2 being due to increased G_i activity (Malbon *et al*, 1988; Milligan and Saggerson, 1990). Hyperthyroidism in rats increases response to ß-agonists and decreases response to adenosine, the latter effect being at least partly due to a decrease in adenosine receptor number (Malbon *et al*, 1988). Hypothyroidism in neonatal lambs prevented a rise in serum fatty acid in response to a catecholamine injection (Wrutniak and Cabello, 1986) showing that these hormones also have a permissive role in sheep and hypothyroidism impaired the ability of glucagon to stimulate lipolysis in chicken adipocytes (Gandarias, Galdiz and Fernandez, 1984).

Insulin has a chronic effect on lipolysis, but for some variables there are apparently conflicting reports. The effects of diabetes in the rat on response to ß-agonists and number of ß-receptors (Lacasa, Agli and Giudicelli, 1983; Chiappe de Cingolani, 1986; Saggerson, 1989) and on response to adenosine and the number of adenosine receptors (Saggerson, 1989; Strassheim, Milligan and

Houslay, 1990; Green and Johnson, 1991) vary from study to study. There is some agreement however, in that diabetes appears to decrease sensitivity to adenosine (Saggerson, 1989; Green and Johnson, 1991), possibly due to a change in G_i activity (Strassheim *et al*, 1990; Green and Johnson, 1991). Diabetes also appears to cause a decrease in maximum adenylate cyclase in the rat (Lacasa *et al*, 1983; Strassheim *et al*, 1990). The effects of diabetes on these variables have not been determined in farm animals, but maintenance of sheep adipose tissue in culture for 48 h decreased response to both adenosine and α_2-agonists (Vernon *et al*, 1991; Watt *et al*, 1991). The effects on response to α_2-agonists could be partly due to an effect in receptor binding but insulin had no effect on ligand binding to the adenosine receptor (Watt *et al*, 1991). The full explanation for these unexpected and paradoxical effects of insulin is not known; preliminary studies suggest that changes in adenylate cyclase and G_i activity are not involved (Vernon, Watt and Finley, unpublished observations).

From the aforesaid, it is clear that the signal transduction systems controlling lipolysis are under complex chronic endocrine control, with individual components of the system being controlled by a different cocktail of hormones (Figure 4). A change of physiological state may thus alter a number of components of systems transmitting stimulatory and inhibitory signals. This will adjust the range of operational cyclic AMP concentrations within the cell and hence the strength of signal transmitted to A-kinase. Lactation provides an example of the sort of range of changes that can occur (Table 1). It seems probable that more detailed studies of lipolysis during growth and fattening will also reveal further changes to the system than are apparent to date.

Table 1 CHANGES IN COMPONENTS OF ADRENERGIC/ADENOSINE SIGNAL TRANSDUCTION SYSTEM DURING LACTATION

Component	Sheep	Rat
ß-receptor	increased	increased
Adenosine receptor	increased	?
G_i activity	?	increased
Adenylate cyclase (max activity)	?	decreased
Cyclic AMP phosphodiesterase	?	decreased
A-kinase (max activity)	no change	no change

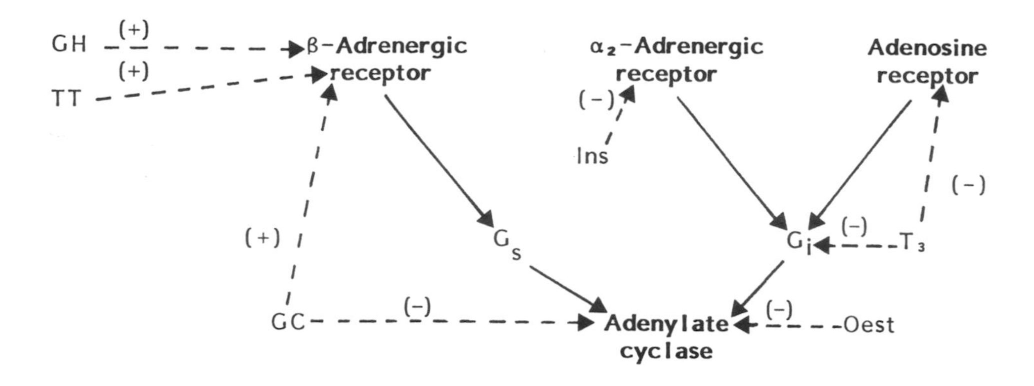

Figure 4 Chronic control of lipolytic signal transduction cascade. (GH: growth hormone; TT, testosterone; GC: glucocorticoid; T_3: tri-iodothyronine; Oest: oestradiol; Ins: insulin; (+): increases; (-) decreases

Adipose tissue metabolism during growth and fattening

Adipose tissue mass can increase through either hyperplasia or hypertrophy, the latter being almost entirely due to lipid accretion in the cells. Both hyperplasia and hypertrophy occur in phases. In lambs for example, there is an apparent rapid increase in cell number over the first hundred days of age in subcutaneous and inter-muscular depots followed by a quiescent period of about one hundred days after which there is a further burst of hyperplasia as animals fatten (see Vernon, 1986); changes in cell size in lambs show a similar pattern (Figure 5). A similar pattern of changes in hypertrophy is also found in pigs (Anderson and Kauffman, 1973; Lee and Kauffman, 1974; Mersmann, Goodman and Brown, 1975). In contrast, in cattle adipocyte volume shows little change until about 150 days of age after which it increases dramatically (see Vernon, 1986).

These studies show that the rate of net lipid deposition varies with age; they also emphasise the importance of expressing metabolic data on a per cell basis

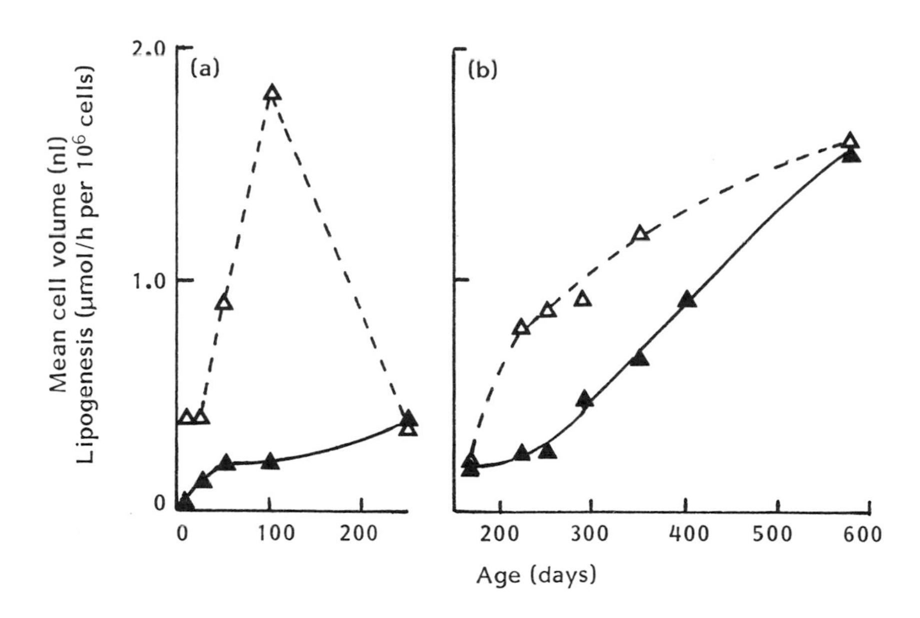

Figure 5 Changes in mean cell volume (▲) and rate of lipogenesis (△) of subcutaneous adipocytes from growing (a) and fattening (b) sheep. Data from Nougues and Vezinhet (1977) and Vezinhet and Nougues (1977) for growing sheep and from Hood and Thornton (1979; 1980) for fattening sheep

rather than per unit weight of tissue when studying developmental changes; failure to do this makes some of the older literature very difficult to interpret.

Lipogenesis has been studied in most detail. In sheep, the rate is very low at birth in perirenal adipose tissue, but increases rapidly after 2 days post-partum until about 5 days post-partum (Vernon, 1986). This is followed by a lag phase of about 3 weeks after which the rate increases steadily until about 100 days after birth (Figure 5). A similar pattern of development between days 10 and 100 post-partum is found in other depots (Vezinhet and Nougues, 1977). Between 100 and 250 days of age the rate falls precipitously in both abdominal and subcutaneous adipose tissue (Vezinhet and Nougues, 1977) and then increases once more in subcutaneous adipose tissue at least as fattening commences (Hood and Thornton, 1980). A similar biphasic pattern was reported by Smith, Jenkins and Prior

(1987), who also showed that changes in flux were paralleled by changes in acetyl CoA carboxylase activity. However, a comparison of these estimates of flux with net rates of lipid deposition derived from changes in fat cell size (Vernon, 1980) shows some curious discrepancies. Net lipid deposition is high between 10 and 25 days of age when the rate of lipogenesis is relatively low. However, at this age lambs would be suckling, and milk is a high fat diet, so much of the lipid deposited probably comes from preformed, dietary fatty acid (lipoprotein lipase activity has not been reported for this period of development) (Vernon, 1980). Paradoxically, the rate of lipogenesis increases between 50 and 100 days, when net lipid accretion is virtually zero and then falls at a time when net lipid accretion begins to rise. Ultimately the increased rate of net lipid deposition after about 200 days of age is paralleled by increases in lipogenesis. Figure 5 shows data for subcutaneous adipose tissue, but a similar picture prevails for perirenal adipose tissue. Presumably the discrepancies between lipogenic flux and net lipid deposition reflect developmental changes in lipoprotein lipase activity and lipolysis, but for these little information is available. Lipoprotein lipase activity does increase during fattening (Haugebak, Hedrick and Asplund, 1974; Vezhinhet *et al*, 1983), while hormone-sensitive lipase activity showed no change between 56 and 112 days of age but increased by about 240 days of age (Sidhu, Emery, Parr and Merkel, 1973).

Little information is available for growing and fattening cattle, but lipogenesis increases between 11 and 19 months of age in several depots (Whitehurst, Beitz, Cianzio and Topel, 1981). The basal rate of lipolysis increased with age between 14 and 19 months but the proportion of fatty acids re-esterified also increased (Smith, Prior, Ferrell and Mersmann, 1984). An increase in basal lipolysis with age can also be inferred from an earlier study by Pothoven, Beitz and Thornton (1975). In contrast, the rate of catecholamine-stimulated lipolysis did not change between 14 and 19 months of age (Smith *et al*, 1984).

Studies with pigs have shown that the increase in fat cell size with fattening is associated with an increase in activity of a variety of lipogenic enzymes including acetyl CoA carboxylase (Anderson and Kauffman, 1973) lipoprotein lipase activity (Lee and Kauffman, 1974) and glycerophosphate acyltransferase activity (a key enzyme of esterification) (Rule, Smith and Mersmann, 1989). As in cattle, it would appear that basal lipolysis also increases with age in pigs (Mersmann *et al*, 1975), rate/g tissue did not vary but number of adipocytes/g decreased with age). In contrast, the maximum rate (*ie* in the presence of catecholamine) decreased between 45 and 100 days of age but may have increased in older animals; there was also a transient decrease in sensitivity to catecholamine between 45 and 100 days of age (Mersmann *et al*, 1975).

The paucity of studies on lipolysis in fattening farm animals is surprising for in other species the response to ß-agonists decreases with age (Dax, Partilla and Gregerman, 1981; Hoffman, Chang, Farahbakhsh and Reaven, 1984) whereas response to adenosine (Rolband, Furth, Staddon, Rogus and Goldberg, 1990) and to α_2-agonists (Arner, Marcus, Karpe, Sonnenfeld and Blome, 1987) increases, in part at least due to an increase in receptor number (Kobatake, Watanabe,

Matsuzawa, Tokunaga, Fujioka, Kawamoto, Keno, Tarui and Yoshida, 1991) suggesting major changes in the signal transduction systems. Curiously a marked correlation between fat cell size and response to α_2-agonists has been found in man (Arner *et al*, 1987).

From these various, albeit limited, studies, a pattern of events begins to emerge. The initial surge of hypertrophy in suckling animals is probably due to deposition of fatty acids derived from milk fat. During the subsequent relatively quiescent phase, in lambs at least there is a paradoxical high rate of lipogenesis (this may reflect a high rate of lipid turnover in growing animals) which perversely appears to decline as fattening commences, before recovering. In general, however fattening is associated with increased lipogenic flux and activities of enzymes of lipogenesis, esterification and also of lipoprotein lipase activity. In addition basal lipolysis, at least as measured *in vitro*, also appears to increase, suggesting increased turnover of triacylglycerol. Fat cells do not continue depositing lipid indefinitely, but reach a maximum size of about 2 nl, when rates of synthesis and hydrolysis must be equal.

Endocrine control during growth and fattening

Our expanding knowledge of the factors and mechanisms involved in the acute and chronic control of adipose tissue metabolism provides some insight into the mechanisms whereby growth-promoting factors such as growth hormone, ß-agonists and, to a lesser extent, steroids decrease adiposity. In contrast the factors involved in the natural switch from muscle growth to fattening have not been defined; there have been no detailed studies of both endocrine changes and changes in lipid accretion and, or, adipocyte metabolism in the same animals during this period. Certainly the packages of hormones promoting muscle growth and fat deposition are not identical, for while insulin promotes both protein synthesis in muscle and lipid accretion in adipocytes, growth hormone, sex steroids and IGF-I, which promote muscle growth, either inhibit or have no effect on adipocyte hypertrophy. Serum growth hormone decreases with age, albeit gradually, and a negative correlation between serum growth hormone and adiposity has been noted (Trenkle and Topel, 1978). In accordance with this serum GH is higher in bulls than in steers and heifers (Plouzek and Trenkle, 1991a). Serum insulin tends to increase with age and a positive correlation between insulin and adiposity has been reported (Trenkel and Topel, 1978). Serum oestradiol increases with age in both male and female cattle (Plouzek and Trenkle, 1991b), whereas triodothyronine falls. The switch to fattening may thus be the result of a number of endocrine changes and while an increase in the serum insulin:GH ratio with age is likely to be an important factor facilitating fattening, it is unlikely to be the only one.

Individual depot differences

The preceding sections have considered adipose tissue metabolism and hypertrophy in general. However fat cells from individual depots grow at different rates (see Vernon, 1986). In general there appears to be little difference in adipocyte mean cell volume (and hence net rate of lipid accretion) in growing animals, but as animals fatten a hierarchy of mean fat cell sizes appears: omental ≥ perirenal > subcutaneous > intermuscular > intramuscular. Several factors contribute to these differences. The rate of lipogenesis within a depot varies with cell size (Hood, 1982; Rule, Beitz and Hood, 1987) and differences in the rate of lipogenesis have been noted between depots (in general omental > perirenal = subcutaneous > intermuscular > intramuscular in ruminants) (*eg* Whitehurst *et al*, 1981; Thornton and Tume, 1984) and pigs (Anderson and Kauffman, 1973; Lee and Kauffman, 1974). However, variations in the rate of lipogenesis alone cannot account for all differences in mean cell size (*eg* Hood and Allen, 1978). Lipoprotein lipase activity and the rate of esterification also vary between depots in ruminants (*eg* Haugebak *et al*, 1974; Thornton and Tume, 1984) and pigs (Lee and Kauffman, 1974; Rule *et al*, 1989) and differences in lipolytic rate have been found in pigs (Benmansour, Demarne, Lecourtier and Lhuillery, 1991). An additional factor is blood flow and hence nutrient supply, which varies between depots (Barnes, Comline and Dobson, 1983; Gregory, Christopherson and Lister, 1986); unfortunately data in both studies was expressed per g of tissue, but it would appear that blood flow per cell ranks perirenal > omental > subcutaneous > intermuscular.

Studies with man and laboratory species have shown the importance of both glucocorticoids and sex steroids in the differential development of adipocyte hypertrophy in different fat depots (Bjorntorp, Ottosson, Rebuffe-Scrive and Xu, 1990). For example ovariectomy in rats decreased the response of abdominal adipose tissue to ß-agonists but had no apparent effect on subcutaneous adipose tissue (Lacasa *et al*, 1991) while differences in the lipoprotein lipase activity in several depots were paralleled by differences in the number of glucocorticoid receptors per cell (Bjorntorp *et al*, 1990). However such studies really only push the key questions further back; we now need to determine why different fat cells have different numbers of steroid receptors.

Summary and Conclusions

The net rate of lipid accumulation by adipocytes varies with age and also with depot. The surge in lipid accumulation during fattening is associated with a general increase in both lipid synthetic and also lipolytic activities, but in growing animals changes in flux and enzyme activities do not always correlate well with changes in lipid accretion and, furthermore, depot specific differences in lipid accretion do not always correlate with differences in synthetic activity. Some discrepancies may be due to differences in blood flow and hence nutrient supply to individual depots.

The importance of insulin, catecholamines and growth hormone in the control of acetyl CoA carboxylase and hence lipogenesis is well defined but the possibility of other factors, such as gut hormones, having a significant role remains to be confirmed. For lipolysis, the key roles of insulin and catecholamines acting as ß-agonists is well defined but the potential importance of factors such as adenosine, prostaglandins and catecholamines acting as α_2-agonist as acute modulators *in vivo* remains to be clarified in farm animals. The chronic endocrine control of lipolysis in ruminants and pigs is still poorly understood. There is growing evidence for a key role of growth hormone, but little information for other factors such as glucocorticoids, thyroid hormones and sex steroids which have been shown to have a role in other species.

While we have some insight into the ways which growth promoting factors such as growth hormone, ß-agonists, and to a lesser extent, sex steroids decrease adiposity, the factors which cause the natural switch from the muscle growth to fattening have not been clearly defined; an increase in the serum insulin:GH ratio with age is likely to be important but it is unlikely to be the only factor involved.

References

Anderson, D.B. and Kauffman, R.G. (1973) *Journal of Lipid Research*, **14**, 160-168

Arner, P., Marcus, C., Karpe, B., Sonnenfeld, T. and Blome, P. (1987) *European Journal of Clinical Investigation*, **17**, 58-62

Barnes, R.J., Comline, R.S. and Dobson, A. (1983) *Quarterly Journal of Experimental Physiology*, **68**, 77-88

Benmansour, N.M.M., Demarne, Y., Lecourtier, M.J. and Lhuillery, C. (1991) *International Journal of Biochemistry*, **23**, 499-506

Berglund, L., Khoo, J.C., Jensen, D. and Steinberg, D. (1980) *Journal of Biological Chemistry*, **255**, 5420-5428

Bjorntorp, P., Ottosson, M., Rebuffe-Scrive, M. and Xu, X. (1990) In *Obesity: Towards a Molecular Approach*, (ed G.A. Bray, D. Ricquier and B.M. Spiegelman), Alan R. Liss Inc, pp 147-157

Butterwith, S.C. (1988) In *Leaness in domestic birds: Genetic metabolic and hormonal aspects*, (ed B. Leclercq and C.C. Whitehead), London, Butterworths, pp 203-222

Campbell, R.M. and Scanes, C.G. (1988) *Proceedings of the Society for Experimental Biology and Medicine*, **188**, 177-184

Chiappe de Cingolani, G.E. (1986) *Diabetes*, **35**, 1229-1232

Chilliard, Y. and Flechet, J. (1988) *Reproduction, Nutrition and Development*, **28**, 195-196

Cordle, S.R., Colbran, R.J. and Yeaman, S.J. (1986) *Biochimica et Biophysica Acta*, **887**, 51-57

Dax, E.M., Partilla, J.S. and Gregerman, R.I. (1981) *Journal of Lipid Research*, **22**, 934-943

DeMazancourt, P., Lacasa, D., Giot, J. and Giudicelli, Y. (1989) *Endocrinology*, **124**, 1131-1139

DiMarco, N.M., Whitehurst, G.B. and Beitz, D.C. (1986) *Journal of Animal Science*, **62**, 363-369

Doglio, A., Dani, C., Fredrikson, G., Grimaldi, P. and Ailhaud, G. (1987) *EMBO Journal*, **6**, 4011-4016

Etherton, T.D. and Evock, C.M. (1986) *Journal of Animal Science*, **62**, 357-362

Fain, J.N., Mohell, N., Wallace, M.A. and Mills, I. (1984) *Metabolism*, **33**, 289-293

Finley, E., Lindsay, S. and Vernon, R.G. (1990) *Biochemical Society Transactions*, **18**, 461-462

Flint, D.J. and Vernon, R.G. (1992) In *The Endocrinology of Growth, Development and Metabolism in Vertebrates*, (ed M.P. Schreibman, C.G. Scanes and P.K.T. Pang), Orlando, Academic Press, pp 469-494

Fredholm, B.B. (1971) *Acta Physiologica Scandanavica*, **81**, 110-123

Gandarias, J.M., Galdiz, B. and Fernandez, B.M. (1984) *Comparative Biochemistry and Physiology*, **77B**, 385-386

Garton, A.J., Campbell, D.G., Carling, D., Hardie, D.G., Colbran, R.J. and Yeaman, S.J. (1989) *European Journal of Biochemistry*, **179**, 249-254

Giudicelli, Y., Lacasa, D., de Mazancourt, P., Pasquier, Y.N. and Pecquery, R. (1989) In *Obesity in Europe 88*, (eds P. Bjorntorp and S. Rossner), London, John Libbey

Green, A. and Johnson, J.L. (1991) *Diabetes*, **40**, 88-94

Gregory, N.G., Christopherson, R.J. and Lister, D. (1986) *Research in Veterinary Science*, **40**, 352-356

Haji Baba, A.S. and Buttery, P.J. (1991) *Biochemical Society Transactions*, (in press)

Hardie, D.G. (1989) *Progress in Lipid Research*, **28**, 117-146

Haugebak, C.D., Hedrick, H.B. and Asplund, J.M. (1974) *Journal of Animal Science*, **39**, 1026-1031

Hoffman, B.B., Chang, H., Farahbakhsh, Z.T. and Reaven, G.M. (1984) *American Journal of Physiology*, **247**, E772-E777

Hood, R.L. (1982) *Federation Proceedings*, **41**, 2555-2561

Hood, R.L. and Allen, C.E. (1978) *Journal of Animal Science*, **46**, 1626-1633

Hood, R.L. and Thornton, R.F. (1979) *Australian Journal of Agricultural Research*, **30**, 153-161

Hood, R.L. and Thornton, R.F. (1980) *Australian Journal of Agricultural Research*, **31**, 155-161

Khachadurian, A.K., Adrouni, B. and Yacoubian, H. (1966) *Journal of Lipid Research*, **7**, 427-436

Kobatake, T., Watanabe, Y., Matsuzawa, Y., Tokunaga, K., Fujioka, S., Kawamoto, T., Keno, Y., Tarui, S. and Yoshida, H. (1991) *Journal of Lipid Research*, **32**, 191-197

Lacasa, D., Agli, B. and Giudicelli, Y. (1983) *European Journal of Biochemistry*, **130**, 457-464

Lacasa, D., Agli, B., Pecquery, R. and Giudicelli, Y. (1991) *Endocrinology*, **128**, 747-753

Larsen, T.S., Nilsson, N.O. and Blix, A.S. (1985) *Acta Physiologica Scandinavica*, **123**, 97-104

Lee, E.T., Yeaman, S.J., Fredrikson, G., Stralfors, P. and Belfrage, P. (1985) *Comparative Biochemistry and Physiology*, **80B**, 609-612

Lee, K.H. and Kim, K.H. (1979) *Journal of Biological Chemistry*, **254**, 1450-1453

Lee, Y.B. and Kauffman, R.G. (1974) *Journal of Animal Science*, **38**, 532-537

Lewis, K.J., Molan, P.C., Bass, J.J. and Gluckman, P.D. (1988) *Endocrinology*, **122**, 2554-2557

Liu, C.Y., Boyer, J.L. and Mills, S.E. (1989) *Journal of Animal Science*, **67**, 2930-2936

Londos, C., Egan, J.J., Greenberg, A.S. and Wek, S.A. (1991) *Journal of Cellular Biochemistry*, (Suppl) **15B**, 7

Lopez-Casillas, F., Bai, D.-H., Luo, X., Kong, I.S., Hermodson, M.A. and Kim, K.-H. (1988) *Proceedings of the National Academy of Science, USA*, **85**, 5784-5788

Magri, K.A., Adamo, M., Leroith, D. and Etherton, T.D. (1990) *Biochemical Journal*, **266**, 107-113

Malbon, C.C., Rapiejko, P.J. and Watkins, D.C. (1988) *Trends in Pharmacological Sciences*, **9**, 33-36

Mersmann, H.J., Goodman, J.R. and Brown, L.J. (1975) *Journal of Lipid Research*, **16**, 269-279

Mersmann, H.J., Phinney, G. and Brown, L.J. (1975) *General Pharmacology*, **6**, 187-191

Mersmann, H.J., Phinney, G. and Brown, L.J. (1976) *Biology of the Neonate*, **29**, 104-111

Milligan, G. and Saggerson, E.D. (1990) *Biochemical Journal*, **270**, 765-769

Moss, J., Yamagishi, M., Kleinschmidt, A.K. and Lane, M.D. (1972) *Biochemistry*, **11**, 3779-3786

Nougues, P.J. and Vezinhet, A. (1977) *Annales de Biologie Animale, Biochimie, Biophysique*, **17**, 799-806

Plouzek, C.A. and Trenkle, A. (1991a) *Domestic Animal Endocrinology*, **8**, 63-72

Plouzek, C.A. and Trenkle, A. (1991b) *Domestic Animal Endocrinology*, **8**, 73-79

Pond, C.M. and Mattacks, C.A. (1985) In *Functional Morphology of Vertebrates*, (eds H.R. Dunker and G. Fleischer), Stuttgart, Springer, pp 485-489

Pothoven, M.A., Beitz, D.C. and Thornton, J.H. (1975) *Journal of Animal Science*, **40**, 957-962

Prior, R.L. (1978) *Journal of Nutrition*, **108**, 926-935

Robertson, J.P., Faulkner, A. and Vernon, R.G. (1982) *Biochemical Journal*, **206**, 577-586

Roche, J.F. and Quirke, J.F. (1986) In *Control and Manipulation of Animal Growth*, (eds P.J. Buttery, N.B. Haynes and D.B. Lindsay), London, Butterworths, pp 39-52

Rolband, G.C., Furth, E.D., Staddon, J.M., Rogus, E.M. and Goldberg, A.P. (1990) *Journal of Gerontology*, **45**, B174-B178

Ros, M., Northup, J.K. and Malbon, C.C. (1989) *Biochemical Journal*, **257**, 737-744

Rule, D.C., Beitz, D.C. and Hood, R.L. (1987) *Animal Production*, **44**, 454-456

Rule, D.C., Smith, S.B. and Mersmann, H.J. (1989) *Journal of Animal Science*, **67**, 364-373

Saggerson, E.D. (1980) *FEBS Letters*, **115**, 127-128

Saggerson, E.D. (1989) *Biochemical Society Transactions*, **17**, 47-48

Sasaki, S. (1990) *Hormone and Metabolic Research*, **22**, 457-461

Saxena, U., Klein, M.G. and Goldberg, I.J. (1991) *Proceedings of the National Academy of Science, USA*, **88**, 2254-2258

Scow, R.O. and Blanchette-Mackie, E.J. (1985) *Progress in Lipid Research*, **24**, 197-241

Sechen, S.J., Dunshea, F.R. and Bauman, D.E. (1990) *American Journal of Physiology*, **258**, E582-E588

Sidhu, K.S., Emery, R.S., Parr, A.F. and Merkel, R.A. (1973) *Journal of Animal Science*, **36**, 658-662

Smith, S.B. (1984) *Journal of Animal Science*, **58**, 1198-1203

Smith, S.B. and Prior, R.L. (1986) *Journal of Nutrition*, **116**, 1279-1286

Smith, S.B., Jenkins, T. and Prior, R.L. (1987) *Journal of Animal Science*, **65**, 1525-1530

Smith, S.B., Prior, R.L. Ferrell, C.L. and Mersmann, H.J. (1984) *Journal of Nutrition*, **114**, 153-162

Snoswell, A.M., Finley, E. and Vernon, R.G. (1990) *Hormone and Metabolic Research*, **22**, 650-651

Stralfors, P. and Honnor, R.C. (1989) *European Journal of Biochemistry*, **182**, 379-385

Strassheim, D., Milligan, G. and Houslay, M.D. (1990) *Biochemical Journal*, **266**, 521-526

Takai, T., Yokoyama, C., Wada, K. and Tanabe, T. (1988) *Journal of Biological Chemistry*, **263**, 2651-2657

Thornton, R.F. and Tume, R.K. (1984) In *Ruminant Physiology - Concepts and Consequences*, University of Western Australia, pp 289-298

Trenkle, A. and Topel, D.G. (1978) *Journal of Animal Science*, **46**, 1604-1609

Vernon, R.G. (1980) *Progress in Lipid Research*, **19**, 23-106

Vernon, R.G. (1986) In *Control and Manipulation of Animal Growth*, (eds P.J. Buttery, N.B. Haynes and D.B. Lindsay), London, Butterworths, pp 67-83

Vernon, R.G. and Clegg, R.A. (1985) In *New Perspectives in Adipose Tissue*, (eds A. Cryer and R.L.R. Van), London, Butterworths, pp 65-86
Vernon, R.G. and Finley, E. (1988) *Biochemical Journal*, **256**, 873-878
Vernon, R.G. and Flint, D.J. (1989) In *Biotechnology in Growth Regulation*, (eds R.B. Heap, C.G. Prosser and G.E. Lamming), London, Butterworths, pp 57-71
Vernon, R.G. and Sasaki, S. (1991) In *Physiological Aspects of Digestion and Metabolism in Ruminants*, (eds T. Tsuda, Y. Sasaki and R. Kawashima), London, Academic Press, pp 155-182
Vernon, R.G. and Taylor, E. (1988) *Biochemical Journal*, **256**, 509-514
Vernon, R.G., Barber, M.C. and Finley, E. (1991) *Biochemical Journal*, **274**, 543-548
Vernon, R.G., Finley, E. and Watt, P.W. (1991) *Journal of Dairy Science*, **74**, 695-705
Vezinhet, A. and Nougues, J. (1977) *Annales de Biologie Animale, Biochimie, Biophysique*, **17**, 851-863
Vezinhet, A., Nougues, J. and Teyssier, J. (1983) *Reproduction, Nutrition and Development*, **23**, 837-846
Volpe, J.J. and Vagelos, P.R. (1976) *Physiological Reviews*, **56**, 339-417
Walton, P.E. and Etherton, T.D. (1986) *Journal of Animal Science*, **62**, 1584-1595
Walton, P.E., Etherton, T.D. and Evock, C.M. (1986) *Endocrinology*, **118**, 2577-2581
Watt, P.W., Madon, R.J., Flint, D.J. and Vernon, R.G. (1990) *Biochemical Society Transactions*, **18**, 486
Watt, P.W., Finley, E., Cork, S., Clegg, R.A. and Vernon, R.G. (1991) *Biochemical Journal*, **273**, 39-42
Whitehurst, G.B., Beitz, D.C., Pothoven, M.A., Ellison, W.R. and Crump, M.H. (1978) *Journal of Nutrition*, **108**, 1806-1811
Whitehurst, G.B., Beitz, D.C., Cianzio, D. and Topel, D.G. (1981) *Journal of Nutrition*, **111**, 1454-1461
Wrutniak, C. and Cabello, G. (1986) *Journal of Endocrinology*, **108**, 451-454
Xu, X., De Pergola, G. and Bjorntorp, P. (1991) *Endocrinology*, **128**, 379-382
Yang, Y.T. and Baldwin, R.L. (1973) *Journal of Dairy Science*, **56**, 350-365
Yang, Y.T., White, L.S. and Muir, L.A. (1982) *Journal of Animal Science*, **55**, 313-320
Yang, S.-D. and Novakofski, J. (1990) *FASEB Journal*, **4**, A916

4
BONE GROWTH

B.M. THOMSON and N. LOVERIDGE
Bone Growth and Metabolism Unit, Division of Biochemical Sciences, Rowett Research Institute, Bucksburn, Aberdeen, U.K. AB2 9SU, UK

Introduction

Bone serves two functions within the body, acting as a support for the musculature and as a reservoir for calcium and phosphorus during mineral homeostasis. It consists of two distinct cell lineages, bone-forming osteoblasts and bone-resorbing osteoclasts, and the extracellular matrix that these cells secrete and remodel.

Bone formation begins in the embryo, either via a membranous intermediate, as in the case of the flat bones of the skull, or via a cartilaginous intermediate, as in the case of the long bones. Continued production of cartilage at specialised sites at the ends of long bones, termed growth plates, and the subsequent conversion of this cartilage into bone, results in longitudinal, post-natal long-bone growth.

Normal bone development requires that osteoclastic bone resorption is co-ordinated with osteoblastic bone formation. Resorption is necessary for the maintenance of proportional thickness and shape during growth and permits bone to contribute to calcium homeostasis by mobilising bone mineral. Furthermore, continued bone formation and resorption throughout life enables bone to remain a dynamic structure capable of micro-repair and readjustment to changing mechanical needs. Current theories suggest that osteoblasts play a pivotal role in controlling osteoclast activity, allowing bone resorption and formation to be integrated to the needs of the individual.

The processes of bone growth and turn-over require a continuous supply of new, mature bone cells. Osteoblasts are derived from resident mesenchymal progenitors located near to the bone's surfaces, whilst osteoclasts are immigrant cells that arise from the haematopoietic lineage and which arrive at resorption sites via the circulation.

Although bone growth is largely pre-determined by genetic programming, bone cell recruitment and maturation are regulated by two sets of soluble factors. Calciotropic hormones maintain calcium homeostasis by triggering mineral release from the skeleton at times of calcium stress (*eg* pregnancy and lactation), whilst numerous systemic (*ie* growth hormone, oestradiol) and locally derived growth factors (IGF-I, TGFβ, FGF) participate in the control of skeletal growth and morphogenesis. There may be considerable overlap between these two groups however as the calciotropic hormone PTH achieves an indirect anabolic effect via the local induction of the growth factor IGF-I. Skeletal development is also influenced by the mechanical loads placed upon it, a relationship termed Wolf's Law.

Bone architecture

Bone matrix is arranged into two types of macroscopic architecture. In cortical or lamellar bone, (*eg* the shafts of long bones), the matrix is deposited in concentric layers around longitudinal vessels within the structure (Haversian canals). It is the stronger of the two forms and fulfils a mainly mechanical role. The second form, trabecular bone, consists of a lattice-work of bony struts. It is more cellular and hence more metabolically active.

Bone-matrix itself is a composite material and derives its strength from interactions between two dissimilar but complementary components. These are a rigid mineral phase that resists compression and a network of organic fibres that are more able to withstand tension and torsion. In general, highly mineralised bone is stiff, whilst less mineralised bone is more flexible and hence more impact resistant (Pritchard, 1956a; Currey, 1984).

Bone: Growth and remodelling

All bones begin in the early embryo with the formation of mesenchymal condensations, regions of increased cell number and density that act as models or formers upon which future development is based. Subsequent growth requires the proliferation and differentiation of skeletogenic precursor cells, the deposition and mineralisation of extracellular matrix by mature bone forming cells and the remodelling of the resulting tissues by bone-resorbing cells and bone-forming cells acting in concert (for review see Ham, 1969).

GROWTH OF FLAT BONES, *eg* THE BONES OF THE CRANIUM

In the case of the flat bones of the cranium, proliferation of the primitive embryonic mesenchyme gives rise to an avascular, multilayered capsule of closely-packed, fusiform cells that lie around the brain. Vascular invasion of this membranous structure coincides with the appearance of the first osteoblasts in the middle of the condensation and it is these cells that secrete the initial spicules of bone (Thompson, Owens and Wilson, 1989). Osteoblasts lining the newly formed bone surfaces enlarge them by synthesising additional matrix, whilst cells proliferating in the surrounding soft tissue increase the pool of synthetically active osteoblastic cells. These events establish a radiating network of bony spicules lying parallel to the surface of the developing skull, termed an ossification centre. There are usually two such centres for each of the cranial bones. The bony network of each ossification centre is expanded by peripheral bone formation, thickened by the addition of perpendicular elements and cross linked by the formation of secondary trabeculae. The spaces enclosed by the growing network become colonised by vascular tissues and the forerunners of the haemopoietic marrow.

Once the growing bones of the skull have expanded to occupy their definitive territories they make contact with neighbouring bones, whereupon their growth rates decline sharply and skeletal remodelling becomes prominent. This reorganisation serves a gross morphological function, reducing the curvature of the vault of the skull as the cranium increases in size, and produces a change in internal architecture. Initially the cranium consists of only a single plate of bone with some limited internal marrow spaces. As growth continues, a combination of resorption from the interior and continued external deposition gradually converts the structure into a double plate of cortical bone with trabecular bone and marrow spaces in between. This "sandwich structure" both optimises the balance between strength and weight, and provides a large surface area of metabolically active trabecular bone for calcium exchange during homeostasis.

GROWTH OF LONG BONES, *eg* FEMUR

A more complex situation prevails in the cylindrical long-bones of the appendicular skeleton, which begin as mesenchymal condensations in the limb-buds of the developing embryo. Cells located in the middle of these aggregates differentiate into chondrocytes and secrete a cartilage-like extracellular matrix, whilst those cells at the periphery give rise to a perichondral sheath that envelops the structure. This periochondrium consists of an outer fibroblast-rich layer of collagenous connective tissue and an inner layer of proliferating, undifferentiated mesenchymal cells.

Initially the aggregate enlarges both by interstitial growth, as cells in the body of the cartilage divide, enlarge and secrete matrix, and by the addition of new chondrocytes derived from the proliferating cells in the inner layers of the perichondrium.

The most rapid growth occurs at the ends of the rudiment however, as small, irregularly-spaced, rapidly proliferating young cells overgrow their predecessors to form columns of flattened, closely packed cells. These columns lie parallel to the long axis of the bone, their arrangement maintained by bundles of collagen fibrils running longitudinally in the partitions of extracellular matrix between adjacent columns of cells. As the cells age and become buried deeper in the growing cartilage-model, they expand, accumulate glycogen and produce additional extracellular matrix (thereby separating the cells from their neighbours and producing longitudinal growth). The oldest cells are therefore found in the central midsection of the model, where they reach their maximum size and become extremely vacuolated. These chondrocytes are termed hypertrophic.

Meanwhile, the progressive development of the embryo's vasculature leads to the invasion of the perichondrium by capillaries, which signals a switch in the differentiation of the perichondral mesenchyme from the chondrocytic to the osteoblastic lineage. Osteoblasts arising from the peripheral mesenchyme secrete a thin layer of bone, termed the "primary bone collar", which lies around the outside of the mid-section of the cartilage model, between the undifferentiated periosteal/ perichondral mesenchyme and that portion of the aggregate containing the hypertrophic cartilage cells. Subsequent generations of osteoblasts expand the primary bone collar to form a multilayered bony cylinder around the cartilage and extend it towards the ends of the model. In the underlying cartilage itself, the hypertrophic cells secrete alkaline phosphatase and their matrix becomes calcified.

Once the primary bone collar is well established, it is penetrated at several points by small cellular masses derived from the periosteum. These cellular masses contain osteoclasts which rapidly erode the internal calcified cartilage to leave only a supportive framework inside the primary bone collar. The spaces thus formed are filled by proliferating cells derived from the invading tissue, some of which differentiate into osteoblasts whilst others give rise to blood-forming marrow tissue. The invasion extends rapidly towards the end of the model, with osteoblasts depositing bone around the remnants of the calcified cartilage to form an irregular trabecular network. As periosteal bone formation increases the bone collar's thickness and strength, many of these internal trabeculae are no longer required to support the bone collar and they are therefore removed, establishing a short marrow cavity.

CONTINUED GROWTH OF LONG BONES: THE GROWTH PLATE

By the time the processes of invasion and remodelling reach the ends of the bone model, the various elements have assumed a more orderly arrangement, termed an epiphyseal growth plate. Chondrocytes at various stages of maturation are arranged in columns that lie at the ends of the model along the long axis of the bone. Proliferating young chondrocytes at the top of these columns increase the number of cells, their more mature descendants enlarge and secrete extracellular matrix (thereby producing longitudinal growth), whilst the most mature chondrocytes at the end of the column adjacent to the marrow cavity hypertrophy and become entrapped within a honeycomb of calcified cartilaginous matrix. Blood vessels, osteoclasts and cells from the osteoblast lineage invade and destroy the calcified ends of the columns, leaving bars of mineralised cartilage projecting into the marrow cavity. Bone is then deposited around these bars, expanding them to form an irregular network of thin bone trabeculae with calcified cartilage cores. Thus the bone of the epiphyses is continuous with the cartilage of the growth plate. Secondary remodelling of this initial honeycomb of calcified matrix by osteoclasts removes many of its elements whilst those that remain are strengthened by additional bone deposition. Formation is greatest at the periphery, where the new trabeculae are ultimately incorporated into the cylindrical shaft of the growing long bone, and least in the centre, where the supportive trabeculae are eventually resorbed.

THE INTERNAL REMODELLING OF BONES

Bone formation and resorption continue after growth is complete, replacing old matrix (thereby preventing the accumulation of micro-damaged or stress-fractured bone) and contributing to calcium homeostasis by mobilising bone mineral. These turn-over events are localised and, in normal bone, occur without any net change in skeletal form or mass. Bone turnover occurs in both trabecular and cortical bone and in all cases occurs via a cycle of osteoclastic resorption, recruitment of new bone forming cells and subsequent repair of the resorption space by new bone formation. Typically, these processes replace 5-11% of cortical bone and 14-44% of trabecular bone per annum.

In cortical bone, turnover involves the excavation of an elongated chamber parallel to the long axis of the bone by osteoclasts, and the subsequent refilling of this space by successive generations of osteoblastic cells. The process forms a cylinder of new bone matrix (termed an osteon), that is aligned with the prevailing loads for optimal mechanical efficiency, and which retains blood vessels in its narrow central lumen for nutrient supply (for review see Lacroix, 1971). Matrix synthesis is most rapid at the beginning of the replacement process (2 μm/day) and very slow at the end (Goss, 1978). In a 30-year-old human male there are approximately 800 osteons/cm^2 of cortical bone of which 10 will be undergoing remodelling. The rate of osteon formation varies within the skeleton (5 times higher in the metaphases than the diaphyses), and with age, being highest in young animals. About 2% of osteons show evidence of either a temporary pause in refilling or more than one cycle of resorption and reformation. Curiously, whereas adult refilling consists of concentric lamellae, replacement in growing individuals is often highly eccentric with formation on one side of the resorption space and continuing resorption on the other side. The reason for this Haversian drift is unknown (Lacroix, 1971).

Bone turnover on the surfaces of trabecular bone involves analogous foci of bone destruction and replacement to those observed in cortical bone.

Cell types involved

OSTEOBLASTS

Osteoblasts were originally defined as non-replicating fully differentiated end cells which synthesised osteoid and participated in mineralisation. More recent studies however, have shown that cells of the osteoblast lineage are heterogeneous, with flexible phenotypes and variable functions. It is now thought that osteoblasts may either form bone, regulate osteoclastic bone resorption or do neither, in response to local and systemic control (Chambers, 1985).

Osteoblast structure

In young animals with growing bones, osteoblasts are arranged on newly-formed bone trabeculae as a single-layered pseudoepithelium with the cells in lateral contact. Active osteoblasts are generally compact and squat (20-30 μm), although cells in areas of particularly high matrix accretion tend to be more columnar and highly packed. Osteoblasts have a single eccentric nucleus located away from the bone surface. The nucleus is large, hypochromatic and either spherical or ovoid with one to three nucleoli. Osteoblast cytoplasm is highly basophilic and, in synthetically active cells, displays prominent endoplasmic reticula and Golgi apparatus. Active osteoblasts are rich in glycogen and alkaline phosphatase, contain many short, thick rod-shaped mitochondria and a large number of small vesicles, especially in the cytoplasm adjacent to the matrix (Pritchard, 1956b, 1971).

Osteoblast cell membranes show an irregular contour, especially at the matrix surface, where many fine cell processes extend into the underlying osteoid and mineralised bone. These processes make contact with adjacent osteoblasts and buried osteocytes via tight and gap junctions (Furseth, 1973) so that osteoblasts and osteocytes form a continuum that penetrates and pervades a volume of bone about 100 μm deep. The osteoblast surface opposing the bone is often in contact with blood capillaries.

Osteoblasts and matrix synthesis

Osteoblasts synthesise bone matrix, depositing 2-3 μm thick layers of unmineralised material via the base of the cell onto pre-existing connective tissue surfaces. Typical cells produce approximately 3 times their own volume of matrix in about 3 days (Owen, 1963), although appositional rates vary considerably, both in time, between skeletal regions and even between adjacent areas on the same trabecular bone surface. The newly-formed matrix (termed osteoid) consists primarily (90%) of a network of highly cross-linked undenatured type I collagen fibrils (60-100 nm diameter; Schenk, Hunziker and Herrmann, 1982; Eastoe, 1956) and these provide the bone's tensile strength (for a review of collagen biochemistry, see Piez, 1984; Robins, 1988). Also present are a number of non-collagenous proteins (*eg* sialoprotein, osteocalcin and osteonectin; Termine, 1990) and, intriguingly, high concentrations of polypeptide growth factors (*eg* transforming growth factor β and insulin-like growth factor-II; Hauschka, Mavrakos, Iafrati, Doleman and Klagsbrun, 1986). Osteoid is highly acidophilic, PAS-positive and orthochromatic. Matrix synthesis produces structures that are much larger than individual osteoblasts, including extensive regions of parallel collagen bundles. Thus "sheets" of osteoblasts must act in concert during bone formation (Boyde, 1972) and must be subject to local control.

The newly-synthesised matrix is then mineralised, a process that requires crystal nucleation and regulated crystal growth. In embryonic bone (and mineralised cartilage), the nucleation sites are provided inside spherical (100-200 nm diameter) membrane vesicles that are shed from the cell's plasma membranes into the extracellular matrix space. Subsequent crystal growth within these vesicles ruptures their membranes, releasing a sphere of tiny (5x20x40 nm) calcium phosphate microcrystalites, which grow by adsorbing calcium and phosphate ions from the tissue fluid. The rate at which these crystals grow is constrained by the release of inhibitory molecules (*eg* pyrophosphate and non-collagenous proteins) by the bone cells. Matrix vesicles are rarely if ever seen in lamellar bone and instead nucleation probably begins at sites along the surface of the collagen fibres that are rich in non-collagenous proteins.

Mineralisation never occurs in contact with the cell surface but advances at about 1-2 μm per day, approximately 10 μm below the osteoblast cell layer (between 7 hours and 10 days after the initial synthesis of the matrix; Schenk *et al*, 1982). It continues briefly after matrix deposition stops, and finishes at the *lamina limitans* several hundred nanometres below the bone surface (Scherft, 1972). All bone is therefore synthesised with a thin covering of unmineralised osteoid separating the cells from the underlying mineralised bone.

The tightly packed mosaic of microcrystalites produced provides a large surface area for ion exchange (Goose and Appleton, 1982) and should delay the spread of fatigue fractures across the bone, as propagating cracks would tend to be arrested at crystalite surfaces (Forsyth, 1969; Currey, 1984). This second point is interesting because hip fractures occur more frequently in regions of bone with higher than average crystal size (Kent, Dodds, Watts, Bitensky and Chayen, 1983).

Osteocytes and bone lining cells

Some osteoblasts become entombed within the matrix that they have synthesised, and are termed osteocytes. Such cells display a reduced Golgi apparatus and endoplasmic reticulum and have fewer mitochondria than actively-secreting cells. The osteocytes in young bone are relatively rich in cytoplasm and are linked to neighbouring cells by a few short processes, whilst, those in mature lamellar bone are flattened and ovoid, with fine, branching processes that emerge roughly every 1.9 μm of the osteocyte perimeter. The processes are packed with intermediate filaments and run inside canaliculae which stain brilliantly with certain basic dyes suggesting they are lined with mucopolysaccharides (Ham, 1969). The canaliculae communicate with other osteocyte lacunae and ultimately with the bone surface or vascular channels, linking the osteocytes with their nutrient supply.

Osteocytes tend to be regularly spaced and aligned along bone lamellae (Pritchard, 1956b). There are about 20,000 osteocytes per cubic mm in cortical bone. That this number is constant suggests entombment in a controlled rather than random event. Although the role of osteocytes is unknown they have been hypothesised to contribute to calcium homeostasis by micro dissolution of bone-mineral. An alternative function is as a sensory array to perceive bone flexion prior to signalling for bone secretion or resorption at the bone surface.

In adults, when bone growth has essentially ceased, plump secretory osteoblasts disappear and are replaced by biosynthetically dormant cells. Inactive osteoblasts are extremely thin, flattened squamous cells, very closely applied to the bone. They have a central flattened nucleus and sparse mitochondria. Precursor cells remain, however, because mechanical or irradiation damage induces the return of synthetically active cells (Pritchard, 1956b, 1971).

Osteoblasts and the control of bone resorption

It has recently been shown that cells from the osteoblast lineage are central to the control of bone resorption (Chambers, 1985; Chambers, McSheehy, Thomson and Fuller, 1985). They possess receptors for the major systemic (*eg* PTH and $1,25(OH)_2D_3$) and local (*eg* IL1, TNF, prostaglandins) bone resorptive agents (Manolagos, Haussler and Deflos, 1980; Partridge, Frampton, Eisman, Michaelangeli, Elms, Bradley and Martin, 1980; Martin and Partridge, 1981; Silve, Hradek, Jones and Arnaud, 1982; Bird and Saklatvala, 1986), and respond to these agents by recruiting osteoclast precursors and activating mature bone resorbing cells (Thomson, Saklatvala and Chambers, 1986, Thomson, Mundy and Chambers, 1987; McSheehy and Chambers, 1986a, 1987). Osteoblasts can also release osteoclasts from calcitonin induced quiescence (Chambers and Magnus, 1982). These findings suggest that by switching from bone formation to the control of bone resorption, osteoblasts may co-ordinate bone turnover in vivo (Chambers, 1980; Rodan and Martin, 1981).

One way by which osteoblasts may contribute to the initiation of bone resorption is by producing matrix degrading enzymes (Heath, Atkinson, Meikle and Reynolds, 1984; Hamilton, Lingelbach, Partridge and Martin, 1984, 1985; Thomson, Atkinson, Reynolds and Meikle, 1987 a, b; Thomson *et al*, 1989; Meikle, McGarrity, Thomson and Reynolds, 1991). These may function *in vivo* to remove the thin layer of unmineralised osteoid that remains on the surface of bones and which appears to shield the underlying calcified matrix from osteoclast attack. (Osteoclasts will only attack mineralised substrates (Chambers, Thomson and Fuller, 1984). It has therefore been hypothesised that osteoid acts as a passive barrier to stray osteoclast action, and that by removing this layer resident osteoblasts provide a permissive mechanism for initiating bone resorption (Chambers, Darby and Fuller, 1985; Chambers and Fuller, 1985).

Osteoblasts also appear to be able to signal directly to mature osteoclasts to increase bone resorption in response to both local and systemic bone resorbing hormones (Thomson *et al*, 1986, 1987; McSheehy and Chambers, 1986a, 1987), an interaction that may involve the production of a soluble mediator (McSheehy and Chambers, 1986b; Perry, Skogen, Chappel, Wilner, Kahn and Teitelbaum, 1987). In addition, cells from the osteoblast lineage may participate in osteoclast recruitment: the invasion of developing bone rudiments by osteoclasts requires cellular activity by non-osteoclastic cells within the embryonic bone (Burger, Thesingh, van der Meer and Nijweide, 1984; Burger, van der Meer and Nijweide, 1984), whilst osteoblast-like MC3T3-E1 cells produce colony-stimulating activity (CSA) both constituitively (Shiina-Ishimi, Abe, Tanaka and Suda, 1986) and in response to recombinant IL-1 α, IL-1 β and TNF α (Sato, Fujii, Asano, Ohtsuki,

Kawakami, Kasono, Tsushima and Shizume, 1986; Sato, Kasono, Fujii, Kawakami, Tsushima and Shizume, 1987).

The origin of the osteoblast

Osteoblasts are members of a super-family of mesenchymal cells that includes chondrocytes, adipocytes and a variety of more fibroblastic cell types (Friedenstein, 1980; Friedenstein, Chailakhyan and Gerasimov, 1987; Owen, 1978, 1980, 1988). They are derived from self-sustaining populations of multipotential stem-cells and mature via a sequence of events that results in a reciprocal gain in function and a loss of proliferative capacity. Thus, during bone growth, a population of stem-cells are thought to give rise to clones of committed osteogenic precursors, which in turn proliferate and begin to express lineage specific markers (pre-osteoblasts) before maturing into fully differentiated, post-mitotic osteoblastic cells.

The earliest recognisable precursor to the osteoblast cell lineage is the "inducible" osteogenic precursor cell. This cell-type is detectable in experimental systems using non-skeletal tissues (*eg* spleen and muscle) and does not form bone unless artificially induced to do so by external agents, *eg* bone matrix extract (Urist, Delange and Fineran, 1983). It may represent a sparse population of highly undifferentiated mesenchymal cells that persist in adult tissues and which may normally function in wound repair. In contrast, "committed" osteogenic stem cells (found in bone marrow stroma and periosteum) spontaneously form bone, cartilage, fat and fibrous tissues in a variety of experimental models. These stem-cells have been estimated to have a short cell cycle time (14.5 h) and are capable of undergoing 13-14 population doublings prior to the expression of differentiation markers *in vitro* (Bab, Ashton, Gazit, Marx, Williamson and Owen, 1986). Maturation of these cells is thought to produce the recognisable pre-osteoblast observed in histological preparations. These are cells with large nuclei that lie in close proximity to forming bone surfaces, which secrete type I collagen and express early markers of osteoblast differentiation, *eg* alkaline phosphatase and osteonectin mRNA, but which lack the intense alkaline phosphatase activity and highly developed endoplasmic reticulum of the mature cell (Yoon, Buenaga and Rodan, 1987; McCulloch, Fair, Tenenbaum, Limeback and Homareau, 1990; Roberts, Mozsary and Klingler, 1982). Terminal differentiation is associated with the display of late markers *eg* osteopontin and osteocalcin (Yoon, *et al*, 1987) and coincides with marked changes in gene transcription and nuclear protein-DNA interaction. There is a decrease in core and H1 histone gene expression, a selective loss of interactions between HiNF-D and the site II region of an H4 histone gene promoter and a dramatic drop in the steady state levels of c-fos and c-myc mRNA. Simultaneously the expression of a structurally distinct set of histone genes (that are not coupled with DNA replication) is increased (Shalhoub, Gerstenfeld, Collart, Lian and Stein, 1989; Owen, Aronow, Shalhoub, Barone, Wilming, Tassinari, Kennedy, Pockwinse, Lian and Stein, 1990). Further changes in nuclear proteins also coincide with the onset of matrix mineralisation (Dworetzky, Fey, Penman, Lian, Stein and Stein, 1990). These events may in part be triggered by changes in c-fos expression (Ruther, Garber, Komitowski, Muller and Wagner, 1987; Closs, Murray, Schmidt, Schon, Erfle and Strauss, 1990) and by interactions between the cells and specific components in the extracellular matrix (Owen *et al*, 1990).

OSTEOCLASTS

Osteoclasts are large (3,000-250,000 μm^3), multinucleate (up to 100 nuclei), bone-resorbing cells that participate in localised skeletal remodelling and contribute to calcium homeostasis by mobilising bone mineral (Hancox, 1956, 1971). Osteoclast action begins in the embryo and continues throughout life, under strict local and systemic control. Osteoclasts can penetrate 50-70 μm into cortical bone in 24 hours and, in a given period of time, are able to destroy a volume of bone equivalent to that formed by 100-1000 osteoblasts (Schenk *et al*, 1982).

Structure

Most osteoclasts have 10-20 nuclei, although the average number of nuclei is species dependent, (cat > human > rabbit > rat) and varies considerably within the individual (Hancox, 1971). The nuclei are generally oval, with 1 or 2 prominent nucleoli, and are clustered at the centre of the cell, away from the pooled centrioles (Marks, 1983). Osteoclasts possess an extensive Golgi apparatus and numerous mitochondria but a sparse endoplasmic reticulum and few ribosomes (Hancox, 1971). Multinuclearity does not appear essential for osteoclast function as calcitonin responsive, bone resorbing, acid phosphate-positive, mononuclear cells have been observed *in vitro* (Chambers and Magnus, 1982; Kaye, 1984).

The osteoclast's bone resorptive apparatus consists of a central "ruffled border", the site at which resorption occurs, and a peripheral "clear zone", which acts as a seal. The ruffled border is a sharply delineated region of cytoplasmic folds and projections which interdigitate with the underlying bone providing a large surface area for the extrusion of lytic cell products and the uptake of degraded material from the resorbing bone (Hancox, 1956). The cytoplasm adjacent to the ruffled border is highly vacuolated and contains large numbers of coated and uncoated vesicles (diameter 0.02-3 μm). These include phagosomes pinched from the ends of cellular invaginations, lysosomes and exocytotic vesicles. Many contain acid phosphatase, an enzyme used as a histochemical marker for osteoclastic cells (Hancox, 1971).

Encircling the ruffled border and forming the marginal rim of the resorbing osteoclast is the clear zone. This structure is believed to anchor osteoclasts to the bone surface, isolating a micro-environment in which extruded cell products can be localised to give a pH and enzyme concentration suitable for osteolysis. The cell membrane in this region is smooth and follows closely the fine contours of the underlying bone. The cytoplasm above it is devoid of vesicles and mitochondria

(Bonucci, 1980) but contains a belt of vimentin-type intermediate filaments (Marchisio, Cirillo, Naldini, Primavera, Teti and Zambonin-Zallone, 1984) and numerous actin fibres, 5-10 nM in diameter (King and Halthrop, 1975). Some of these actin fibres show a "random" orientation, whilst others are arranged as bands perpendicular to the surface of the bone. The perpendicular fibres often end in short processes that contain vinculin, α-actinin and fimbrin (termed podosomes). They make contact with indentations in the bone surface and may constitute an osteoclast specific adhesion device (Marchisio *et al*, 1984). Adhesion may involve interactions between integrin receptors on the cell surface and specific matrix components (Teti, Blair, Schlesinger, Grano, Zambonin-Zallone, Kahn, Teitelbaum and Hruska, 1989; Glowacki, Rey, Cox and Lian, 1990).

Function

Osteoclasts resorb both the mineral and organic phases in bone (Chambers, Thomson and Fuller, 1984; Chambers, 1985). The initial degradative processes appear to be predominantly extracellular and to occur adjacent to the ruffled border, within the micro-environment delineated by the clear zone (Vaes, 1988). Scanning electron microscopy shows that dissolution of the mineral phase precedes the removal of the organic matrix, producing a transient 1-2 μm thick layer of demineralised matrix below the resorbing osteoclast (Parfitt, 1984).

The available evidence suggests that dissolution of bone mineral is effected by the secretion of hydrogen ions by proton pumping ATPases located on the osteoclast's ruffled border. This is consistent with morphological similarities between resorbing osteoclasts and H^+ secreting parietal cells (Ham, 1969), and accords with the observation that stimulation of bone resorption increases osteoclast acidity (pH 4.5-4.8), whilst calcitonin diminishes it (Vaes, 1988). H^+/K^+ ATPases , "kidney type" proton pumping ATPases and Na^+/H^+ antiports have all been reported in osteoclasts (Baron, Neff, Louvard and Courtoy, 1985; Ghiselli, Blair, Teitelbaum and Gluck, 1987; Hall and Chambers, 1990), whilst the acidification of active osteoclasts can be demonstrated with pH sensitive dyes (Baron *et al*, 1985) and direct micro-puncture techniques (Baron *et al*, 1985). Osteoclast function also requires the activity of both chloride/bicarbonate exchangers and Na^+/K^+ ATPase, although these may function to drive secondary active transport systems (Baron *et al*, 1985; Baron, Neff, Lippincott-Schwarz, Louvard, Mellman, Helenius and Marsh, 1986; Anderson, Woodbury and Jee, 1986; Tuukkanen and Vaananen, 1986). In addition, osteoclast activity requires carbonic anhydrase, an enzyme that is also required for acid production in parietal cells (Hall and Kenny, 1985a, 1985b).

The dissolution of bone mineral is followed by the degradation of the exposed organic matrix components. At the acidic pH found adjacent to the ruffled border, collagen can be efficiently degraded by lysosomal enzymes and it seems likely that these enzymes are exocytosed across the ruffled border into the sub- osteoclastic space. At least three distinct cysteine proteases from bone are able both to depolymerise fibrillar collagen at acid pH, (causing the collagen monomers to denature) and to degrade the resultant gelatin. The action of these enzymes should be enhanced by the high calcium concentration that presumably exists beyond the ruffled border (Vaes, 1988). The suggestion that osteoclast thiol-proteinases might act as collagenolytic enzymes during the destruction of mineralised matrix is supported by the observation that the inhibition of thiol proteases blocks osteoclastic bone resorption (Delaisse, Eekhout and Vaes, 1984).

The origin of the osteoclast

Multinucleate osteoclasts arise from the fusion of mononuclear precursor cells. The precursors are derived from the haemopoietic lineage and arrive at sites of bone resorption via the circulation (Gothlin and Ericsson, 1973; Marks, 1983; Schneider, Relfson and Nicholas, 1986). The earliest osteoclast precursor is therefore the IL-3 dependent haemopoietic stem-cell (Scheven, Visser and Nijweide, 1986). Later pre-osteoclastic intermediates are thought to include granulocyte colony forming cells (Mundy, 1990).

In mammals, primitive osteoclast precursors exist before the appearance of the embryo's long bones (Scheven, Kawilarang-De Haas, Wassenaar and Nijweide, 1986). They originate in the yolk sac and from there migrate via the vasculature to the embryo's tissues and haematopoietic organs (Thesingh, 1986; Burger, Thesingh, van der Meer and Nijweide, 1984; Burger, van der Meer and Nijweide, 1984).

Elegant histochemical studies on mouse foetal metatarsals show that after 15 days gestation, tartrate-resistant acid phosphatase negative (TRAP-ve) osteoclast precursor cells arrive and proliferate in the metatarsal periosteum. These cells have become post-mitotic on day 16 and subsequently differentiate, expressing TRAP and calcitonin receptors on day 17. On day 18 the pre-osteoclasts fuse to form TRAP+ve multi-nucleate cells. These multinucleate cells are active, mature osteoclasts which resorb calcified cartilage to form a marrow cavity (Scheven, Burger, Kawilarang-De Haas, Wassenaar and Nijweide, 1985; Scheven, Visser and Nijweide, 1986; Taylor, Tertineggi, Okuda and Heersche, 1989).

The developmental sequence requires cellular activity by non-osteoclastic cells within the embryonic bone supporting the concept that resident bone cells support osteoclast recruitment and activation (Burger, Thesingh, van der Meer and Nijweide, 1984; Burger, van der Meer and Nijweide, 1984).

Osteoclasts have a half-life of 6-10 days *in vivo*, leaving the endosteal surface upon the cessation of resorption and migrating into the adjacent marrow space. There they may degenerate and die (Chambers, 1985).

Regulatory factors involved

Although bone growth is largely pre-determined by genetic programming, bone cell recruitment and maturation are regulated by mechanical influences and two sets of soluble factors. Calciotropic hormones maintain calcium homoeostasis by triggering mineral release from the skeleton at times of calcium stress (*eg* pregnancy and lactation), whilst numerous systemic (*ie* growth hormone, oestradiol) and locally-derived growth factors (IGF-I, TGFβ, FGF) participate in the control of skeletal growth and morphogenesis. Bone matrix itself contains a complex mixture of polypeptide growth factors and osteogenic substances and these are presumed to play an important role in the development and maintenance of bone tissue.

GENETIC

Embryonic bone growth is a remarkably autonomous process. Immature bones and foetal femoral-heads continue to grow in organ culture, indicating that they contain the necessary developmental information for morphogenesis, cellular maturation, matrix synthesis and calcification. Thus intrinsic, presumably genetic, control mechanisms lie at the heart of skeletal development. Furthermore, the changing proportions of the skeleton during post-natal life (*eg* the relative decrease in the size of the head) suggest that these genetic control mechanisms continue beyond birth. Skeletal growth mechanisms contain some flexibility however, as if one bone from the developing cranium is destroyed, its neighbours will expand to occupy its usual territory (Dickson, Koenig and Silbermann, 1991; for review see Goss, 1978).

MECHANICAL

A wide variety of clinical and physiological observations testify to the influence of physical forces on skeletal development in growing and adult oganisms. In general, an increase in physical activity leads to an increase in bone mass (but not length), whilst a reduction in loading results in bone loss. Thus the cortical width of the humerus in a professional tennis player's serving arm is increased by 35% relative to his other arm (Jones, Priest and Hayes, 1977), whilst paralysis, weightlessness or prolonged bed-rest all lead to bone loss. This relationship between mechanical loading and bone formation is known as Wolf's Law (for review see Rubin and Hausman, 1988).

At the level of a small volume of bone, the mechanical loads imposed upon the skeleton by gravity and locomotion result in "strain" (defined as change in length/original length). It is remarkable therefore that the size and architecture of the bones from many species restricts the peak functional strain experienced by an individual volume of bone to within very narrow limits (0.003-0.004 strain). This suggests that the skeleton monitors its mechanical environment and adjusts local bone formation and resorption until an optimal strain environment is produced (Rubin, 1984). Although the mechanism(s) by which bone perceives mechanical loading remain totally unknown, one candidate may be the osteocyte network. These cells penetrate and pervade a volume of bone 100 μm deep and therefore are in the correct position to sense bone flexion. Furthermore they show increased [^{3}H] uridine incorporation and glucose-6-phosphate dehydrogenase activity following bone loading (Lanyon, Rawlinson and Ali, 1991), indicating that they are capable of responding to mechanical stimulation. Bone cells appear to be preferentially sensitive to mechanical stimuli at frequencies much higher than those experienced during locomotion (10-20 Hz) suggesting that it may be the amplitude of specific resonant frequencies or harmonics that initiate cellular responses (Rubin and Hausman, 1988).

Mechanical influences appear to be quantitative rather than qualitative however as removal of mechanical influences though reducing the size of skeletal structures rarely result in their total absence or loss. This suggests that genetic influences are responsible for dictating the basic structure of the skeleton, whilst responses to mechanical loading adjust the bones to their functional environment.

CALCIOTROPIC HORMONES

The calciotropic hormones maintain plasma calcium within tight limits. As the skeleton contains around 99% of the body calcium it has to be accessed during times of calcium stress such as pregnancy and lactation. The major calciotropic hormones are parathyroid hormone, 1,25 dihydroxyvitamin D and calcitonin.

Parathyroid hormone (PTH)

PTH is an 84-amino acid polypeptide released in response to systemic hypocalcaemia. It promotes calcium retention by the kidney and uptake from the gut (via increased $1,25(OH)_2D_3$ synthesis), and mobilises skeletal mineral by inhibiting calcium influx into bone and promoting bone resorption.

Skeletal response to PTH is concentration dependent. High concentrations of PTH produce an initial resorptive response and a secondary increase in bone formation (Selye, 1932; Herrman-Erlee, Heersche and Hekkelman, 1976), whilst lower concentrations of PTH stimulate osteoblastic proliferation and markers of bone formation without eliciting the bone-resorptive effect. The exact physiological concentrations at which these skeletal responses occur are difficult to determine, but the anabolic effects require intermittent rather than continuous exposure to PTH (Tam, Heersche, Murray and Parsons, 1982).

PTH binds to receptors on cells from the osteoblastic lineage, activating both the cAMP and phosphoinositide second messenger systems. Elegant *in vivo* studies indicate that the principal target cell is a pre-osteoblastic cell-type located just above the cells lining the bone surfaces (Rouleau, Mitchell and Goltzman, 1988). Osteoblast responses to PTH *in vitro* include increased alkaline phosphatase expression (a marker enzyme of bone formation; McPartlin, Skrabanek and Powell, 1978; Loveridge, Dean, Goltzman and Hendy, 1991) and enhanced mitosis, an effect linked to increased glucose-6-phosphate dehydrogenase activity (Bradbeer, Dunham, Fischer, Nagant De Deuxchaisnes and Loveridge, 1988) and the induction of IGF-1 production (Canalis, Centrella, Burch and McCarthy, 1989). The mature osteoclast does not have receptors for PTH and instead the action of this hormone on bone resorption appears to be mediated through the osteoblast, presumably through the release of factors that stimulate osteoclast activity (McSheehy and Chambers, 1986 a, b).

The role of a PTH-like peptide is currently under investigation (Martin, Allan, Caple, Care, Danks, Diefenbachjagger, Ebeling, Gillespie, Hammonds, Heath and Hudson, 1989). This peptide, which shares amino-terminal homologies with PTH, acts in an analogous manner to PTH in a number of *in vivo* and *in vitro* systems (Yates, Guiterrez, Smolens, Travis, Katz, Aufdemorte, Boyce, Hymer, Poser and Mundy, 1988; Loveridge, Dean, Goltzman and Hendy, 1991). It may contribute to calcium homeostasis during pregnancy and lactation as well as contributing to the aetiology of hypercalcaemia of malignancy.

Vitamin D metabolites

Vitamin D_3 is essential for normal skeletal growth and development. It is made in the skin by the action of sunlight on 7-dehydrocholesterol and is available from the diet in fish oils and margarines. Vitamin D_3 is hydroxylated at the 25 position in the liver to form the major circulating metabolite ($25(OH)D_3$), and at the 1 position in the kidney to form the most potent form, $1{,}25(OH)_2D_3$. This second hydroxylation is controlled by a number of factors including hypo- and hypercalcaemia, PTH and circulating $1{,}25(OH)_2D_3$. $1{,}25(OH)_2D_3$ has a number of biological functions which include increasing intestinal calcium absorption and promoting bone-matrix mineralisation (for review see Henry and Norman, 1990). It is becoming increasingly apparent however that $1{,}25(OH)_2D_3$ plays a major role throughout the body in the control of cell proliferation and differentiation, possibly through regulation of cellular oncogenes (Minghetti and Norman, 1988). Thus $1{,}25(OH)_2D_3$ influences chondrocyte differentiation within the growth-plate and exerts effects on osteoclast precursors (Bar-Shavit, Teitelbaum, Reitsma, Hall, Pegg, Trail and Kahn, 1983).

$1{,}25(OH)_2D_3$ acts in a conventional manner for steroid hormones, binding to intracellular receptors which then translocate to the nucleus and influence the transcription of specific genes. In the osteoblast, $1{,}25(OH)_2D_3$ regulates a number of genes associated with matrix production (collagen I and fibronectin) and calcification (alkaline phosphatase and osteocalcin), whilst both $25(OHD_3$ and $1{,}25(OH)_2D_3$ appear to enhance the response of chondrocytes and bone cells to PTH (Endo, Kiyoki, Kawashima, Naruchi and Hashimoto, 1980; Bradbeer, Mehdizadeh, Fraher and Loveridge, 1988). A second dihydroxylated metabolite, $24{,}25(OH)_2D_3$, also produced by the kidney, may have a specific role in matrix mineralisation (Ornoy, Goodwin, Noft and Edelstein, 1978).

Calcitonin (CT)

Calcitonin is a 32-amino acid polypeptide that is produced by the thyroidal C-cells in response to hypercalcaemia and which acts to inhibit bone resorption. Unlike the other calciotropic hormones which affect osteoclasts indirectly, CT binds directly to mammalian osteoclasts (Tashjian, Wright, Ivey and Pont, 1978), inhibiting their motility and resorptive function (Chambers and Dunn, 1983; Chambers *et al*, 1985). The physiological role of CT is uncertain however, as skeletal disease does not follow from abnormalities in CT production.

OTHER SYSTEMIC HORMONES

There are a number of other systemic hormones that are known to influence skeletal physiology. These include growth hormone, thyroid hormones, oestradiol and vitamin A.

Growth hormone

Growth hormone stimulates longitudinal bone growth, increasing mitosis amongst the proliferating chondrocytes of the growth plate and increasing chondrogenesis and osteogenesis in in vitro systems (Isaksson, Lindahl, Nilsson and Isgaard, 1987; Maor, Hockberg, von-der-Mark, Heinegard and Silbermann, 1989). It may either act directly or via increased local and systemic production of IGF-1. Growth hormone also affects carbohydrate, protein and lipid metabolism, producing short term insulin-like effects, especially in GH-deficient individuals. In most species foetal and early post-natal growth proceed normally in the absence of growth hormone, until shortly after birth, whereupon continued growth becomes growth hormone dependent (for review see Isaksson *et al*, 1987).

Oestradiol

Whilst there is no doubt that the removal of oestradiol following normal or artificially induced menopause leads to increased bone loss, oestradiol's role in skeletal biology and target cell-type remains obscure. Although for many years its actions were believed to be mediated via indirect mechanisms involving either systemic (calcitonin; Stevenson, White, Joplin and MacIntyre, 1982) or local mediators (IGF-1; Gray, Mohan, Linkhart and Baylink, 1989), recent work has indicated the presence of oestradiol receptors in osteoblastic cells, suggesting some direct effect (Eriksen, Colvard, Berg, Graham, Mann, Spelsberg and Riggs, 1988).

Vitamin A

In vivo effects of vitamin A were first demonstrated by Mellanby (1947) who showed its importance in the maintenance of osteoclast populations and bone remodelling. It may also act as a morphogen in limb bud development.

Thyroid hormones

Osteoblastic cells have receptors for thyroid hormones (Rizzoli, Poser and Burgi, 1986) and respond to them with increased expression of differentiated function, *eg* enhanced alkaline phosphatase expression and increased PTH responsiveness

(Sato, Han, Fujii, Tsushima and Shizume, 1987; Schmid, Steiner and Froesch, 1986). Thyroid hormones also decrease osteoblast-like cell proliferation (Kasono, Sato, Han, Fujii, Tsushima and Shizume, 1988).

POLYPEPTIDE GROWTH FACTORS AND BONE

Transforming growth factor beta

Transforming growth factor beta (TGFβ) belongs to a class of local hormones that regulate cell growth, control phenotype expression and supply positional information in organisms ranging from drosophila to man (Sporn, Roberts, Wakefield and Assoian, 1986). Expression is associated with morphogenic changes in tissues derived from embryonic mesenchyme, (*eg* the emergence of the digits from the limb buds and the formation of the vertebrae, jaws and palette), and with developmental events where mesenchyme and epithelia interact (*eg* tooth bud formation; Heine, Munoz, Flanders, Ellingsworth, Lam, Thompson, Roberts and Sporn, 1987).

Although essentially ubiquitous, TGFβ is particularly abundant in bone (200 μg/kg; Noda and Rodan, 1986); 100 times higher than in soft tissue (Sporn *et al*, 1986). TGFβ localises to osteoblasts at sites of intramembranous and endochondral ossification (Roberts and Sporn, 1990), and, apparently by modulating the phenotype and activity of skeletogenic precursor cells, ultimately increases bone-matrix deposition (Pfeilshifter, Oechsner, Naumann, Gronwald, Minne and Ziegler, 1990; Hock, Canalis and Centrella, 1990; Joyce, Roberts, Sporn and Bolander, 1990; Carrington and Reddi, 1990).

In common with most cells, osteoblasts have receptors for TGFβ (Sporn *et al*, 1986). They also produce it, suggesting an autocrine role (Robey, Young, Flanders, Roche, Kondaiah, Reddi, Termine, Sporn and Roberts, 1987). Bone resorbing hormones (PTH, $1,25(OH)_2D_3$ and IL1) all increase TGFβ release from organ cultured bone and it has been suggested that active TGFβ is liberated from bone matrix during osteoclast action (Pfeilschifter and Mundy, 1987). This may indicate a role for TGFβ in the coupling of bone formation and resorption. Interestingly, TGFβ is also self-inducing (Van Obberghen-Schilling, Roche, Flanders, Sporn and Roberts, 1988) and does not down regulate its own receptor (Sporn *et al*, 1986).

Unfortunately, the nature of *in vitro* responses to TGFβ appears to depend upon the exact dose, time and differentiated state of the responding cells and the literature frequently conflicts. TGFβ is mitogenic for cells in organ cultured calvaria, and can thereby lead to a subsequent increase in alkaline phosphatase expression and collagen synthesis (Centrella, Massague and Canalis, 1986). TGFβ inhibits phenotype expression in monolayer cultures of osteoblast-like cells however, reducing PTH responsiveness (Guenther, Cecchini, Elford and Fleisch, 1988), diminishing osteocalcin production in response to bFGF (Globus, Plouet and Gospodarowicz, 1989) and causing an almost total suppression of alkaline phosphatase content (Centrella, McCarthy and Canalis, 1987a; Guenther *et al*,

1988). Furthermore, although there is little doubt that TGFβ can modulate osteoblastic cell proliferation, its effects appear to be dose, time and cell-density dependent, and no consensus has yet emerged as to its precise effects.

Fibroblast growth factors

Acidic and basic fibroblast growth factors (a/bFGF) are related mitogenic and angiogenic factors that were initially believed to be involved in wound healing. They promote osteoblast proliferation and appear to be particularly effective on less differentiated cell populations, suggesting a role in progenitor cell recruitment (Shen, Kohler, Huang, Huang and Peck, 1989). In general, FGFs diminish markers of osteoblast differentiation, reducing the expression of alkaline phosphatase, diminishing collagen I and osteocalcin synthesis and inhibiting PTH responsiveness (Rodan, Wesolowski, Thomas, Yoon and Rodan, 1989; Rodan Wesolowski, Yoon and Rodan, 1989). Bovine bone cells produce FGF and store it in their matrix, suggesting that it may have an autocrine or paracrine role. FGF action may be altered by interactions with other growth factors. Thus FGFs stimulate TGFβ production but may be inhibited by it (Centrella, McCarthy and Canalis, 1987b; Globus, Patterson-Buckendahl and Gospodarowicz, 1988; Chen, Mallory and Chang, 1989; Noda and Vogel 1989).

Insulin-like growth factors

Insulin-like growth factors exert an anabolic effect upon osteoblastic cells, increasing DNA synthesis (Canalis and Lian, 1988; Hickman and McElduff, 1989; Kozawa, Takatsuki, Kotake, Yoneda, Oiso and Saito, 1989), altering oncogene expression (*eg* c-fos; Slootweg, van-Genesen, Otte, Duursma and Kruijer, 1990) and stimulating collagen, osteonectin and osteocalcin production (McCarthy, Centrella and Canalis, 1987; Canalis and Lian, 1988). IGFs are synthesised by bone cells and deposited in their extracellular matrix, suggesting a possible autocrine role (Canalis, McCarthy and Centrella, 1988; Mohan, Jennings, Linkhart and Baylink, 1988). IGF production is increased by PTH (McCarthy, Centrella and Canalis, 1989), estradiol, $1,25(OH)_2D_3$ (Gray *et al*, 1989), growth hormone (Ernst and Froesch, 1988) and TGFβ (Elford and Lamberts, 1990) but blocked by cortisol (perhaps contributing to the osteopenic effect of glutocorticoids; McCarthy, Centrella and Canalis, 1990).

Osteoblastic responses to IGFs are influenced by interactions with other growth factors. Their mitogenic effects are strongly synergised by FGF, EGF and PDGF, and whilst PDGF blocks IGF enhanced collagen synthesis (Slootweg *et al*, 1990; Canalis, McCarthy and Centrella, 1989), TGFβ increases it (Pfeilschifter *et al*, 1990).

Transforming growth factor α and epidermal growth factor

Transforming growth factor α and epidermal growth factor are polypeptide mitogens that act via a common receptor to increase cell division amongst osteoblastic cells (Roberts, Frolik, Anzano and Sporn, 1983; Massague 1983 a, b). Both factors appear to increase osteoprogenitor cell proliferation, and thereby produce a secondary increase in osteoblast activity when these newly-formed cells mature (Maric, Hott and Perheentupa, 1990; Antosz, Bellows and Aubin, 1987). TGFα and EGF have also been reported to increase bone resorption, possibly via increased PGE_2 production (Stern, Kriegèr, Nissenson, Williams, Winkler, Derynck and Strewler, 1985; Ibbotson, Harrod, Gowen, D'Souza, Smith, Winkler, Derynck and Mundy, 1986; Yokota, Kusaka, Ohshima, Yamamoto, Kurihara, Yoshino and Kumagawa, 1986).

Platelet-derived growth factor

Platelet-derived growth factor increases DNA synthesis amongst the fibroblast and progenitor rich cell layers of the periosteum (Canalis, McCarthy and Centrella, 1989). It has been reported to increase the rate of bone matrix apposition (Pfeilschifter *et al*, 1990) and has been isolated from bone matrix, suggesting an autocrine role (Hauschka, Chen and Maurakos, 1988).

Cytokines and haemopoietic growth factors

Osteoclast recruitment and maturation appears to be regulated by a number of haemopoictic growth factors (for review see Testa, 1988). Of particular interest are the multifunctional cytokines, interleukin-1 and tumour necrosis factors α and β. These hormone-like molecules are released during disease and promote host survival by triggering defensive responses. The beneficial effects of IL1 and TNF occur at considerable metabolic cost to the host however, so that prolonged or unregulated cytokine production may itself be pathogenic.

Both IL1 and the TNFs promote bone resorption (Gowen, Wood, Ihrie, McGuire and Russell, 1983; Bertolini, Nedwin, Bringman, Smith and Mundy, 1986), increasing osteoclast numbers and activity following an initial interaction with osteoblastic cells (Thomson *et al*, 1986; Thomson, Mundy and Chambers, 1987). The resulting bone loss appears to be part of the systemic catabolic state induced by these agents, a state that is thought to increase nutrient supply during disease.

IL1 and TNF also inhibit alkaline phosphatase expression and matrix synthesis by osteoblastic cells, generally reducing collagen synthesis (Canalis, 1986), blockading $1,25(OH)_2D_3$ induced osteocalcin production (Beresford, Gallagher, Gowen, Couch, Poser, Wood and Russell, 1984; Gowen, MacDonald and Russell, 1988) and reducing bone nodule formation (Stashenko, Dewhirst, Rooney, Desjardins and Heeley, 1987). Both IL1 and TNF are, however, bone cell mitogens, stimulating osteoprogenitor cell division and thereby producing a secondary increase in osteoblastic activity (Canalis, 1986, 1987). This suggests that IL1 and TNF may contribute to subsequent repair processes.

Concluding remarks

Bone cell biology is currently advancing rapidly following the introduction of molecular biology techniques. Future research should therefore lead to an increased understanding of the interplay between the genetic, mechanical and hormonal factors that regulate bone-cell development. This knowledge could form the basis for the rational development of pharmacological agents to influence such diverse processes as growth, wound healing and the accumulation of lean body mass.

References

Anderson, R.E., Woodbury, D.M. and Jee, W.S.S. (1986) *Calcified Tissue International*, **39**, 252-258

Antosz, M.E., Bellows, C.G. and Aubin, J.E. (1987) *Journal of Bone Mineral Research*, **2**, 385-393

Bab, I., Ashton, B.A., Gazit, D., Marx, G., Williamson, M.C. and Owen, M.E. (1986) *Journal of Cell Science*, **84**, 139-151

Bar-Shavit, Z., Teitelbaum, S.L., Reitsma, P., Hall, A., Pegg, L.E., Trail, J. and Kahn, A.J. (1983) *Proceedings of National Academy of Science USA*, **80**, 5907-5911

Baron, R., Neff, L., Louvard, D. and Courtoy, P.J. (1985) *Journal of Cell Biology*, **101**, 2210-2222

Baron, R., Neff, L., Lippincott-Schwartz, J., Louvard, D., Mellman, I., Helenius, A. and Marsh, M. (1986) *Journal of Bone Mineral Research*, **1**, Suppl 1, 32

Beresford, J.N., Gallagher, J.A., Gowen, M., Couch, M., Poser, J., Wood, D.D. and Russell, R.G.G. (1984) *Biochimica Biophysica Acta*, **801**, 58-65

Bertolini, D.R., Nedwin, G.E., Bringman, T.S., Smith, D.D. and Mundy, G.R. (1986) *Nature*, **319**, 516-518

Bird, T.A. and Saklatvala, J. (1986) *Nature*, **324**, 263-266

Bonucci, E. (1980) *Clinical Orthopaedics and Related Research*, **158**, 252-269

Boyde, A. (1972) In *The biochemistry and physiology of bone*, (ed G.H. Bourne), New York, Academic Press Inc, vol 1, pp 258-310

Bradbeer, J.N., Dunham, J., Fischer, J.A, Nagant De Deuxchaisnes, C. and Loveridge, N. (1988) *Journal of Clinical Endocrinology and Metabolism*, **67**, 1237-1243

Bradbeer, J.N., Mehdizadeh, S., Fraher, L.J. and Loveridge, N. (1988) *Journal of Bone Mineral Research*, **3**, 47-52

Burger, E.H., Thesingh, C.W., van der Meer, J.W.M. and Nijweide, P.J. (1984) In *Endocrine control of bone and calcium metabolism*, (ed D.V. Cohn, T. Fujita, J.T. Potts, and R.V. Talmage), Elsevier Science Publishers BV, pp 125-130

Burger, E.H., van der Meer, J.W.M. and Nijweide, P.J. (1984) *Journal of Cell Biology*, **99**, 1901-1906

Canalis, E. (1986) *Endocrinology*, **118**, 74-81

Canalis, E. (1987) *Endocrinology*, **121**, 1596-1604

Canalis, E. and Lian, J.B. (1988) *Bone*, **9**, 243-246

Canalis, E., McCarthy, T. and Centrella, M. (1988) *Calcified Tissue International*, **43**, 346-351

Canalis, E., McCarthy, T. and Centrella, M. (1989) *Journal of Cell Physiology*, **140**, 530-537

Canalis, E., Centrella, M., Burch, W. and McCarthy, T. (1989) *Journal of Clinical Investigation*, **83**, 60-65

Carrington, J.L. and Reddi, A.H. (1990) *Experimental Cell Research*, **186**, 368-373

Centrella, M., Massague, J. and Canalis, E. (1986) *Endocrinology*, **119**, 2306-2312

Centrella, M., McCarthy, T.L. and Canalis, E. (1987a) *Journal of Biological Chemistry*, **262**, 2869-2874

Centrella, M., McCarthy, T.L. and Canalis, E. (1987b) *FASEB Journal*, **1**, 312-317

Chambers, T.J. (1980) *Clinical Orthopaedics*, **151**, 283

Chambers, T.J. (1982) *Journal of Cell Science* **57**, 247-260

Chambers, T.J. (1985) *Journal of Clinical Pathology*, **38**, 241-252

Chambers, T.J. and Magnus, C.J. (1982) *Journal of Pathology*, **136**, 27-39

Chambers, T.J. and Dunn, C.J. (1983) *Calcified Tissue International*, **35**, 566-570

Chambers, T.J., Thomson, B.M. and Fuller, K. (1984) *Journal of Cell Science*, **70**, 61-71

Chambers, T.J., Darby, J.A. and Fuller, K. (1985) *Cell Tissue Research*, **241**, 671-675

Chambers, T.J. and Fuller, K. (1985) *Journal of Cell Science*, **76**, 155-165

Chambers, T.J., McSheehy, P.M.J., Thomson, B.M. and Fuller, K. (1985) *Endocrinology*, **116**, 234-239

Chen, T.L., Mallory, J.B. and Chang, S.L. (1989) *Growth Factors*, **1**, 335-345

Closs, E.I., Murray, A.B., Schmidt, J., Schon, A., Erfle, V. and Strauss, P.G. (1990) *Journal of Cell Biology*, **111**, 1313-1323

Currey, J. (1984) In *A short course on biomaterials*, Department of Mechanical Engineering, University of Leeds

Delaisse, J-M., Eekhout, Y and Vaes, G. (1984) *Biochemical and Biophysical Research Communication*, **125**, 441-447

Dickson, G.R., Koenig, E. and Silbermann, M. (1991) *Calcified Tissue*, **48**, (Suppl) A164

Dworetzky, S.I., Fey, E.G., Penman, S., Lian, J.B., Stein, J.L. and Stein, G.S. (1990) *Proceedings of the National Academy of Science USA*, **87**, 4605-4609

Eastoe, J.E. (1956) In *The biochemistry and physiology of bone*, (ed G.H. Bourne), New York, Academic Press Inc, pp 81-105

Elford, P.R. and Lamberts, S.W. (1990) *Endocrinology*, **127**, 1635-1639

Endo, H., Kiyoki, M., Kawashima, K., Naruchi, T. and Hashimoto, Y. (1980) *Nature*, **286** 262-264

Eriksen, E.F., Colvard, D.S., Berg, N.J., Graham, M.L., Mann, K.G., Spelsberg, T.C. and Riggs, B.L. (1988) *Science*, **214**, 84-86

Ernst, M. and Froesch, E.R. (1988) *Biochemical and Biophysical Research Communications*, **4151**, 142-147

Forsyth, P.J.E. (1969) *The physical basis of metal fatigue*, Blackie and Sons Ltd

Friedenstein, A.J. (1980) In *Immunobiology of bone marrow transplantation*, (ed S. Thienfelder), Berlin, Springer-Verlag, pp 19-29

Friedenstein, A.J., Chailakhyan, R.K. and Gerasimov, U.V. (1987) *Cell and Tissue Kinetics*, **20**, 263-272

Furseth, R. (1973) *Scandinavian Journal of Dental Research*, **81**, 339-341

Ghiselli, R., Blair, H., Teitelbaum, S. and Gluck, S. (1987) *Journal of Bone and Mineral Research*, **2** (Suppl 1), 295

Globus, R.K., Patterson-Buckendahl, P. and Gospodarowicz, D. (1988) *Endocrinology*, **123**, 98-105

Globus, R.K., Plouet, J. and Gospodarowicz, D. (1989) *Endocrinology*, **124**, 1539-1547

Glowacki, J., Rey, C., Cox, K and Lian, J. (1990) *Connective Tissue Research*, **20**, 121-129

Goose, D.H. and Appleton, J. (1982) *Human dentofacial growth*, Oxford, Pergamon Press Ltd

Goss, R.J. (1978) *The physiology of growth*, New York, Academic Press

Gothlin, G. and Ericsson, A.J. (1973) *Virchows Archiv B Cell Pathology*, **12**, 318-329

Gowen, M., Wood, D.D., Ihrie, E.J., McGuire, M.K.B. and Russell, R.G. (1983) *Nature*, **306**, 378-380

Gowen, M., MacDonald, B.R. and Russell, R.G. (1988) *Arthritis and Rheumatism*, **31**, 1500-1507

Gray, T.K., Mohan, S., Linkhart, T.A. and Baylink, D.J. (1989) *Biochemical and Biophysical Research Communications*, **158**, 407-412

Guenther, H.L., Cecchini, M.G., Elford, P.R. and Fleisch, H. (1988) *Journal of Bone and Mineral Research*, **3**, 211-218

Hall, G.E. and Kenny, A.D. (1985a) *Calcified Tissue International*, **37**, 134-142

Hall, G.E. and Kenny, A.D. (1985b) *Pharmacology*, **30**, 339-347

Hall, T.J. and Chambers, T.J. (1990) *Journal of Cell Physiology*, **142**, 420-424

Ham, A.W. (1969) *Histology* (6th edition), Philadelphia and Toronto, Lippincott Company

Hamilton, J.A., Lingelbach, S.R., Partridge, N.C. and Martin, T.J. (1984) *Biochemical and Biophysical Research Communications*, **122**, 230-236

Hamilton, J.A., Lingelbach, S., Partridge, N.C. and Martin, T.J. (1985) *Endocrinology*, **116**, 2186-2191
Hancox, N.M. (1956) The osteoclast. In *The biochemistry and physiology of bone*, (ed G.H. Bourne), New York, Academic Press Inc, pp 213-247
Hancox, N.M. (1971) The osteoclast. In *The biochemistry and physiology of bone*, (ed G.H. Bourne), New York, Academic Press Inc, pp 45-69
Hauschka, P.V., Mavrakos, A.E., Iafrati, M.D., Doleman, S.E. and Klagsbrun, M. (1986) *Journal of Biological Chemistry*, **261**, 12665-12674
Hauschka, P.V., Chen, T.L. and Maurakos, A.E. (1988) *Ciba Foundation Symposium*, **136**, 207-225
Heath, J.K., Atkinson, S.J., Meikle, M.C. and Reynolds, J.J. (1984) *Biochimica Biophysica Acta*, **802**, 151-154
Heine, U., Munoz, E.F., Flanders, K.C., Ellingsworth, L.R., Lam, H.Y., Thompson, N.L. Roberts, A.B. and Sporn, M.B. (1987) *Journal of Cell Biology*, **105**, 2861-2876
Henry, H.L. and Norman, A.W. (1990) Vitamin D: Metabolism and mechanism of action. In *Primer on the metabolic bone diseases and disorders of mineral metabolism*, American Society for Bone and Mineral Research, pp 47-52
Herrman-Erlee, M.P.M., Heersche, J.N.M. and Hekkelman, J.W. (1976) *Endocrine Research Communications*, **3**, 21-27
Hickman, J. and McElduff, A. (1989) *Endocrinology*, **124**, 701-706
Hock, J.M., Canalis, E. and Centrella, M. (1990) *Endocrinology*, **126**, 421-426
Ibbotson, K.J., Harrod, J., Gowen, M., D'Souza, S., Smith, D.D., Winkler, M.E., Derynck, R. and Mundy, G.R. (1986) *Proceedings of National Academy of Science USA*, **83**, 2228-2232
Isaksson, O.G.P., Lindahl, A., Nilsson, A. and Isgaard, J. (1987) *Endocrine Reviews*, **8**, 426-437
Jones, H.H., Priest, J.D. and Hayes, W.C. (1977) *Journal of Bone and Joint Surgery*, **59**, 204-208
Joyce, M.E., Roberts, A.B., Sporn, M.B. and Bolander, M.E. (1990) *Journal of Cell Biology*, **110**, 2195-2207
Kasono, K., Sato, K., Han, D.C., Fujii, Y., Tsushima, T. and Shizume, K. (1988) *Bone and Mineral*, **4**, 355-363
Kaye, M. (1984) *Journal of Clinical Pathology*, **37**, 398-400
Kent, G.N., Dodds, R.A., Watts, R.W.E., Bitensky, L., Chayen, J. (1983) *Journal of Bone and Joint Surgery*, **65**, 189-194
King, G.J. and Halthrop, M.E. (1975) *Journal of Cell Biology*, **66**, 445
Kozawa, O., Takatsuki, K., Kotake, K., Yoneda, M., Oiso, Y. and Saito, H. (1989) *FEBS Letters*, **243**, 183-185
Lacroix, P. (1971) Internal remodelling in bone. In *The biochemistry and physiology of bone*, (ed G.H. Bourne), New York, Academic Press Inc, p 119
Lanyon, L.E., Rawlinson, S. and Ali, N.N. (1991) *Calcified Tissue*, **48** (Suppl), A330
Loveridge, N., Dean, V., Goltzman, D. and Hendy, G.N. (1991) *Endocrinology*, **128**, (in press)

Manolagos, S.C., Haussler, M.R. and Deflos, L.I. (1980) *Journal of Cell Biology*, **255**, 4414-4417
Maor, G., Hockberg, Z., von-der-Mark, K., Heinegard, D. and Silbermann, M. (1989) *Endocrinology*, **125**, 1239-1245
Marchisio, P.C., Cirillo, D., Naldini, L., Primavera, M.V., Teti, A. and Zambonin-Zallone, A. (1984) *Journal of Cell Biology*, **99**, 1696-1705
Marie, P.J., Hott, M. and Perheentupa, J. (1990) *American Journal of Physiology*, **258**, E275-281
Marks, S.C. (1983) *Journal of Oral Pathology*, **12**, 226-256
Martin, T.J. and Partridge, N.C. (1981) In *Hormonal control of calcium metabolism*, (eds D.V. Cohn, R.V. Talmage and J.L. Matthews), Amsterdam, Excerpta Medica, pp 147-156
Martin, T.J., Allan, E.H., Caple, I.W., Care, A.D., Danks, J.A., Diefenbachjagger, H., Ebeling, P.R., Gillespie, M.T., Hammonds, G., Heath, J.A. and Hudson, P.J. (1989) *Recent Progress in Hormone Research*, **45**, 467-506
Massague, J. (1983a) *Journal of Biological Chemistry*, **258**, 13606-13613
Massague, J. (1983b) *Journal of Biological Chemistry*, **258**, 13614-13620
McCarthy, T.L., Centrella, M. and Canalis, E. (1987) *Bone and Mineral*, **2**, 185-192
McCarthy, T.L., Centrella, M. and Canalis, E. (1989) *Endocrinology*, **125**, 2118-2126
McCarthy, T.L., Centrella, M. and Canalis, E. (1990) *Endocrinology*, **126**, 1569-1575
McCulloch, C.A., Fair, C.A., Tenenbaum, H.C., Limeback, H. and Homareau, R. (1990) *Developmental Biology*, **140**, 352-361
McPartlin, J., Skrabanek, P. and Powell, D. (1978) *Endocrinology*. **103**, 1573-1578
McSheehy, P.M.J. and Chambers, T.J. (1986a) *Endocrinology*, **118**, 824-828
McSheehy, P.M.J. and Chambers, T.J. (1986b) *Endocrinology*, **119**, 1654-1659
McSheehy, P.M.J. and Chambers, T.J. (1987) *Journal of Clinical Investment*, **80**, 425-429
Meikle, M.C., McGarrity, A.M., Thomson, B.M. and Reynolds, J.J. (1991) *Bone and Mineral*, **12**, 41-55
Mellanby, E. (1947) *Journal of Physiology*, **105**, 382-399
Minghetti, P.P. and Norman, A.W. (1988) *FASEB Journal*, **2**, 3043-3054
Mohan, S., Jennings, J.C., Linkhart, T.A. and Baylink, D.J. (1988) *Biochimica Biophysica Acta*, **966**, 44-55
Mundy, G.R. (1990) In *Primer on the metabolic bone diseases and disorders of mineral metabolism*, American Society for Bone and Mineral Research, pp 18-22
Noda, M. and Rodan, G.A. (1986) *Biochemical and Biophysical Research Communications*, **140**, 56-65
Noda, M. and Rodan, G.A. (1989) *Connective Tissue Research*, **21**, 71-75
Noda, M. and Vogel, R. (1989) *Journal of Cell Biology*, **109**, 2529-2535

Ornoy, A., Goodwin, D., Noff, D. and Edelstein, S. (1978) *Nature*, **276**, 517-519
Owen, M. (1963) *Journal of Cell Biology*, **19**, 19-32
Owen, M. (1978) *Calcified Tissue International*, **25**, 205-207
Owen, M. (1980) *Arthritis and Rheumatism*, **23**, 1073-1080
Owen, M. (1988) *Journal of Cell Science*, **10**, 63-76
Owen, T.A., Aronow, M., Shalhoub, V., Barone, L.M., Wilming, L., Tassinari, M.S., Kennedy, M.B., Pockwinse, S., Lian, J.B. and Stein, G.S. (1990) *Journal of Cell Physiology*, **143**, 420-430
Parfitt, A.M. (1984) *Calcified Tissue International*, **36**, 37
Partridge, N.C., Frampton, R.J., Eisman, J.A., Michaelangeli, V.P., Elms, E., Bradley, T.R. and Martin, T.J. (1980) *FEBS Letters*, **115**, 139-142
Piez, K.A. (1984) *Extracellular matrix biochemistry*, (eds K.A. Piez and A.H. Reddi), New York, Elsevier
Perry, H.M., Skogen, W., Chappel, J.C., Wilner, G.D., Kahn, A.J. and Teitelbaum, S.L. (1987) *Calcified Tissue International*, **40**, 298-300
Pfeilschifter, J. and Mundy, G.R. (1987) *Proceedings of National Academy of Science USA*, **84**, 2024-2028
Pfeilschifter, J., Oechsner, M., Naumann, A., Gronwald, R.G., Minne, H.W. and Ziegler, R. (1990) *Endocrinology*, **127**, 69-75
Pritchard, J.J. (1956a) General anatomy and physiology of bone. In *The biochemistry and physiology of bone*, (ed G.H. Bourne), New York, Academic Press Inc, pp 1-25
Pritchard, J.J. (1956b) The osteoblast. In *The biochemistry and physiology of bone*, (ed G.H. Bourne), New York, Academic Press Inc, pp 179-211
Pritchard, J.J. (1971) The osteoblast. In *The biochemistry and physiology of bone*, (ed G.H. Bourne), New York, Academic Press Inc, pp 21-44
Rizzoli, R., Poser, J. and Burgi, U. (1986) *Metabolism*, **35**, 71-74
Roberts, A.B., Frolik, C.A., Anzano, M.A. and Sporn, M.B. (1983) *Federation Proceedings*, **42**, 2621- 2625
Roberts, A.B. and Sporn, M.B. (1990) In *Cell biology of cartilage and bone*, Oxford, British Society for Cell Biology meeting
Roberts, W.E., Mozsary, P.G. and Klingler, E. (1982) *American Journal of Anatomy*, **165**, 373-384
Robey, P.G., Young, M.F., Flanders, K.C., Roche, N.S., Kondaiah, P., Reddi, A.H., Termine, J.D., Sporn, M.B. and Roberts, A.B. (1987) *Journal of Cell Biology*, **105**, 457-463
Robins, S.P. (1988) *Balliere's Clinical Rheumatology*, **2**, 1-36
Rodan, G.A. and Martin, T.J. (1981) *Calcified Tissue International*, **3**, 349-351
Rodan, S.B., Wesolowski, G., Thomas, K.A., Yoon, K. and Rodan, G.A. (1989) *Connective Tissue Research*, **20**, 283-288
Rodan, S.B., Wesolowski, G., Yoon, K. and Rodan, G.A. (1989) *Journal of Biological Chemistry*, **264**, 19934-19941
Rouleau, M.F., Mitchell, J. and Goltzman, D. (1988) *Endocrinology*, **118**, 919-931

Rubin, C.T. (1984) *Calcified Tissue International*, **36**, 11-18
Rubin, C.T. and Hausman, M.R. (1988) *Rheumatic Disease Clinics of North America*, **14**, 503-517
Ruther, U., Garber, C., Komitowski, D., Muller, R. and Wagner, E.F. (1987) *Nature*, **325**, 412-416
Sato, K., Fujii, Y., Asano, S., Ohtsuki, T., Kawakami, M., Kasono, K., Tsushima, T. and Shizume, K. (1986) *Biochemical and Biophysical Research Communications*, **141**, 285-291
Sato, K., Kasono, K., Fujii, Y., Kawakami, M., Tsushima, T. and Shizume, K. (1987) *Biochemical and Biophysical Research Communications*, **145**, 323-329
Sato, K., Han, D.C., Fujii, Y., Tsushima, T. and Shizume, K. (1987) *Endocrinology*, **120**, 1873-1881
Schenk, R.K., Hunziker, E. and Herrmann, W. (1982) Structural properties of cells related to tissue mineralisation. In *Biological mineralisation and demineralisation*, (ed G.H. Nancollas), Berlin, Heidelberg, New York Springer-Verlag, pp 143-160
Scherft, J.P. (1972) *Journal of Ultrastructure Research*, **38**, 318-331
Scheven, B.A.A., Burger, E.H., Kawilarang-De Haas, W.M., Wassenaar, A.M. and Nijweide, P.J. (1985) *Laboratory Investigations*, **53**, 72-79
Scheven, B.A.A., Visser, J.W.M. and Nijweide, P.J. (1986) *Nature*, **321**, 79-81
Scheven, B.A.A., Kawilarang-De Haas, W.M., Wassenaar, A,M. and Nijweide, P.J. (1986) *Anatomical Record*, **214**, 418-423
Schmid, C., Steiner, T. and Froesch, E.R. (1986) *Acta Endocrinologica Copenhagen*, **111**, 213-216
Schneider, G.B., Relfson, M. and Nicolas, J. (1986) *American Journal of Anatomy*, **177**, 505-511
Selye, H. (1932) *Endocrinology*, **16**, 547-558
Shalhoub, V., Gerstenfeld, L.C., Collart, D., Lian, J.B. and Stein, G.S. (1989) *Biochemistry*, **28** 5318-5322
Shen, V., Kohler, G., Huang, J., Huang, S.S. and Peck, W.A. (1989) *Bone and Mineral*, **7**, 205-219
Shiina-Ishimi, Y., Abe, E., Tanaka, H. and Suda, T. (1986) *Biochemical and Biophysical Research Communications*, **134**, 400-406
Silve, C.M., Hradek, G.T., Jones, A.L. and Arnaud, C.D. (1982) *Journal of Cell Biology*, **94**, 379-386
Slootweg, M.C., van-Genesen, S.T., Otte, A.P., Duursma, S.A. and Kruijer, W. (1990) *Journal of Molecular Endocrinology*, **4**, 265-274
Sporn, M.B., Roberts, A.B., Wakefield, L.M. and Assoian, R.K. (1986) *Science*, **233**, 532-534
Stashenko, P., Dewhirst, F.E., Rooney, M.L., Desjardins, L.A. and Heeley, J.D. (1987) *Journal of Bone and Mineral Research*, **2**, 559-565
Stern, P.H., Krieger, N.S., Nissenson, R.A., Williams, R.D., Winkler, M.E., Derynck, R. and Strewler, G.J. (1985) *Journal of Clinical Investigation*, **76**, 2016-2019

Stevenson, J.C., White, M.C., Joplin, G.F. and MacIntyre, I. (1982) *British Medical Journal*, **285**, 1010-1011
Tam, C.S., Heersche, J.N.M., Murray, T.M. and Parsons, J.A. (1982) *Endocrinology*, **110**, 506-512
Tashjian, A.H.Jr., Wright, D.R., Ivey, J.L. and Pont, A. (1978) *Rec. Prog. Hormone Research*, **34**, 285-334
Taylor, L.M., Tertinegg, I., Okuda, A. and Heersche, J.N. (1989) *Journal of Bone and Mineral Research*, **4**, 751-758
Termine, J.D. (1990) In *Primer on the metabolic bone diseases and disorders of mineral metabolism*, pp 16-17
Testa, N.G. (1988) *Ciba Foundation Symposium*, **136**, 257-274
Teti, A., Blair, H.C., Schlesinger, P., Grano, J., Zambonin-Zallone, A., Kahn, A.J., Teitelbaum, S.L. and Hruska, K.A. (1989) *Journal of Clinical Investigation*, **84**, 773-780
Thesingh, C.W. (1986) *Developmental Biology*, **117**, 127-134
Thompson, T.J., Owens, P.D. and Wilson, D.J. (1989) *Journal of Anatomy*, **166**, 55-65
Thomson, B.M., Saklatvala, J. and Chambers, T.J. (1986) *Journal of Experimental Medicine*, **164**, 104-112
Thomson, B.M., Atkinson, S., Reynolds, J.J. and Meikle, M.C. (1987a) *Bone*, **8**, 265-266
Thomson, B.M., Atkinson, S.J., Reynolds, J.J. and Meikle, M.C. (1987b) *Biochemical and Biophysical Research Communications*, **148**, 596-602
Thomson, B.M., Mundy, G.R. and Chambers, T.J. (1987) *Journal of Immunology*, **138**, 775-779
Thomson, B.M., Atkinson, S.J., McGarrity, A.M., Hembry, R.M., Reynolds, J.J. and Meikle, M.C. (1989) *Biochimica Biophysica Acta*, **1014**, 125-132
Tuukkanen, J. and Vaananen, H.K. (1986) *Calcified Tissue International*, **38**, 123-125
Urist, M.R., Delange, R.J. and Fineran, G.A.M. (1983) *Science*, **220**, 680-686
Vaes, G. (1988) *Clinical Orthopaedics*, **231**, 239-271
Van Obberghen-Schilling, E., Roche, N.S., Flanders, K.C., Sporn, M.B. and Roberts, A.B. (1988) *Journal of Biological Chemistry*, **263**, 7741-7746
Yates, A.J.P., Guiterrez, G.E., Smolens, P., Travis, P.S., Katz, M.S., Aufdemorte, T.B., Boyce, B.F., Hymer, T.K., Poser, J.W. and Mundy, G.R. (1988) *Journal of Clinical Investigation*, **81**, 932-938
Yokota, K., Kusaka, M., Ohshima, T., Yamamoto, S., Kurihara, N., Yoshino, T. and Kumagawa, M. (1986) *Journal of Biological Chemistry*, **261**, 15410-15415
Yoon, K., Buenaga, R. and Rodan, G.A. (1987) *Biochemical and Biophysical Research Communications*, **148**, 1129-1136

5

FOETAL GROWTH AND ITS INFLUENCE ON POSTNATAL GROWTH AND DEVELOPMENT

A.W. BELL
Department of Animal Science, Cornell University, Ithaca, New York, USA

Introduction

More than two decades ago, at the 14th Easter School on Growth and Development of Mammals, Graham Everitt (1968) complained that "prenatal growth of domesticated farm animals has received scant attention as compared with postnatal studies." He also noted that "the extent to which [events of later life] may be modified by factors operating during the intra-uterine formative stages appears to be insufficiently appreciated." Since then, with the advent of techniques for studying the conceptus *in utero*, much has been learned about the regulation of foetal metabolism and growth (see Gluckman, 1986; Bell, Bauman and Currie, 1987). However, the second of Everitt's complaints appears to be as relevant now as it was then.

Foetal growth accounts for only a small fraction (5 to 8% in ruminants, < 1% in pigs) of mature body size in meat animals. However, the vital importance of this developmental phase is indicated by the fact that initiation and differentiation of all cell types is completed and maturation of most is well under way before birth. This raises two questions central to the theme of this chapter: firstly, how responsive is the prenatal development of key tissues, such as skeletal muscle and adipose, to non-genetic influences; and secondly, how does this influence rate and composition of postnatal growth?

Before addressing these questions, an overview of the changing patterns of pre- and postnatal growth and possible modes of regulation is provided, with special emphasis on foetal growth.

Patterns of foetal and postnatal growth

Division of the continuum of growth from conception to maturity into pre- and postnatal phases must be somewhat arbitrary. However, birth results in a particularly dramatic alteration in nutritional and environmental influences on gene expression and the periods of growth before and after birth can be further categorized according to changes in nutrient supply, endocrine and paracrine/autocrine influences and relative growth rate. Four readily discernible stages of prepubertal growth are early foetal, late foetal, neonatal and postweaning.

EARLY FOETAL GROWTH

Specific teratogenic influences can cause disastrous disruption of cell differentiation and organogenesis during embryonic life. Such congenitally malformed animals usually die or are destroyed at birth and are not relevant to this discussion. The possibility that embryonic development may be subjected to more subtle influences with consequences for later foetal and postnatal growth is worth considering but has been little studied in meat animal species. An exception is the work of Parr, Williams, Campbell, Witcombe and Roberts (1986) who showed that restriction of post-implantation embryonic growth in sheep by severe maternal undernutrition in early pregnancy had no residual effect on lamb birth weight if ewes were later fed normally.

Postembryonic foetal growth through mid-pregnancy proceeds at a rapid but slowly diminishing relative rate which in sheep varies from about 15%/d at 40 days to about 6%/d at 100 days gestation. Until then foetal growth appears to be largely insensitive to variations in maternal nutrition or other potential constraints, suggesting that it occurs at rates close to genetic potential. Thus, Ferrell (1991a) found that variation in bovine foetal growth from Brahman or Charolais embryos cross-transferred between Brahman and Charolais dams was entirely explained by foetal genotype until about 230 days gestation. Such observations need not mean that early foetal growth is unregulated by extracellular factors, as implied by some reviewers (Gluckman, 1986; Milner, 1988). However, it is true that the factors responsible for coordination of the orderly, allometric development of foetal organ systems through this extremely dynamic phase remain to be identified.

Consistent with the notion that foetal growth in mid-pregnancy is not restricted by nutrient supply, weight-specific rates of umbilical uptake of oxygen, glucose (Bell, Kennaugh, Battaglia, Makowski, and Meschia, 1986) and amino acids (Bell, Kennaugh, Battaglia and Meschia, 1989), and of whole-body protein synthesis (Kennaugh, Bell, Teng, Meschia and Battaglia, 1987) in foetal sheep are much greater in mid than in late gestation. Similar gestational declines in foetal metabolic rates occur in cattle (Reynolds, Ferrell, Robertson and Ford, 1986). Reduced foetal growth before 90 to 100 days has been observed in ewes which were severely undernourished (Everitt, 1968) or heat-stressed during mid-pregnancy (Alexander and Williams, 1971; Alexander, Hales, Stevens and Donnelly, 1987). The degree to which placental factors mediated these early reductions in foetal growth is unknown. However, in both underfed and heat-stressed ewes, placental growth was retarded more than foetal growth. The idea that heat-induced reduction of placental size and functional capacity precedes, then mediates a slowing of foetal growth is supported by our observation that chronic heat exposure of ewes between days 50 and 75 of pregnancy reduced placental weight by 20% but did not affect foetal weight at 75 days (I. Vatnick and A.W. Bell, unpublished).

LATE FOETAL GROWTH

As noted by Gluckman (1986), the major characteristic of foetal growth in late gestation is that it is constrained. Among the numerous factors which may act to prevent the foetus from achieving its genetic capacity for growth, placental restriction of foetal nutrient supply, even in very well-fed dams, appears to be predominant (Gluckman, 1986; Bell, Bauman, and Currie, 1987).

The specific aspect(s) of placental functional capacity which may be limiting for foetal growth are not known. Intrauterine growth retardation induced experimentally by chronic heat stress (Bell, Wilkening and Meschia, 1987) or premating caruncIectomy (Owens, Falconer and Robinson, 1987a; 1987b) of pregnant ewes is associated with reduced placental transport of oxygen and glucose, and foetal hypoxaemia and hypoglycaemia, each of which is well correlated with reduction in placental weight. In animals with more normal placentae it seems unlikely that oxygen supply is a primary limit to foetal growth, although it is intriguing that above-normal growth of chick embryos can be stimulated by incubating eggs under hyperoxic conditions (Stock, Francisco, McCutcheon and Metcalfe, 1982). Placental capacity for glucose transport is a more likely limiting factor although convincing evidence for this notion in well-nourished animals has been lacking. However, chronic infusion of singleton sheep foetuses with glucose throughout the final month of gestation caused an 18% increase in birth weight, even though their dams were fed above predicted nutrient requirements during this period (Table 1; Stevens, Alexander and Bell, 1990).

Table 1 EFFECT OF PROLONGED GLUCOSE INFUSION INTO FOETAL SHEEP ON BODY GROWTH AND FAT DEPOSITION
(from Stevens *et al*, 1990)

Infusate	Body weight (kg)	Internal fat (g/kg)	Subcutaneous fat (g/kg)
Saline	3.28±0.24(11)	6.73±0.37(10)	0.27±0.13(11)
Glucose	3.86±0.16(11)*	9.91±0.65(9)**	1.25±0.44(11)*

Values are means ± SEM with number of observations in parentheses
Foetuses were infused from about 115 days to about 145 days gestation
*Significantly different from saline control, $P<0.05$
**Significantly different from saline control, $P<0.01$

Placental glucose transport occurs by facilitated diffusion (see Bell, Bauman and Currie, 1987). Specific glucose carriers are probably located in maternal and foetal placental membranes but these proteins have yet to be isolated and characterized. *In vivo* measurements in pregnant ewes suggest that glucose transport capacity is greater on the foetal than on the maternal surface of the placenta (Hay, Molina, Di Giacomo and Meschia, 1990). This raises the possibility that maternal capacity for uterine uptake and placental transport of glucose may be a component of placental insufficiency during late pregnancy. Restriction of growth during late gestation in Charolais foetuses carried by Brahman cows, compared with those in Charolais cows, was associated with substantially reduced growth in maternal (caruncular) but not foetal (cotyledonary) tissues of the placenta (Ferrell, 1991a). Data for umbilical glucose uptake and maternal-foetal arterial glucose concentration differences were also collected from these animals at 232 days gestation (Ferrell, 1991b). From these it can be calculated that placental glucose transport capacity in the Charolais foetus/Brahman cows was only about 50% of that in the Charolais foetus/Charolais cows.

There is little evidence that placental capacity for active transport of amino acids is a specific limitation on foetal growth. However, Robinson, McDonald, Brown and Fraser (1985) estimated that even in well-fed pregnant ewes, maternal supply of cystine and histidine from microbial protein was insufficient to account for observed rates of foetal accretion of these amino acids. Whether these or any other amino acids are limiting for foetal growth remains to be investigated. It is, however, pertinent that in contrast to the considerable excess of umbilical uptake over foetal accretion of most amino acids, the uptake of sulphur amino acids, histidine and lysine barely matches their rates of foetal accretion in sheep (Meier, Teng, Battaglia and Meschia, 1981).

Mediation of the declining rate of foetal growth during late pregnancy must involve interplay between nutrient supply and both extracellular (endocrine) and local (autocrine/paracrine) trophic factors. The relative importance of these factors in the regulation of foetal growth has been a matter of some argument (see Gluckman, 1986; Milner, 1988; Browne and Thorburn, 1989). This issue is far from being resolved, but some recent developments deserve comment.

Local mediation of the growth restriction in foetal tissues during late gestation almost certainly involves the somatomedins, particularly insulin-like growth factor I (IGF-I). Plasma levels, presumably reflecting tissue synthesis, of IGF-I are consistently decreased by a variety of treatments which result in decreased foetal nutrient availability and growth (Table 2). Gene expression for IGF-I in foetal liver is also dramatically reduced by maternal fasting in rats (Straus, Ooi, Orlowski and Rechler, 1991). In contrast, high plasma levels and abundant tissue mRNA for IGF-II are either unchanged or increased by foetal nutrient restriction. This seeming paradox might be explained in terms of a role for IGF-II in the mediation of placental adaptive responses to foetal nutrient demand, but this is entirely speculative.

Table 2 EFFECTS OF FOETAL NUTRIENT DEPRIVATION ON FOETAL PLASMA CONCENTRATIONS OF IGF-I AND IGF-II

Species	Treatment	IGF-I	IGF-II	Source
Sheep	Pancreatectomy	↓	↑	Gluckman *et al* (1987)
	Carunclectomy	↓	↑	Jones *et al* (1988)
	Uterine artery ligation (acute)	NC	↑	Jones *et al* (1988)
Guinea pig	Uterine artery ligation (chronic)	↓	↑	Jones *et al* (1987)
Rat	Maternal protein restriction	↓	↓	Pilistine *et al* (1984)
	Maternal fast (late gestation)	↓	NC	Davenport *et al* (1990)
		↓	Small ↓	Straus *et al* (1991)
	Uterine artery ligation (chronic)	↓	--	Vileisis and D'Ercole (1986)

NC, no change

Putative endocrine regulators of the expression and local activity of the somatomedins include growth hormone (GH), placental lactogen, thyroid hormones, cortisol and insulin, while the possibility of a direct influence of specific nutrient availability cannot be ignored.

The prevailing view that GH has little effect on foetal IGF-I or skeletal growth (see Gluckman, 1986; Milner, 1988) should be revised. Certainly, this influence is less profound than in postnatal life. However, recent, careful experiments have shown that hypophysectomy causes retardation of bone growth and a substantial decrease in plasma IGF-I in foetal lambs (Mesiano, Young, Baxter, Hintz, Browne and Thorburn, 1987; Mesiano, Young, Hey, Browne and Thorburn, 1989). Hypophysectomy also retards muscle fibre differentiation in foetal pigs (Hausman, Hentges and Thomas, 1987). In terms of whole body weight, this reduction in skeletal growth is masked by increased body fatness and water retention (Liggins and Kennedy, 1968; Stevens and Alexander, 1986). Replacement of thyroxine results in only minor improvement in plasma IGF-I and bone growth, implying a more important role for GH (Mesiano *et al*, 1987; 1989), while replacement of GH, but not triiodothyronine, abolishes the increased fat deposition in hypophysectomized sheep foetuses (Stevens and Alexander, 1986).

Direct evidence is lacking for the relative importance of insulin, glucose or other nutrients as influences on somatomedin-regulated foetal tissue growth. Insulin plays an important permissive role, judging from the negative effects of foetal streptozotocin treatment or pancreatectomy on plasma IGF-I and rate of growth (see Fowden, 1989). However, chronic administration of upper physiological doses of insulin had little effect on late-gestation foetal growth in pigs (Garssen, Spencer, Colenbrander, Macdonald and Hill, 1983) or sheep (Milley, 1986; Fowden, Hughes and Comline, 1989; Schoknecht, Slepetis and Bell, 1990). Additional infusion of glucose to maintain euglycaemia resulted in increased fat deposition but did not affect lean growth (Schoknecht *et al*, 1990).

Increasing foetal secretion of corticosteroids may directly or indirectly influence foetal growth in late pregnancy. Treatment of pregnant rats with synthetic glucocorticoids decreased foetal weight and plasma IGF-I (Mosier, Dearden, Jansons, Roberts and Biggs, 1982; Mosier, Spencer, Dearden and Jansons, 1987) and foetal weight was increased in adrenalectomized foetal sheep (Trahair, Perry, Silver and Robinson, 1987). Also, plasma cortisol levels were increased in foetal lambs with growth retardation caused by maternal undernutrition (Mellor, Matheson and Small, 1977) or placental insufficiency (Robinson, Hart, Kingston, Jones and Thorburn, 1980).

NEONATAL GROWTH

The slowing of growth during late gestation is perhaps best illustrated by comparison with its rapid acceleration almost immediately after birth in well-fed animals. Daily measurement of heart girth before and after birth in the same lambs elegantly demonstrated that growth rate during the first 10 days of neonatal life was more than 3-fold greater than that during the final 10 days of gestation (Mellor and Murray, 1982). The marked increase in growth rate of milk-fed lambs is accompanied by an equally impressive increase in metabolic rate, independent of the effects of environmental temperature (Young, Bell and Hardin, 1989). This is probably related to a similar ontogenic pattern in protein synthetic rate of skeletal muscle and other tissues (see Bell, Bauman and Currie, 1987).

Rapid postnatal growth of skeletal muscle is characterized by steady accumulation of DNA, presumably through incorporation of satellite cells which continue to differentiate and proliferate long after the prenatal cessation of primary fibre hyperplasia (see Allen, Merkel and Young, 1979). In contrast, hypertrophic growth, indicated by muscle protein/DNA ratio, ceases relatively soon after birth (Black, 1983). Although the rate of lean tissue growth in well-fed neonates is impressive, it is outstripped by even faster growth of adipose tissue through a combination of cellular hyperplasia and hypertrophy (Vernon, 1986). This is associated with a rapid rate of deposition of pre-formed fatty acids in adipose triglycerides which is especially dramatic in the young pig (Mersmann, Houk, Phinney, Underwood and Brown, 1973). The composition of the accelerated weight gain of the neonatal animal thus becomes increasingly fat as weaning approaches.

The regulatory roles of specific hormones and growth factors in early postnatal growth are poorly defined. Plasma levels and presumably, tissue synthesis of IGF-I increase immediately after birth in lambs (Gluckman and Butler, 1983; Mesiano *et al*, 1989), and a few days later in pigs (Breier, Gluckman, Blair and McCutcheon, 1989). This, with other local factors, is probably involved in the mediation of differentiation and proliferation of satellite cells as shown *in vitro* (Allen and Rankin, 1990), leading to rapid "hyperplastic" growth in skeletal muscle. IGF-I may also stimulate postnatal proliferation of differentiated preadipocytes, thereby increasing the capacity for adipose tissue growth and fat accumulation (Hausman, Jewell and Hentges, 1989). However, to the extent that IGF-I synthesis is controlled by GH, such a general influence on adipose hyperplasia is difficult to reconcile with the potent inhibitory effects of GH on adipose growth (Boyd and Bauman, 1989).

The early postnatal induction of IGF-I expression and synthesis may be substantially influenced by extracellular factors other than GH. In sheep, cattle and pigs, specific binding of GH by hepatocyte membranes is low at birth and increases markedly during the early postnatal period (Gluckman, Butler and Elliott, 1983; Breier, Gluckman and Bass, 1988; Breier *et al*, 1989). The time at which GH receptors become fully functional in neonatal liver and other tissues probably varies with species, tissue and nutrition. Hypophysectomy and GH replacement studies suggest that in sheep, the dominant influence of GH on postnatal growth of skeletal and adipose tissues is established within 3 to 4 weeks of birth (Vézinhet, 1973). Treatment of unweaned baby pigs with exogenous GH between days 7 and 21 had no effect on average daily gain but body composition was not measured (Scamurra, 1986).

Alternative influences during this period, when postnatal relative growth is maximal, include nutrient availability and insulin (Breier *et al*, 1988; 1989). Evidence for the influence of nutrition is circumstantial, in that the postnatal elevation in plasma IGF-I coincides with a major increase in availability of glucose, long-chain fatty acids and amino acids in well-fed neonates. Levels then decline upon weaning, at least in ruminants (Breier *et al*, 1988). As in the foetus, the influence of insulin may be permissive, but definitive experiments have not been reported. Oddy, Lindsay, Barker and Northrop (1987) saw no effect of exogenous administration of insulin on whole-body or muscle growth, or protein synthesis or deposition in young, milk-fed lambs.

POSTWEANING GROWTH

Relative growth rate slows progressively with time after weaning. This is associated with declining rates of fractional synthesis and deposition of muscle protein (see Bell, Bauman and Currie, 1987) and increased partitioning of nutrients towards lipid deposition in adipose tissue. Absolute rate and to a lesser extent, composition of gain is affected by energy and protein intake but it is becoming increasingly clear that most meat animal species are operating considerably below their genetic potential for lean tissue growth (Campbell, 1988). In particular, it appears that endogenous secretion of GH is a major constraint on growth performance as indicated by the often dramatic improvement of lean growth in pigs and ruminants treated with exogenous GH (Beermann and Boyd, 1992).

Muscle DNA content continues to increase after weaning as a linear function of muscle weight and protein content, probably through continuing satellite cell proliferation and incorporation (Campion, Richardson, Kraeling and Reagen, 1981). This "hyperplastic" growth phase ceases at a body weight which is well below mature weight but is equal to or greater than usual market weight in lambs (Johns and Bergen, 1976) and pigs (Campbell and Dunkin, 1983). The relative slowing of muscle growth in finishing animals may be due to the waning influence of the GH - IGF-I axis on satellite cell mitogenesis. The capacity for amplification of this aspect of muscle growth is illustrated by substantial increases in muscle fibre diameter and protein content, paralleled by a proportionate increase in DNA content in GH-treated pigs (Beermann, Fishell, Roneker, Boyd, Armbruster and Souza, 1990).

Adipose tissue also continues to undergo hyperplastic growth after weaning, but in general this contributes less than cellular hypertrophy to the increasing mass of this tissue in finishing meat animals. The pattern of cellular growth is complex and appears to vary with species, anatomical location, physiological age and plane of nutrition (see Vernon, 1986). Much of the increase in fat deposition during this phase may also be due to the declining influence of GH. Specific inhibitory influences and mechanisms of action of GH on adipose lipogenesis are discussed in detail elsewhere in this volume (Vernon, 1992).

Prenatal influences on postnatal tissue growth and body composition

Gestational nutrient deprivation influences not only growth of the whole foetus but relative growth of specific organs and tissues in a manner which is reasonably consistent across non-altricial species and with different causes of nutrient restriction (see Alexander, 1974; Widdowson, 1977). This section firstly reviews the cellular basis for some of these observations, especially in relation to likely key stages of gestational development. The degree to which altered prenatal tissue development can influence rate and composition of postnatal growth is then examined.

FOETAL NUTRIENT SUPPLY AND TISSUE DEVELOPMENT

Alexander (1974) reviewed data for organ composition of newborn lambs which were growth-retarded because of maternal undernutrition or placental insufficiency caused by maternal heat stress, premating carunclectomy or uterine vascular embolization. In these small lambs, relative weight of the brain was consistently greater, while skeletal muscles, liver, spleen, thymus and thyroid were always disproportionately smaller than in normally grown lambs. Effects on adipose tissue were more variable, but small lambs tended to have relatively smaller dissectible depots and less chemically-extractable fat. Bone weights were closely proportional to birth weight.

These observations generally agree with effects of natural or experimental intrauterine growth retardation in other species, including the runt pig (Widdowson, 1977). With the notable exception of the liver, they also fit Hammond's (1932) theory of allometric growth, in that early-maturing organs and tissues would be expected to be least affected by prenatal nutritional insult.

The mediation of these responses probably also varies with the gestational pattern of cellular growth, especially in late pregnancy when foetal nutrient deprivation is most likely. Foetal hepatic hyperplasia normally slows appreciably or ceases in late gestation in sheep (Dick, 1956) and cattle (Godfredson, Holland, Odde and Hossner, 1991). Thus, much of the marked trophic response to undernutrition must be via decreased hypertrophy and protein deposition, consistent with observations of a dramatic reduction in hepatic protein synthetic rate in late-gestation foetal lambs whose dams had been starved for 48 h (Krishnamurti and Schaefer, 1984). However, earlier, chronic nutrient restriction may also inhibit cell proliferation in the foetal liver, as indicated by a marked reduction in hepatic DNA content of severely growth-retarded newborn pigs (Widdowson, 1971).

It is generally accepted that formation of primary myotubes and thus, "true" hyperplasia in skeletal muscle ceases after about two-thirds of gestation in domestic meat animals (see Swatland, 1984). However, DNA content of foetal muscle continues to increase through late gestation in sheep (Rattray, Robinson, Garrett and Ashmore, 1975), cattle (Godfredson *et al*, 1991) and pigs (Widdowson, 1971). This is apparently due to continuing proliferation and organization of myoblasts into secondary myotubes, which then develop into type II fibres (see Swatland, 1984). Much of the disproportionate reduction in muscle weight of growth-retarded foetuses may occur through disruption of this process, as judged by decreased muscle DNA content in runt piglets (Widdowson, 1971) and less severely retarded twin or triplet lambs (Rattray *et al*, 1975). More specific evidence that secondary but not primary fibre number is reduced in runt piglets was recently obtained (Handel and Stickland, 1987). Suppression of foetal muscle hyperplasia by inadequate nutrient supply is indicated by a comparison of changes in DNA content of *semitendinosus* muscle between days 232 and 271 in Charolais foetuses carried in Brahman versus Charolais dams (Ferrell, 1991a). The increase in DNA content during late gestation was much greater in the Charolais/Charolais animals, possibly related to their greater placental size and capacity for nutrient transport (Figure 1).

The meagre growth of foetal adipose tissue during late gestation in sheep is characterized by modest increases in DNA and lipid content at relative rates which vary between depots (Broad and Davies, 1980). The variable decrease in relative weight of adipose tissue in growth-retarded newborn lambs (Alexander, 1974) may involve reduced rates of adipocyte proliferation, as well as of lipogenesis and fat deposition but this has not been specifically examined. The issue is complicated by the different developmental characteristics of the predominantly brown versus more sparse white adipose tissues in this species (see Vernon, 1986). Limited evidence suggests that adipocyte number may be reduced at birth in growth-retarded piglets (Powell and Aberle, 1981).

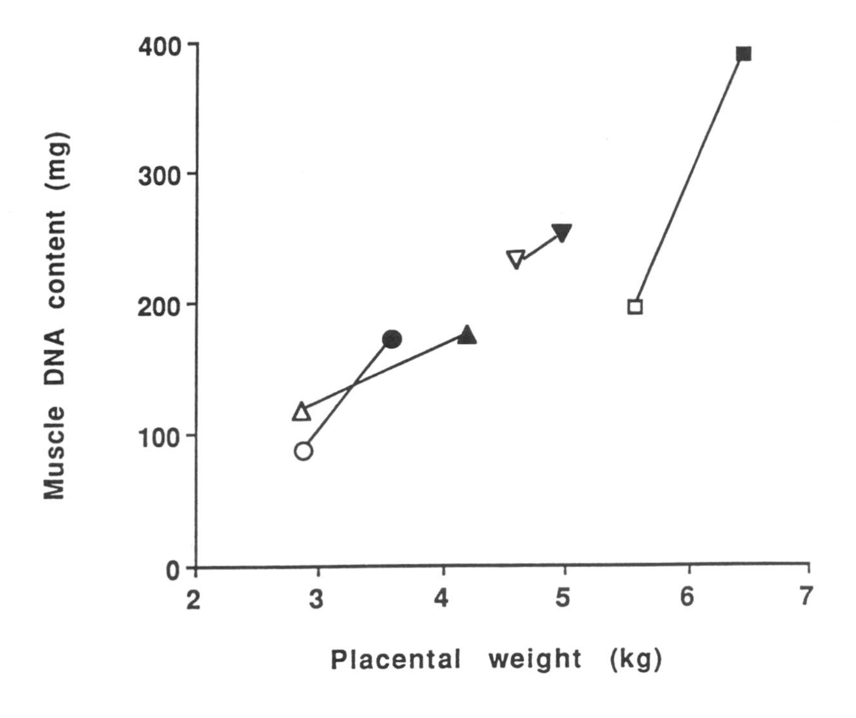

Figure 1 Relation between DNA content of *M. semitendinosus* and placental weight in foetal calves at 232 days (open symbols) and 271 days (closed symbols) of gestation. Symbols denote combinations of foetal and maternal genotypes achieved by reciprocal embryo transfer between Brahman and Charolais. Brahman foetus/Brahman dam (○); Brahman foetus/Charolais dam (△); Charolais foetus/Brahman dam (▽); Charolais foetus/Charolais dam (□). Drawn from data of Ferrell (1991a).

As suggested earlier in this review, a common mediating factor for reduction in foetal cell proliferation may be the reduced expression and synthesis of IGF-I in liver and presumably, other foetal tissues during nutrient deprivation caused by maternal undernutrition or placental insufficiency. Other extracellular and local factors are undoubtedly involved in the mediation and modulation of foetal tissue responses to nutrient availability but these are presently obscure.

POSTNATAL CONSEQUENCES

Studies on the consequences of altered foetal tissue development for postnatal growth rate and composition in meat animals have been mainly confined to pigs. Despite the very rapid postnatal growth and large ratio of mature weight to birth weight in this species, permanent effects on skeletal and vital organ capacity for growth have been observed in natural runts (Widdowson, 1971) and in the growth-retarded offspring of severely protein-deficient sows (Pond, Yen, Mersmann and Maurer, 1990). These included reduced carcass length, implying an effect on bone growth. This was supported by observation of shorter long bones in natural runts but not in the less prenatally-retarded progeny of protein-deficient dams. Reduction in the mass of soft tissues and organs, including skeletal muscles, was generally proportional to that in body weight (16% in natural runts, 15% in protein-deficient progeny), except for the disproportionate stunting of the mature liver in natural runts. As might be predicted from the lack of effect of foetal undernutrition on the brain and early attainment of mature size in this organ, absolute weights at maturity of the cerebrum, cerebellum and pituitary were unaffected by prenatal nutrition (Pond *et al*, 1990). In the same study, both maternal protein and energy restriction during pregnancy caused permanent reduction in backfat depth. Interestingly, this effect was greater in the progeny of energy-restricted sows (Pond *et al*, 1990).

The failure of prenatally-retarded pigs to attain the mature size of normal animals inevitably means that their rate of growth to market weight (usually about 30% of mature weight) must be reduced (Atinmo, Pond and Barnes, 1974; Hegarty and Allen, 1978; Powell and Aberle, 1980; Pond, Mersmann and Yen, 1985). This can be illustrated by simple correlation of average daily gain with birth weight (Figure 2).

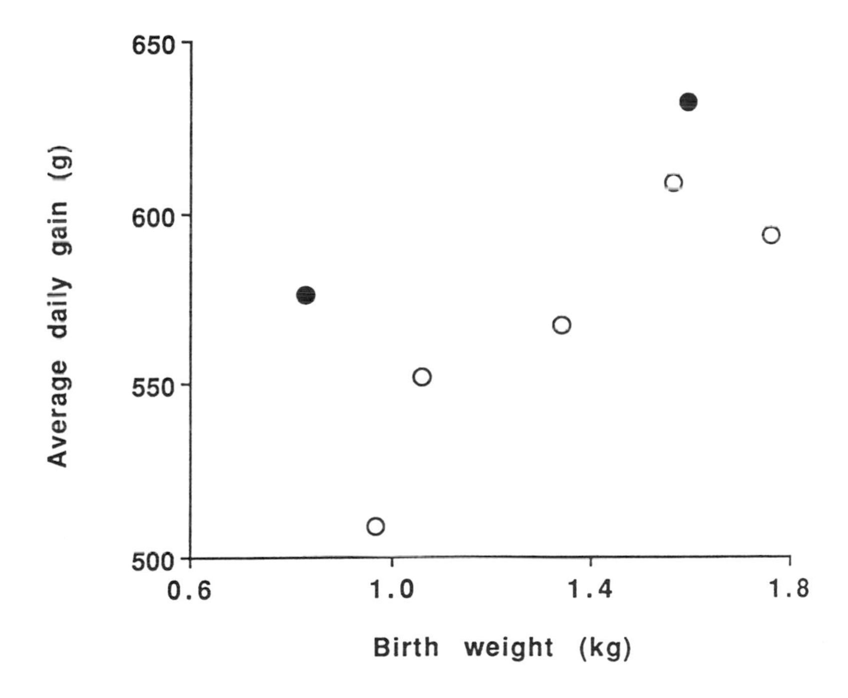

Figure 2 Relation between average daily gain from birth to slaughter at 108 kg and birth weight in naturally-reared (○) and artificially-reared (●) pigs. Drawn from data of Powell and Aberle (1980).

It is not surprising that in naturally-reared pigs, such correlations are strongest before weaning, partly because small neonates suffer from competition with their larger litter mates. Also, inadequate maternal nutrition during pregnancy may affect both birth weight and lactation performance of the dam. Nevertheless, in neither pigs (Powell and Aberle, 1980) nor sheep (Penning, Corcuera and Treacher, 1980) could the handicap of low birth weight on preweaning growth be fully overcome by artificially feeding milk *ad libitum*. Artificially-reared runt pigs were also slower to reach slaughter weight (108 kg), but much of this effect may have been related to the severity of their prenatal growth retardation (Figure 2; Powell and Aberle, 1980).

Limited evidence suggests that not only rate but also composition and efficiency of postnatal gain may be inferior in pigs suffering intrauterine growth retardation. For example, both naturally- and artificially-reared runts had less carcass muscle and muscle protein, smaller *longissimus* muscle area, more intramuscular lipid and higher feed:gain ratios than large littermates slaughtered at the same weight (108 kg) (Hegarty and Allen, 1978; Powell and Aberle, 1980). Reduced muscle weight was associated with fewer type II fibres of similar or greater diameter than those of larger birth-weight pigs (Powell and Aberle, 1980; Handel and Stickland, 1987). This predominant fibre type is thought to be derived from foetal secondary myotubes, formation of which ceases before birth (see Swatland, 1984). As discussed in the previous section, the number of these present at birth may be especially susceptible to prenatal nutrient restriction. In severely retarded pigs, it seems that this deficit in fibre number at birth cannot be fully overcome by a compensatory increase in the normal rate of satellite cell proliferation and inclusion during postnatal growth. Thus, even in mature pigs, Widdowson (1971) and Pond *et al* (1990) found that the smaller muscles of runts had a similar concentration and therefore, a smaller content of DNA.

Links between pre- and postnatal cellular growth of adipose tissue are less easy to discern, not least because patterns of adipose growth seem to vary considerably between depots. In cases where increased fatness of slower-growing runt pigs has been observed, an apparent increase in recruitment of preadipocytes occurred, especially in the perirenal depot (Powell and Aberle, 1981). It is not clear why this occurs in natural runts but not in small pigs born to undernourished sows; in both cases, adipocyte number appears to be reduced at birth, but only in the latter does this reduced fatness persist to market weight (Atinmo *et al*, 1974) and beyond (Pond *et al*, 1990).

POSTNATAL COMPENSATION FOR PRENATAL GROWTH RETARDATION

Many of the longer term effects of prenatal growth retardation on postnatal tissue growth are consistently observed only when the retardation is relatively severe. This implies an ability of many less extremely affected newborn animals to undergo postnatal compensation. The physiological or cytological bases for this phenomenon are unknown. However, Handel and Stickland (1988) found that the ability of low birth-weight pigs to catch up to their larger littermates was related to their muscle fibre number at slaughter. They interpreted this to mean that a low muscle fibre number is not an inevitable consequence of being small at birth and suggested that early identification of small pigs with adequate versus low fibre numbers could aid in the management of such animals. It is also possible that small neonates vary in their ability to compensate by increased muscle growth through satellite cell inclusion. Given the likely involvement of IGF-I and other local factors in such a process (Allen and Rankin, 1990) it is interesting to note that the ability of human infants suffering intrauterine growth retardation to undergo catch-up growth by 12 months was related to their serum IGF-I concentrations through the first postnatal year (Theriot-Prevost, Boccara, Francoual, Badoual and Job, 1988).

Conclusions

Despite an amazing increase in knowledge of foetal metabolism and growth physiology in farm animals since Everitt (1968) bemoaned the lack of such information, much remains to be learned about the regulation of foetal growth. It is now clear that foetal metabolism and growth are normally constrained during late pregnancy, probably by placental capacity to supply nutrients. Normal and abnormal growth restriction during this phase may be locally mediated by suppression of expression and synthesis of IGF-I in foetal tissues. However, the extracellular regulation and specific roles of this factor, its enigmatic cousin IGF-II and their binding proteins, as well as the many other factors likely to be influencing foetal growth, remain to be defined.

Restriction of hyperplastic growth of foetal muscle appears to influence postnatal growth rate, and in severe cases, carcass composition at market weight. If energy intake remains high in these animals, increased adiposity and carcass fatness leads to decreased efficiency of feed conversion. Postnatal ability to compensate for prenatal growth restriction is inversely related to birth weight, but the ability of some small neonates to catch up to their larger peers is impressive. Elucidation of the cellular basis for catch-up growth, and its regulation, especially in muscle, might do much to explain this individual variation.

References

Alexander, G. (1974) In *Size at Birth*, (eds K. Elliott and J. Knight), Amsterdam, Elsevier, pp 215-245

Alexander, G., Hales, J.R.S., Stevens, D. and Donnelly, J.B. (1987) *Journal of Developmental Physiology*, **9**, 1-15

Alexander, G. and Williams D. (1971) *Journal of Agricultural Science,* **76**, 53-72

Allen, R.E., Merkel, R.A. and Young, R.D. (1979) *Journal of Animal Science*, **49**, 115-127

Allen, R.E. and Rankin, L.L. (1990) *Proceedings of the Society for Experimental Biology and Medicine*, **194**, 81-86

Atinmo, T., Pond, W.G. and Barnes, R.H. (1974) *Journal of Animal Science*, **39**, 703-711

Beermann, D.H. and Boyd, R.D. (1992) In *The Control of Fat and Lean Deposition*, (eds P.J. Buttery, K.N. Boorman and D.B. Lindsay), London, Butterworths, (in press)

Beermann, D.H., Fishell, V.K., Roneker, K., Boyd, R.D., Armbruster, G. and Souza, L. (1990) *Journal of Animal Science*, **68**, 2690-2697

Bell, A.W., Kennaugh, J.M., Battaglia, F.C., Makowski, E.L. and Meschia, G. (1986) *American Journal of Physiology*, **250**, E538-E544

Bell, A.W., Bauman, D.E. and Currie, W.B. (1987) *Journal of Animal Science*, **65**, (Supplement 2), 186-212

Bell, A.W., Wilkening, R.B. and Meschia, G. (1987) *Journal of Developmental Physiology*, **9**, 17-29

Bell, A.W., Kennaugh, J.M., Battaglia, F.C. and Meschia, G. (1989) *Quarterly Journal of Experimental Physiology*, **74**, 635-643

Black, J.L. (1983) In *Sheep Production*, (ed W. Haresign), London, Butterworths, pp 21-58

Boyd, R.D. and Bauman, D.E. (1989) In *Animal Growth Regulation*, (eds D.R. Campion, G.J. Hausman and R.J. Martin), New York, Plenum, pp 257-293

Breier, B.H., Gluckman, P.D. and Bass, J.J. (1988) *Journal of Endocrinology*, **119**, 43-50

Breier, B.H., Gluckman, P.D., Blair, H.T. and McCutcheon, S.N. (1989) *Journal of Endocrinology*, **123**, 25-31

Broad, T.E. and Davies, A.S. (1980) *Animal Production*, **31**, 63-71

Browne, C.A. and Thorburn, G.D. (1989) *Biology of the Neonate*, **55**, 331-346

Campbell, R.G. (1988) *Nutrition Research Reviews*, **1**, 233-253

Campbell, R.G. and Dunkin, A.C. (1983) *British Journal of Nutrition*, **50**, 619-626

Campion, D.R., Richardson, R.L., Kraeling, R.R. and Reagen, J.O. (1981) *Journal of Animal Science*, **52**, 1014-1018

Davenport, M.L., D'Ercole, A.J. and Underwood, L.E. (1990) *Endocrinology*, **126**, 2062-2067

Dick, D.A.T. (1956) *Journal of Embryology and Experimental Morphology*, **4**, 97-109

Everitt, G.C. (1968) In *Growth and Development of Mammals*, (eds G.A. Lodge and G.E. Lamming), London, Butterworths, pp 131-157

Ferrell, C.L. (1991a) *Journal of Animal Science*, **69**, 1945-1953

Ferrell, C.L. (1991b) *Journal of Animal Science*, **69**, 1954-1965

Fowden, A.L. (1989) *Journal of Developmental Physiology*, **12**, 173-182

Fowden, A.L., Hughes, P. and Comline, R.S. (1989) *Quarterly Journal of Experimental Physiology*, **74**, 703-714

Garssen, G.J., Spencer, G.S.G., Colenbrander, B., Macdonald, A.A. and Hill, D.J. (1983) *Biology of the Neonate*, **44**, 234-242

Gluckman, P.D. (1986) In *Control and Manipulation of Animal Growth*, (cds P.J. Buttery, D.B. Lindsay and N.B. Haynes), London, Butterworths, pp 85-104

Gluckman, P.D. and Butler, J.H. (1983) *Journal of Endocrinology*, **99**, 223-232
Gluckman, P.D., Butler, J.H. and Elliott, T.B. (1983) *Endocrinology*, **112**, 1607-1612
Gluckman, P.D., Butler, J., Comline, R.S. and Fowden, A.L. (1987) *Journal of Developmental Physiology*, **9**, 79-88
Godfredson, J.A., Holland, M.D., Odde, K.G. and Hossner, K.L. (1991) *Journal of Animal Science*, **69**, 1074-1081
Hammond, J. (1932) Growth and Development of Mutton Qualities in the Sheep. London, Oliver and Boyd
Handel, S.E. and Stickland, N.C. (1987) *Animal Production*, **44**, 311-317
Handel, S.E. and Stickland, N.C. (1988) *Animal Production*, **47**, 291-295
Hausman, G.J., Hentges, E.J. and Thomas, G.B. (1987) *Journal of Animal Science*, **64**, 1255-1261
Hausman, G.J., Jewell, D.E. and Hentges, E.J. (1989) In *Animal Growth Regulation*, (eds D.R. Campion, G.J. Hausman and R.J. Martin), New York, Plenum, pp 49-68
Hay, W.W., Jr., Molina, R.A., Di Giacomo, J.E. and Meschia, G. (1990) *American Journal of Physiology*, **258**, R569-R577
Hegarty, P.V.J. and Allen, C.E. (1978) *Journal of Animal Science*, **46**, 1634-1640
Johns, J.T. and Bergen, W.G. (1976) *Journal of Animal Science*, **43**, 192-200
Jones, C.T., Gu, W., Harding, J.E., Price, D.A. and Parer, J.T. (1988) *Journal of Developmental Physiology*, **10**, 179-189
Jones, C.T., Lafeber, H.N., Price, D.A. and Parer, J.T. (1987) *Journal of Developmental Physiology*, **9**, 181-201
Kennaugh, J.M., Bell, A.W., Teng, C., Meschia, G. and Battaglia, F.C. (1987) *Pediatric Research*, **22**, 688-692
Krishnamurti, C.R. and Schaefer, A.L. (1984) *Growth*, **48**, 391-403
Liggins, G.C. and Kennedy, P.C. (1968) *Journal of Endocrinology*, **40**, 371-381
Meier, P., Teng, C., Battaglia, F.C. and Meschia, G. (1981) *Proceedings of the Society for Experimental Biology and Medicine*, **167**, 463-468
Mellor, D.J., Matheson, I.C. and Small, J. (1977) *Research in Veterinary Science*, **23**, 119-121
Mellor, D.J. and Murray, L. (1982) *Research in Veterinary Science*, **32**, 377-382
Mersmann, H.J., Houk, J.M., Phinney, G., Underwood, M.C. and Brown, L.J. (1973) *American Journal of Physiology*, **224**, 1123-1129
Mesiano, S., Young, I.R., Baxter, R.C., Hintz, R.L., Browne, C.A. and Thorburn, G.D. (1987) *Endocrinology*, **120**, 1821-1830
Mesiano, S., Young, I.R., Hey, A.W., Browne, C.A. and Thorburn, G.D. (1989) *Endocrinology*, **124**, 1485-1491
Milley, J.R. (1986) *Growth*, **50**, 390-401
Milner, R.D.G. (1988) In *Perinatal Nutrition*, (ed B.S. Lindblad), San Diego, Academic Press, pp 45-62
Mosier, H.D., Jr., Dearden, L.C., Jansons, R.A., Roberts, R.C. and Biggs, C.S. (1982) *Developmental Pharmacology and Therapeutics*, **4**, 89-105

Mosier, H.D., Jr., Spencer, E.M., Dearden, L.C. and Jansons, R.A. (1987) *Pediatric Research*, **22**, 92-95

Oddy, V.H., Lindsay, D.B., Barker, P.J. and Northrop, A.J. (1987) *British Journal of Nutrition*, **58**, 437-452

Owens, J.A., Falconer, J. and Robinson, J.S. (1987a) *Journal of Developmental Physiology*, **9**, 137-150

Owens, J.A., Falconer, J. and Robinson, J.S. (1987b) *Journal of Developmental Physiology*, **9**, 225-238

Parr, R.A., Williams, A.H., Campbell, I.P., Witcombe, G.F. and Roberts, A.M. (1986) *Journal of Agricultural Science*, **106**, 81-87

Penning, P.D., Corcuera, P. and Treacher, T.T. (1980) *Animal Feed Science and Technology*, **5**, 321-336

Pilistine, S.J., Moses, A.C. and Munro, H.N. (1984) *Proceedings of the National Academy of Sciences*, USA, **81**, 5853-5857

Pond, W.G., Mersmann, H.J. and Yen, J.-T. (1985) *Journal of Nutrition*, **115**, 179-189

Pond, W.G., Yen, J.-T., Mersmann, H.J. and Maurer, R.R. (1990) *Growth, Development and Aging*, **54**, 77-84

Powell, S.E. and Aberle, E.D. (1980) *Journal of Animal Science*, **50**, 860-868

Powell, S.E. and Aberle, E.E. (1981) *Journal of Animal Science*, **52**, 748-756

Rattray, P.V., Robinson, D.W., Garrett, W.N. and Ashmore, R.C. (1975) *Journal of Animal Science*, **40**, 783-788

Reynolds, L.P., Ferrell, C.L., Robertson, D.A. and Ford, S.P. (1986) *Journal of Agricultural Science*, **106**, 437-444

Robinson, J.J., McDonald, I., Brown, D.S. and Fraser, C. (1985) *Journal of Agricultural Science*, **105**, 21-26

Robinson, J.S., Hart, I.C., Kingston, E.J., Jones, C.T. and Thorburn, G.D. (1980) *Journal of Developmental Physiology*, **2**, 239-248

Scamurra, R.W. (1986) MS Thesis, Virginia Polytechnic Institute and State University

Schoknecht, P.A., Slepetis, R. and Bell, A.W. (1990) *FASEB Journal*, **4**, A1079

Stevens, D. and Alexander, G. (1986) *Journal of Developmental Physiology,* **8**, 139-145

Stevens, D., Alexander, G. and Bell, A.W. (1990) *Journal of Developmental Physiology*, **13**, 277-281

Stock, M.K., Francisco, D.L., McCutcheon, I.E. and Metcalfe, J. (1982) *Federation Proceedings*, **41**, 1491

Straus, D.S., Ooi, G.T., Orlowski, C.C. and Rechler, M.M. (1991) *Endocrinology*, **128**, 518-525

Swatland, H.J. (1984) Structure and Development of Meat Animals. Englewood Cliffs, NJ, Prentice-Hall

Theriot-Prevost, G., Boccara, J.F., Francoual, C., Badoual, J. and Job, J.C. (1988) *Pediatric Research*, **24**, 380-383

Trahair, J.F., Perry, R.A., Silver, M. and Robinson, P.M. (1987) *Quarterly Journal of Experimental Physiology*, **72**, 61-69
Vernon, R.G. (1986) In *Control and Manipulation of Animal Growth*, (eds P.J.Buttery, N.B. Haynes and D.B. Lindsay), London, Butterworths, pp 67-83
Vernon, R.G. (1992) In *The Control of Fat and Lean Deposition*, (eds P.J. Buttery, K.N. Boorman and D.B. Lindsay), London, Butterworths, (in press)
Vézinhet, A. (1973) *Annales de biologie animale et Biochimie et Biophysique*, **13**, 51-73
Vileisis, R.A. and D'Ercole, A.J. (1986) *Pediatric Research*, **20**, 126-130
Widdowson, E.M. (1971) *Biology of the Neonate*, **19**, 329-340
Widdowson, E.M. (1977) *Nutrition and Metabolism*, **21**, 76-87
Young, B.A., Bell, A.W. and Hardin, R.T. (1989) In *Energy Metabolism of Farm Animals*, compiled by Y. van Honing and W.H. Close, Wageningen, Pudoc, pp 155-158

6
THE NEUROPHYSIOLOGICAL REGULATION OF GROWTH AND GROWTH HORMONE SECRETION

F.C. BUONOMO and C.A. BAILE
Monsanto Company, St Louis, Missouri 63198, USA

Introduction

With the advent of genetic engineering, the importance of growth hormone (GH) in the regulation of growth and metabolism in domestic species has been clearly demonstrated. Ample evidence of an integral role for GH in the processes of growth and lactation exists in dairy cattle (Machlin, 1973; Bauman, Eppard, DeGeeter and Lanza, 1985), sheep (Zainur, Tassell, Kellaway and Dodemaide, 1989), beef cattle (Eisemann, Hammond, Bauman, Reynolds, McCutcheon, Tyrrell and Haaland, 1979) and swine (Machlin, 1972). Endogenous GH secretion is primarily controlled by the central nervous system (CNS) via two specific hypothalamic neurohormones: growth hormone-releasing factor (GRF) and somatostatin (SRIF), an inhibitor of GH release. The secretion of GRF and SRIF is governed by a host of neuropeptides and neurotransmitters which provide a functional link between higher CNS centres and hypophysiotropic neurons. This review will focus on the CNS regulation of GH secretion and circulating factors which feedback either to stimulate or to inhibit its release. In addition, limited data demonstrating improvements in growth and/or production efficiency through stimulation of endogenous GH secretion will be presented.

Neuroendocrine regulation of GH secretion

In many domestic species GH is released in a pulsatile manner (Klindt and Stone, 1984; Dubreuil, Lapierre, Pelletier, Petitclerc, Couture, Gaudreau, Monisset and Brazeau, 1988; Buonomo, Lauterio and Scanes, 1984), the pattern of which is regulated by the sequence of GRF and SRIF secretion. For example, measurements of GRF and SRIF collected from the hypophysial portal system of sheep indicated that SRIF (Frohman, Down, Clarke and Thomas, 1990) and GRF (Frohman *et al*, 1990; Thomas, Cummins, Francis, Sudbury, McCloud and Clarke, 1991) are secreted episodically. A mean pulse interval of 71 min with peaks of 25-40 pg/ml were observed for GRF (Frohman *et al*, 1990). In the case of SRIF, peaks ranged from 65 to 160 pg/ml with intervals of approximately 54 min (Frohman *et al*, 1990). The GH pulse interval for these sheep was 62 min. Interestingly, a significant association was observed between peaks of GRF simultaneous with or immediately preceding GH secretory peaks (approximately 60%), but not between SRIF troughs and GH peaks or between GRF and SRIF peaks in portal plasma. These observations suggest that in sheep, GRF plays a primary role in dictating the pulsatile secretory pattern of GH, while SRIF plays a minor role.

Central regulation of GRF and SRIF secretion

It is now evident that the secretion of GRF and SRIF from the hypothalamus is regulated by a number of neurotransmitters and peptides present both within and outside the CNS. The biogenic amines act as neurotransmitters to process information from areas higher within the CNS to adjacent hypothalamic neurons which either stimulate or inhibit the release of GRF and SRIF. In general, dopamine inhibits GH secretion (Collu, Fraschini, Visconti and Martini, 1972; Kato, Dupre and Beck, 1973; Davis and Borgen, 1973), presumably by stimulating SRIF release. For example, L-DOPA administration has been reported to inhibit GH release in sheep (Davis and Borgen, 1973). In addition, the intraventricular injection of dopamine has been reported to stimulate SRIF release into the hypophyseal portal system of anaesthetized rats *in vivo* (Chihara, Arimura and Schally, 1979), hypothalamic tissue *in vitro* (Wakabayshi, Myazawa, Kondon, Miki, Demura, Demura and Shizume, 1977) and perfused hypothalamic cells (Richardson, Naguyen and Hollander, 1983). Moreover, cytochemical studies have indicated that neural fibres containing dopamine are located close to the end terminals of SRIF-containing hypothalamic neurons (Hokfelt, Elde, Fuxe, Johansson, Ljungda, Goldstein, Luft, Efendic, Nilsson, Terenius, Ganten, Jeffcoate, Rehfeld, Said, Perez, Posain, Taia, Teran and Palacios, 1978). However, it should be noted that in conscious rats, dopaminergic agonists, such as apomorphine and piribedil, cause a significant increase in GH levels (Mueller, Simpkins, Meites and Moore, 1976), while chlorpromazine and haloperidol (dopamine antagonists) inhibit episodic GH secretion (Eden, Bolle and Modigh, 1979).

There is evidence that the adrenergic neurotransmitters are also involved in the regulation of both GRF and SRIF. Although the administration of phenoxybenzamine (PBZ), an alpha$_1$-adrenergic antagonist, has been reported to stimulate GH secretion in sheep (Davis and Borgen, 1973), in general the alpha-adrenergic system appears to exert a permissive role on GH secretion in most species. For example, the administration of alpha-adrenergic antagonists, attenuates GH secretion in rats (Durand, Martin and Brazeau, 1977), rabbits (Durand *et al*, 1977) and poultry (Buonomo *et al*, 1984). In rats and rabbits, pretreatment with either PBZ, yohimbine (YOH, alpha$_2$-adrenergic antagonist) or isoproterenol (ISO, beta-adrenergic agonist) blocked GRF-induced GH secretion, while the beta-adrenergic antagonist, propranolol, enhanced the GRF effect (Durand *et al*, 1977; Minamitani, Chihara, Kaji, Kodama, Kita and Fujita, 1989; Krieg, Perkins, Johnson, Rogers, Arimura and Cronin, 1988). Passive immunization against SRIF restored GRF-induced GH secretion in the PBZ- or

ISO-treated rats, but was not effective in the YOH-treated rats (Minamitani *et al*, 1989; Krieg *et al*, 1988). Alternatively, clonidine (an alpha$_2$-agonist) reportedly increased GH levels even further than that already achieved by pretreatment with antisomatostatin (Durand *et al*, 1977). These data indicate that depending on the species, the noradrenergic system is in part responsible for the stimulation of GRF release and that central alpha$_2$-adrenergic mechanisms play a more important role in the regulation of GH secretion than alpha$_1$-receptors. Blockade of alpha$_2$-adrenergic receptors results in the suppression of hypothalamic GRF release and enhanced release of endogenous SRIF, thereby suppressing GH secretion. In addition, the inactivation of the beta-adrenergic system appears to enhance GH secretion by inhibiting SRIF release.

Other neurotransmitters including histamine, acetylcholine, serotonin and gamma-aminobutyric acid (GABA) have also been implicated in the regulation of SRIF and GRF release. Evidence of a permissive role for GABA has been suggested in that picrotoxin, a GABA antagonist, was observed to stimulate SRIF release from cultured cortical cells, while GABA itself was inhibitory (Robbins and Landon, 1983). Dual roles for histamine have been proposed due to the presence of H_1 and H_2 receptors, while acetylcholine has been shown to stimulate GH secretion in both rats as well as humans (Robbins and Landon, 1983).

Serotonin has been reported to stimulate GH secretion following intraventricular administration in rats (Muller, 1987) possibly through stimulation of GRF secretion, as portal levels of SRIF are unaffected following serotonin injection in anaesthetized rats (Kukucska and Makara, 1983) and passive immunization against GRF attenuates serotonin-induced GH secretion (Murakami, Kato, Kabayama, Tojo, Inque and Imura, 1986). However, the reduction in GH secretion following the administration of the serotonin receptor antagonist, metergoline, was partially attenuated by the simultaneous infusion of SRIF antisera (Arnold and Fernstrom, 1981), suggesting increased SRIF levels following serotonin blockade. In steers, serotonin receptor blockade by cyproheptadine decreased GH secretion (Figure 1), while activation of serotonin receptors by quipazine produced an increase in circulating GH levels in steers (Sartin, Kemppainen, Marple, Carnes, Dieberg and Oliver, 1987).

In addition to and possibly in concert with the brain monamines, endogenous opioid peptides stimulate GH secretion in several mammalian species. Opioid peptides have been reported to stimulate GH secretion in primates (Stubbs, Jones, Edwards, Delitata, Jeffcoate, Ratler, Besser, Bloom and Alberti, 1978), rats (Rivier, Vale, Ling, Brown and Guillemin, 1977), sheep (Bolton, Roud, Redekopp, Livesay, Nicholls and Donald, 1983; Gluckman, Marti-Henneberg, Kaplan, Li and Grumbach, 1980; Buonomo, Tou and Kaempfe, 1991), cattle (Leshin, Rund, Thompson, Mahaffey, Chang, Byerley and Kiser, 1990) and swine (Buonomo *et al*, 1991; Barb, Kraeling and Rampacek, 1989; Trudeau, Meijer, van de Wiel and Erkens, 1988). For example, circulating GH levels are increased in sheep, pigs and rats within five minutes following intravenous injection of the mu-opioid

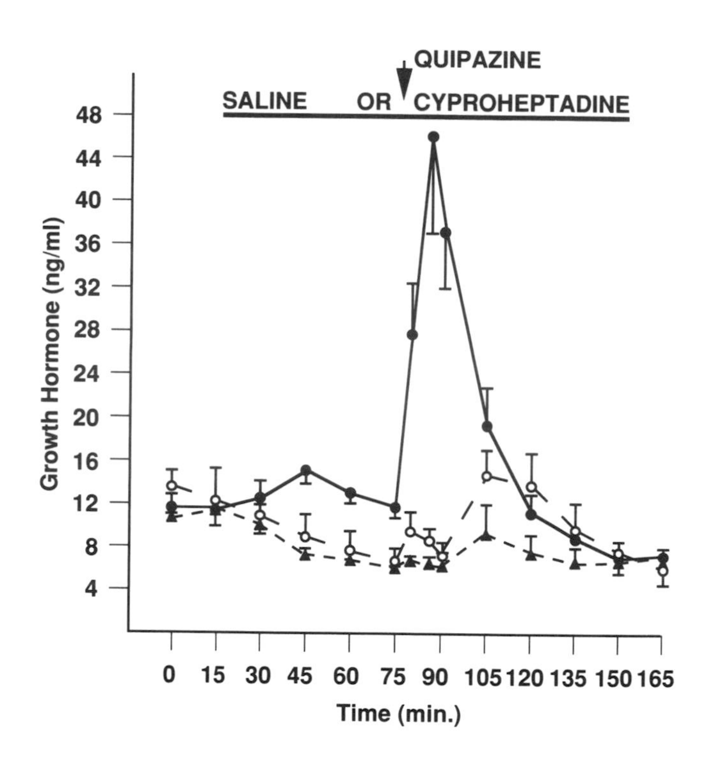

Figure 1 The effect of saline (•) or cyproheptadine (∘, 0.06 mg/kg; ▴, 0.6 mg/kg) pretreatment on quipazine (0.5 mg/kg)-stimulated GH secretion in steers. Reprinted with permission from Sartin *et al*, 1987.

receptor agonist, syndyphalin 33 (Buonomo *et al*, 1991). Sheep continued to respond with significant elevations in plasma GH levels following repeated injections of syndyphalin 33 administered two hours apart (Figure 2). Moreover, this tripeptide agonist maintained bioactivity following oral administration in swine. Evidence exists to suggest that opioid-stimulated GH secretion may be mediated in part by increased GRF secretion. In pigs, GRF immunoneutralization resulted in an attenuation of the GH-stimulatory effects of the opioid agonist,

FK33-824 (Trout and Schanbacher, 1990). In rats, the development of antibodies against GRF has been shown to depress metenkephalin-induced GH secretion, while SRIF depletion has been reported to enhance this response (Wehrenberg, Bloch and Ling, 1985). Moreover, passive immunization of rats using a monoclonal antibody against rat GRF attenuated the GH secretory response to both alpha-adrenergic stimulation by clonidine, as well as to the opiates morphine and beta-endorphin (Wehrenberg *et al*, 1985) indicating a common GH-stimulatory mechanism.

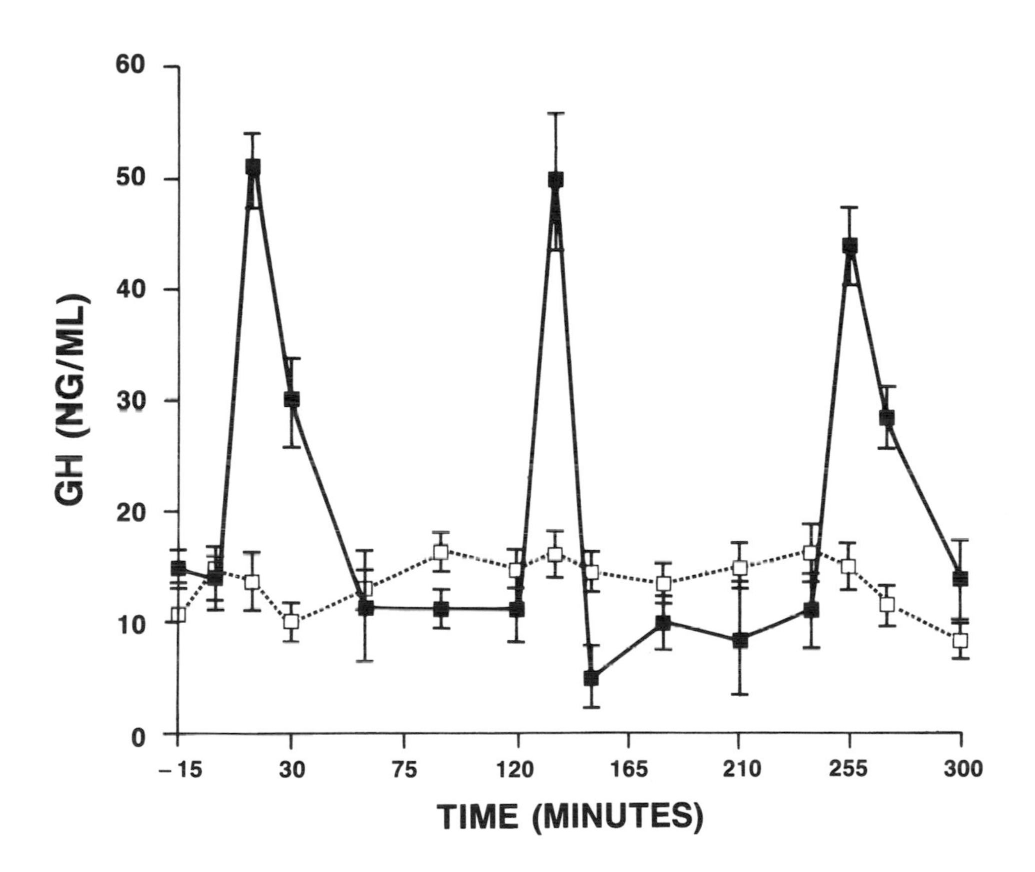

Figure 2 Peak GH levels after i.v. syndyphalin 33 injection (■, 5 nmol/kg) given 2 hours apart were similar after each injection and higher (P<.05) than that of control sheep (□). Reprinted with permission from Buonomo *et al*, 1991.

A number of neuropeptides have also been implicated in the regulation of GH secretion. These include vasoactive intestinal peptide (VIP), neuropeptide Y (NPY), motilin, galanin, neurotensin and substance P, which in general have been reported to stimulate GH secretion (McCann, 1982). Although few data are available in domestic species, these peptides, with some exceptions, appear to act both within the CNS (Muller, 1987) and at the level of the pituitary gland. For example, galanin, a peptide first isolated from the porcine intestine (Tatemoto, Rokaeus, Joruvall, McDonald and Mutt, 1983) has been reported to stimulate GH secretion in rats following both intravenous (Murakami, Kato, Shimatsu, Koshiyama, Hattori, Yanaihara and Imura, 1989) and intraventricular administration in rats (Murakami, Kato, Koshiyama, Inoue, Yanaihara and Imura, 1987). Galanin-immunoreactivity is widely distributed throughout the CNS, especially in the hypothalamus (Ching, Christofides, Anand, Gibson, Allen, Su and Tatemoto, 1985) and in the external layer of the median eminence (Nordstron, Melander, Hokfelt, Bartfai and Goldstein, 1986). Recent studies have shown that the GH-stimulatory effects of centrally administered galanin are attenuated by alpha-adrenergic antagonists, as well as passive immunization against GRF. These data suggest that galanin may play a neuroregulatory role in the control of GH secretion through modulation of GRF release. Vasoactive intestinal peptide is another peptide originally isolated from the gastrointestinal tract which has been reported to exert GH stimulatory effects and for which data are available in a domestic species. Hashizume and Kanematsu (1990) have demonstrated that VIP stimulates GH secretion from bovine pituitary glands using an *in vitro* perfusion system; albeit at higher molar concentrations than GRF. Ogwuegbu, Hashizume and Kanematsu (1991) have further demonstrated that VIP-stimulated GH secretion from caprine pituitary fragments is attenuated by a hypothalamic factor, most likely SRIF. Motilin, a gastrointestinal peptide originally isolated from porcine duodenal extracts has also been reported to stimulate GH secretion both *in vitro* (Samson, Koenig and McCann, 1982) and *in vivo* (Samson, Lumpkin, Nilaver and McCann, 1984) in rats. Motilin-like immunoreactivity has been identified in the hypothalamus of several mammalian species (Jacobowitz, O'Donohue, Chey and Chang, 1981; O'Donohue, Beinfeld, Chey, Chang, Nilaver, Zimmerman, Yajima, Adachi, Poth, McDevitt and Jacobowitz, 1981; Yanaihara, Sato, Yanaihara, Naruse, Foresmann, Helmstaedter, Fujita, Yamaguchi and Abe, 1978) suggesting that a possible neuroendocrine role for motilin may exist. Limited data exist in support of this contention in that intraventricular administration of motilin results in a significant depression of circulating GH levels in conscious rats (Samson *et al*, 1982), similar to the ultrashort feedback action of GRF. In addition, immunohistochemical data indicate that motilin coexists in axon terminals containing SRIF (Niswender, Midgley, Monroe and Reichert, 1968) and thus motilin may exert its effects on GH secretion via neuromodulation of hypothalamic SRIF.

Recent evidence suggests that a large proportion of CNS neurons contain peptide-peptide and amine-peptide co-localizations. For example, NPY is found in a variety of hypothalamic loci frequently co-localized with catecholamines and its SRIF-stimulating activity can be attenuated by either alpha$_1$- or beta-adrenergic receptor blockade (Rettori, Milenkovic, Aguila and McCann, 1990). More recently, co-localization studies have indicated that GRF neurons may also contain neurotensin (Hokfelt, Johansson and Goldstein, 1984), NPY (Ciofi, Croix and Tramu, 1987) and galanin (Rokaeus, 1987). The functional significance of co-localization of peptides in GRF neurons is presently unknown, however theoretically, the peptide which is co-released from the GRF neuron may act presynaptically on adjacent SRIF receptors to inhibit or stimulate SRIF, or may act directly at the pituitary level to modulate GRF-induced GH secretion following transport via the hypophysial portal system.

Recently a new class of peptides having GH regulatory activities has been identified. Cytokines are immunomodulator peptides which are secreted by monocytes and T-cells in response to infectious challenge or immunologic activation (Beutler and Cerami, 1988). In addition, these peptides are produced in and also affect a number of non-immune cell types, including fibroblasts, endothelial and neural cells (Spangelo, Judd, Isakson and MacLeod, 1989). The cytokines, tumor necrosis factor alpha (TNF-α), interleukin-1 beta (IL-1β) and interleukin-6 (IL-6) have all been reported to modulate GH secretion at both the hypothalamic and pituitary levels. Recent neuroanatomical studies demonstrating the presence of immunoreactive IL-1β neuronal processes in the hypothalamus (Breder, Dinarello and Saper, 1988) and brain (Lechan, Toni, Clark, Cannon, Shaw, Dinarello and Reichlin, 1990), as well as IL-1β receptors (Katsuura, 1989) suggest that IL-1β may be a hypophysiotropic factor. *In vitro*, IL-1β increases SRIF mRNA and peptide content, and stimulates SRIF release from rat diencephalic cells in culture (Scarborough and Dinarello, 1989). Moreover, intraventricular administration of IL-1β results in the attenuation of pulsatile GH secretion in rats *in vivo* (Lumpkin and Hartmann, 1989). However, immunoreactive IL-1β also has been localized in cytoplasmic granules of pituicytes suggesting that IL-1β may exert a paracrine or autocrine influence directly at the level of the pituitary gland (Koenig, Snow, Clark, Toni, Cannon, Shaw, Dinarello, Reichlin, Lee and Lechan, 1990). Evidence of direct inhibition by other cytokines at the level of the pituitary gland exists, as TNF-α has been shown to depress both basal GH secretion, as well as GRF-stimulated GH release from rat (Walton and Cronin, 1989; Gaillard, Turnill, Sappino and Muller, 1990) and bovine (Elsasser, Caperna, Kenison and Fayer, 1989) pituitary cells in culture. Alternatively, the cytokine IL-6 stimulates GH secretion from dispersed rat pituitary cells, and potentiates both GRF- and TRH-induced GH release (Spangelo *et al*, 1989). We have recently shown that systemic administration of IL-6 stimulates GH secretion in chickens (Figure 3) and that IL-6 stimulated GH release can be attenuated by simultaneous SRIF administration (Buonomo *et al*, 1991). It has also been reported that IL-6 is synthesized and secreted by the folliculo-stellate cells of the rat anterior pituitary gland (Vankelecom, Carmeliet, Damme, Billiau and Denef, 1989) and is spontaneously released in significant quantities *in vitro* and thus may function as an intrapituitary releasing factor (Spangelo *et al*, 1990). Cytokines have also been implicated in the regulation of other hormones such as ACTH

(Sharp, Matta, Peterson, Newton, Chao and McAllen, 1989; Kohler, O'Malley, Rayford, Lipsett and Odell, 1967) and TSH (Pang, Hershman, Mirell and Pekary, 1989; Dubuis, Dayer, Siegrist-Kaiser and Burger, 1988), thus these immunoregulatory peptides appear also to possess neuroendocrine modulatory functions.

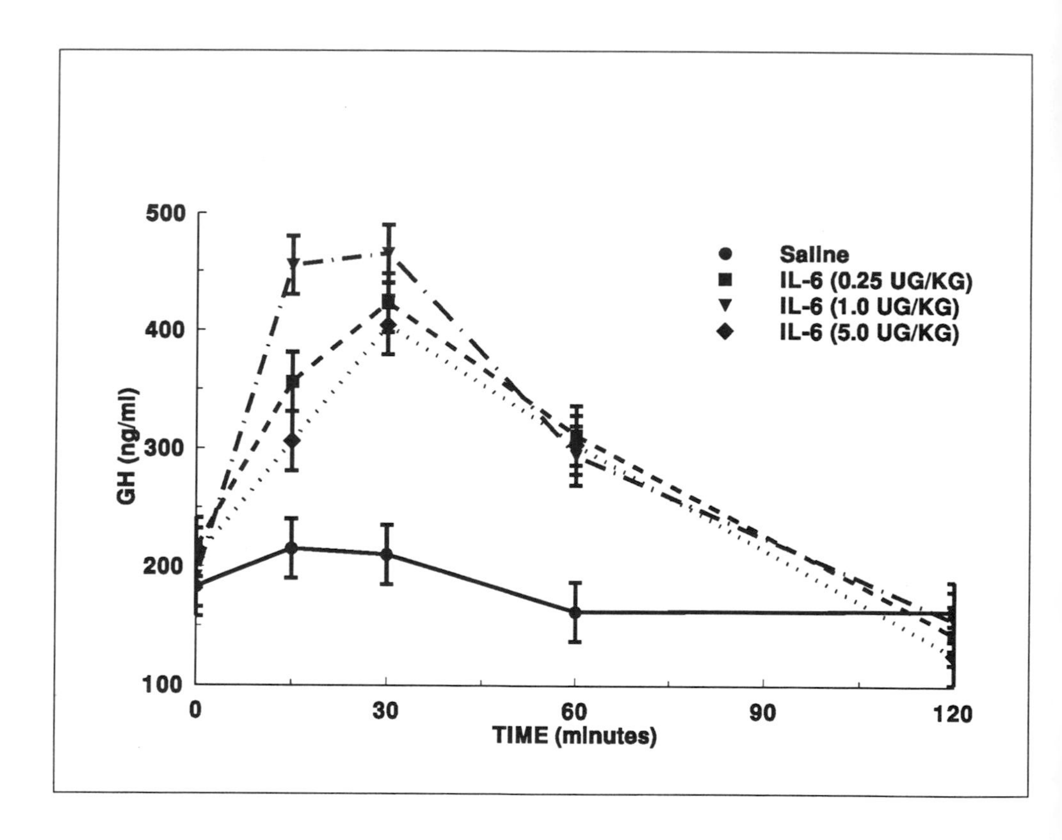

Figure 3 Intravenous administration of IL-6 stimulates GH secretion in conscious chicks at doses between 0.25 and 5.0 μg/kg. Reprinted with permission from Buonomo (unpublished).

Negative feedback effects on the GH regulatory system

The presence of ultra-short, short and long-loop negative feedback systems by GRF, GH and IGF-I on GH secretion are evident. Intraventricular administration of GRF inhibits basal and spontaneous GH secretion in rats (Tannenbaum, 1984; Lumpkin, Samson and McCann, 1985; Katakami, Arimura and Frohman, 1986) and attenuates TRH-stimulated GH secretion in chickens (Ahene, Lea and Harvey, 1991). The suppressive effect of centrally administered GRF on GH secretion was blocked by pretreatment with a specific SRIF antisera suggesting the presence of an ultra-short loop inhibitory feedback system within the CNS for GH which is mediated by SRIF (Katakami *et al*, 1986). In rats, GRF-containing neurons in the arcuate nucleus innervate SRIF fibres in the preoptic area of the hypothalamus (Smith, Howe, Oliver and Willoughby, 1984). Moreover, GRF has been reported to stimulate SRIF release from chicken (Ahene *et al*, 1991) and rat (Aguila and McCann, 1987) hypothalamic tissue, and increases the secretion of SRIF into the hypophysial portal circulation (Mitsugi, Arita and Kimura, 1990).

Considerable evidence exists which suggests that GH can regulate its own secretion via a negative feedback mechanism, *ie* short-loop feedback. This phenomenon was first demonstrated in rats by Krulich and McCann (1966) utilizing bovine GH and was confirmed by others with the homologous peptide (Groesbeck, Parlow and Daughaday, 1987). Growth hormone has been shown to suppress the amplitude of the spontaneous GH peaks in rats regardless of whether injected by the intraventricular (Tannenbaum, 1980; Clark, Corlsson and Robinson, 1988) or the intravenous (Clark *et al*, 1988; Willoughby, Menadue, Zeegers, Wise and Oliver, 1980) route. Recently, the CNS injection of avian GH has been reported to suppress circulating GH concentrations in chickens (Lea and Harvey, 1990).

Although the evidence for a GH feedback loop is strong, the mechanism by which it occurs has not been elucidated. It is possible that GH may inhibit its own secretion at the hypothalamic level or by direct action on the somatotroph. The demonstration of increased SRIF content in the hypothalami of GH-treated hypophysectomized rats (Kanatsuka, Makin, Matsushina, Osegawa, Yamamoto and Kumagai, 1983), the increase in hypothalamic SRIF mRNA following GH treatment (Rogers, Vician, Steiner and Clifton, 1988) and the stimulatory effect of GH on hypothalamic SRIF release (Chihara, Minamitani, Kaji, Arimura and Fujita, 1981; Robbins, Leidy and Landon, 1985) suggest that GH autoregulation involves an increase in inhibitory hypothalamic tone. Moreover, experiments in anaesthetized rats have shown that GH stimulates the secretion of SRIF into the hypophyseal portal system *in vivo* (Chihara *et al*, 1981) and from rat hypothalamic fragments *in vitro* (Berelowitz, Szabo, Frohman, Firestone and Chu, 1981). Alternatively, there is evidence to suggest that GH autoregulatory feedback cannot be entirely explained by increased SRIF secretion. For example, immunoneutralization against SRIF was not capable of restoring either endogenous pulsatile GH secretion (Tannenbaum, 1985) or clonidine-induced GH release (Conway, McCann and Krulich, 1985) following inhibition by intraventricular GH administration.

The observation that the GH-induced suppression was reversed by GRF administration and not by SRIF immunoneutralization suggests that GH may be producing its negative feedback effects by inhibiting GRF release. Moreover, GH administration has been reported to decrease both hypothalamic GRF content (Ganzetti, Petraglia, Capuano, Rosi, Wehrenberg, Muller and Cocchi, 1987), secretion (Miki, Ono, Miyoshi, Tsushima and Shizume, 1989) and GRF mRNA levels (De Gennaro Colonna, Cattaneo, Cocchi, Muller and Maggi, 1988). These data provide evidence of a GH negative feedback mechanism via modulation of hypothalamic neuroendocrine regulatory peptides. It is unlikely that the GH negative feedback effects observed *in vivo* are the result of direct GH action at the level of the pituitary since GH does not alter GH secretion by pituitary cells *in vitro* (Richman, Weiss, Hochberg and Florini, 1981) or block GRF-stimulated GH release *in vivo* (Abe, Molitch, Van Wyk and Underwood, 1983).

In addition, evidence exists for modulation of circulating GH concentrations via GH-induced alterations in insulin-like growth factor (IGF) levels. IGF receptors are present on both the pituitary gland and hypothalamus (Goodyer, De Stephano, Lai, Guyda and Posner, 1984) and IGF-I has been shown to inhibit spontaneous, as well as GRF-induced GH release both *in vitro* and *in vivo* in rats (Berelowitz *et al*, 1981; Abe *et al*, 1983; Goodyer, De Stephano, Guyda and Posner, 1984; Brazeau, Guillemin, Ling, Van Wyk and Humbel, 1982; Tannenbaum, Guyda and Posner, 1983) and chickens (Perez, Malamed and Scanes, 1985; Buonomo, Lauterio, Baile and Daughaday, 1987). More recent studies indicate that central administration of IGF-I depresses circulating GH concentrations in the foetal pig (Spencer, Macdonald, Buttle and Carlyle, 1989), while peripheral IGF-I administration reduces GH secretion in foetal (De Zegher, Bettendorf, Kappan and Grumbach, 1988), but not mature sheep (Spencer, Bass, Hodgkinson, Edgley and Moore, 1991). While long-term negative feedback effects in GH-treated animals may be mediated by IGF peptides, the acute effects observed following exogenous GH administration are too rapid to be produced by changes in IGF levels.

Nutritional status

In several domestic species, the secretion of GH is influenced by changes in the nutritional or metabolic status of the animal. In sheep, circulating GH concentrations change with time after feeding (Bassett, 1974; Driver and Forbes, 1978). For example, plasma GH levels were reduced for 1 to 3 h following feeding in wether lambs exposed to a once daily feeding regimen (Bassett *et al*, 1974). Alternatively, feed restriction results in elevated circulating GH concentrations in pigs (Antinmo, Baldijao, Houpt, Pond and Barnes, 1978; Buonomo and Baile, 1991), sheep (Driver and Forbes, 1978; Foster, Ebling, Micka, Vannerson, Bucholtz, Wood, Suttie and Fenner, 1989; Thomas, Mercer,

Karalis, Rao, Cummins and Clarke, 1990), cattle (Blum, Schnyder, Kunz, Blom, Bickel and Schurch, 1985; Breier, Bass, Butler and Gluckman, 1986), rabbits (McIntyne and Odel, 1974) and chickens (Scanes, Griminger and Buonomo, 1981; Lauterio and Scanes, 1987). Recently, Thomas *et al* (1990) reported that feed restriction was associated with increased mean plasma GH concentrations as well as GH pulse amplitude, but did not affect GH pulse frequency in ovariectomized ewes. It was shown in a subsequent study that the level of feeding had no effect on mean portal concentrations of GRF, GRF pulse amplitude or GRF pulse frequency and that portal concentrations of SRIF in the restricted sheep were significantly reduced (Thomas *et al*, 1991). Alternatively, *ad libitum* feeding following nutritional restriction reduced circulating GH concentrations in sheep (Foster *et al*, 1989; Landefeld, Ebling, Suttie, Vannerson, Padmanabhan, Beitins and Foster, 1989).

Alterations in nutritional status appear to influence GH secretion at both the pituitary and hypothalamic levels. For example, pituitary levels of GH mRNA were increased by restricted feeding (Thomas *et al*, 1990) and conversely decreased following *ad libitum* feeding in sheep (Landefeld *et al*, 1989). Since pituitary GH synthesis is in part under the control of GRF, changes in nutritional status may affect GH synthesis via a change in GRF secretion. There is also evidence that nutritional status alters pituitary sensitivity to hypothalamic regulatory factors such as GRF. For example, Moseley, Alaniz, Claflin and Krabill (1988) reported that GRF-induced GH secretion was reduced immediately after feeding in cattle. Alternatively, fasting increased the GH response to arginine in steers (Trenkle, 1978) and to GRF in sheep (Hart, Chadwick, Coert, James and Simmonds, 1986) and chickens (Lauterio and Scanes, 1988).

Since GH exerts profound effects on intermediary metabolism, it is expected that changes in specific plasma metabolites such as glucose and free fatty acids (FFA) affect somatotropin secretion. Insulin-induced hypoglycaemia in sheep results in decreased plasma GH concentrations (Frohman, Down, Clark and Thomas, 1990), an effect which is in part mediated at the CNS level. Early studies in the rat indicated that neuroglucocytopoenia induced by 2-deoxyglucose stimulated SRIF release from hypothalamic fragments *in vitro* (Berelowitz, Dudlak and Frohman, 1982). More recently however, Frohman *et al* (1990) demonstrated that insulin-induced hypoglycaemia in sheep resulted in a rapid and brief stimulation of SRIF secretion into the hypophysial portal system, followed by a decline in circulating GH levels. Thus increased SRIF hypothalamic secretion may account for the attenuation of GRF-stimulated GH release observed following either central glucopoenia or peripheral hypoglycaemia in sheep (Sartin, Bartol, Kemppainen, Dieberg, Buxton and Soyoola, 1988). A similar observation has been reported in the rat as GRF-induced secretion of GH is attenuated following insulin-induced hypoglycaemia (Painson and Tannenbaum, 1985).

On the basis of the lipolytic activity of GH, Hertelendy and Kipnis (1973) first theorized that elevated FFA levels may act as a negative feedback signal to suppress GH secretion. Support for this concept was obtained in studies demonstrating that pharmacological suppression of FFA release enhanced GH secretion in sheep (Hertelendy and Kipnis, 1973) and cattle (Reynart, De Paepe, Marcus and Peeters, 1975). It was subsequently demonstrated that elevated FFA concentrations produced by lipid infusion were associated with decreased circulating GH levels in sheep (Hertelendy and Kipnis, 1973; Estienne, Schillo, Green and Boling, 1989). This report was recently extended to show that lipid-induced reductions in mean GH concentrations could be attributed to a decrease in GH pulse frequency following lipid infusion in ovariectomized ewes (Estienne *et al*, 1989) and lambs (Estienne, Schillo, Hileman, Green, Hayes and Boling, 1990).

The suppressive effects of FFA infusion on GH secretion appears to act via altered release of and sensitivity to hypothalamic GH release and inhibiting hormones. In sheep, infusion of FFA attenuates GRF-stimulated GH secretion (Sartin *et al*, 1988; Estienne *et al*, 1989) similar to observations in humans (Imaki, Shibasaki, Shizume, Masuda, Hotta, Kiuosawa, Jibiki, Demura, Tsushima and Ling, 1985; Casanueva, Villanueva, Dieguez, Diaz, Cabranes, Szoke, Scanion, Schally and Fernandez-Cruz, 1987) and rats (Imaki, Shibaski, Masuda, Hotta, Yamauchi, Demura, Shizume, Wakabayashi and Ling, 1986). Imaki *et al* (1986) reported that the suppressive effect of FFA on GRF-induced GH release in rats could be abolished by pretreatment with antisomatostatin serum. Since it is known that SRIF is capable of inhibiting the GH response to GRF in dairy cows (Sartin, Pierce, Kemppainen, Bartol, Cummins, Marple and Williams, 1989) and in bovine pituitary cells (Padmanabhan, Enright, Zinn, Convey and Tucker, 1987), it is possible that FFA infusion blunts GRF-stimulated GH secretion by increasing SRIF release from the hypothalamus in sheep.

Neuroendocrine regulation of growth

Somatostatin and GRF are active in a number of species including primates (Besser, 1974), rats (Martin, Renaud and Brazeau, 1974), swine, sheep, cows (Della-Fera, Buonomo and Baile, 1986) and poultry (Scanes, Carsia, Lauterio, Huybrecht, Rivier and Vale, 1984). The administration of SRIF blocks not only endogenous, but GRF-stimulated GH secretion as well (Wehrenberg, Ling, Bohlen, Esch, Brazeau and Guillemin, 1982). Conversely, plasma GH concentrations are elevated following SRIF immunoneutralization in several species such as sheep (Varner, Davis and Reeves, 1980), rats (Terry and Martin, 1981) and chickens (Buonomo, Sabacky, Della-Fera and Baile, 1987). Immunoneutralization against GRF results in an inhibition of pulsatile GH secretion in rats (Wehrenberg, Brazeau, Luben, Bohlen and Guillemin, 1982) and pigs (Armstrong, Esbenshade, Johnson, Coffey, Heiner, Campbell, Mowles and Felix, 1990), and decreased mean GH concentrations in cattle (Armstrong *et al*, 1990). Moreover, circulating IGF-I concentrations were reduced in the GRF-immunized steers (Trout and Schanbacher, 1990) and swine (Armstrong *et al*, 1990); and a decrease in somatic growth was observed in GRF-immunized rats (Wehrenberg, Bloch and Phillips, 1984) and cattle (Trout and Schanbacher, 1990).

Recently, Cella, Locatelli, Mennini, Zanini, Bendotti, Forloni, Fumagalli, Arce, De Gennaro Colonna, Wehrenberg and Muller (1990) demonstrated that passive immunization against GRF in rats during the early postnatal period (days 1 to 10) reduced both basal and GRF-stimulated GH secretion as long as 60 days after treatment and induced a marked and chronic impairment of growth rate. These data emphasize the importance of maintenance of a normal GH regulatory system during the period of "GH-independent" growth.

Exogenous GRF administration has been shown to stimulate growth in both normal and GH-deficient rats (Clark and Robinson, 1985), GH-deficient children (Smith, Brook, Rivier, Vale and Thorner, 1986), low-birth-weight lambs (Pastoureau, Charrier, Blanchard, Boivin, Dulor, Theriez and Barenton, 1989) and swine (Etherton, Wiggins, Chungs, Evock, Rebun and Walton, 1986; Dubreuil, Petitclerc, Pelletier, Gaudreau, Farmer, Mowles and Brazeau, 1990). However neither daily injections of GRF, thyrotropin-releasing factor (TRH) or the simultaneous administration of GRF and TRH were effective in stimulating rate of gain or efficiency in dairy calves (Lapierre, Pelletier, Petitclerc, Dubreuil, Morisset, Gaudreau, Couture and Brazeau, 1991) and poultry (Buonomo and Baile, 1986). In contrast, several studies have demonstrated the ability of exogenous GRF administration to stimulate milk yield in ewes (Hart, Chadwick, James and Simmonds, 1985), and dairy cattle (Enright, Chaplin, Moseley and Tucker, 1985; Lapierre, Pelletier and Petitclerc, 1988) over a short-term period. Although milk yield was not reported, a subsequent study reported that GRF was still effective in stimulating GH secretion in dairy cattle following long-term (182 days) GRF treatment (Lacasse, Petitclerc, Pelletier, Delorme, Morisset, Gaudreau and Brazeau, 1991).

Summary

The mechanism regulating growth involves the interaction of a myriad of hormones, but most notably, GH. The pattern of endogenous GH secretion is regulated by the CNS via two specific hypothalamic neurohormones, GRF and SRIF, an inhibitor of GH release. In turn, the secretion of GRF and SRIF is governed by a host of neuropeptides and neurotransmitters which provide a functional link between higher CNS centres and hypophysiotropic neurons. Factors that have been implicated include the biogenic amines, *eg*, dopamine, histamine, acetylcholine, serotonin, and GABA, the opioid peptides, galanin and NPY. The cytokines, TNF-α, IL1-ß and IL-6 apparently can modulate GH secretion by acting at both hypothalamic and pituitary levels. Clearly the hypothalamo-pituitary axis is responsive to many factors associated with metabolic, behavioral, and immune system alterations and in turn the axis orchestrates a multitude of coordinated activities to meet the physiological demands of the individual animal in its internal and external environment.

References

Abe, H., Molitch, M.E., Van Wyk, J.J. and Underwood, L.E. (1983) *Endocrinology*, **113**, 1319-1324
Aguila, M.C. and McCann, S.M. (1987) *Endocrinology*, **117**, 762-765
Ahene, C.A., Lea, R.W. and Harvey, S. (1991) *Journal of Endocrinology*, **128**, 13-19
Antinmo, T., Baldijao, K.A., Houpt, K.A., Pond, W.G. and Barnes, R.H. (1978) *Journal of Animal Science*, **46**, 409-413
Armstrong, J.D., Esbenshade, K.L., Johnson, J.L., Coffey, M.T., Heiner, E., Campbell, R.M., Mowles, T.and Felix, A. (1990) *Journal of Animal Science*, **68**, 427-434
Arnold, M.A. and Fernstrom, J.D. (1981) *Neuroendocrinology*, **31**
Barb, C.R., Kraeling, R.R. and Rampacek, G.B. (1989) *Journal of Dairy Science* **72** (Suppl 1), 330-331
Bassett, J.M. (1974) *Australian Journal of Biological Sciences*, **27**, 167-173
Bauman, D.E., Eppard, P.J., De Geeter, M.J. and Lanza, G.M. (1985) *Journal of Dairy Science*, **68**, 1352-1362
Berelowitz, M., Szabo, M., Frohman, L.A., Firestone, S. and Chu, L. (1981) *Science*, **212**, 1279-1281
Berelowitz, M., Dudlak, D. and Frohman, L.A. (1982) *Journal of Clinical Investigation*, **69**, 1293-1301
Besser, G.M. (1974) *British Medical Journal*, **4**, 622-624
Beutler, B. and Cerami, A. (1988) *Endocrine Review*, **9**, 57-67
Blum, J.W., Schnyder, W., Kunz, P.L., Blom, A.K., Bickel, H. and Schurch, A. (1985) *Journal of Nutrition*, **115**, 417-424
Bolton, J.E., Roud, H.K., Redekopp, C., Livesay, J.H., Nicholls, M.G. and Donald, R.A. (1983) *Hormone and Metabolic Research*, **15**, 165-176
Brazeau, P., Guillemin, R., Ling, N., Van Wyk, J.J. and Humbel, R. (1982) *Comptes Rendu Academie Sciences (Paris)*, **295**, 651-654
Breder, C.D., Dinarello, C.A. and Saper, C.B. (1988) *Science*, **240**, 321-323
Breier, H.H., Bass, J.J., Butler, J.H. and Gluckman, P.D. (1986) *Journal of Endocrinology*, **111**, 209-216
Buonomo, F.C. (1991) *Life Science* (unpublished)
Buonomo, F.C. and Baile, C.A. (1986) *Domestic Animal Endocrinology*, **3**, 269-276
Buonomo, F.C. and Baile, C.A. (1991) *Journal of Animal Science*, **69**, 755-760
Buonomo, F.C., Lauterio, T.J., Baile, C.A. and Daughaday, W.H. (1987) *General and Comparative Endocrinology*, **66**, 274-279
Buonomo, F.C., Lauterio, T.J. and Scanes, C.G. (1984) *Comparative Biochemistry and Physiology*, **78C**, 409-413
Buonomo, F.C., Sabacky, M.J., Della-Fera, M.A. and Baile, C.A. (1987) *Domestic Animal Endocrinology*, **4**, 191-200
Buonomo, F.C., Tou, J.S. and Kaempfe, L.A. (1991) *Life Sciences*, **48**, 1953-1961

Casanueva, F.F., Villanueva, L., Dieguez, C., Diaz, Y., Cabranes, J.A., Szoke, B., Scanion, M.F., Schally, A.V. and Fernandez-Cruz, A. (1987) *Journal of Clinical Endocrinology Metabolism*, **65**, 634-638

Cella, S.G., Locatelli, V., Mennini, T., Zanini, A., Bendotti, C., Forloni, G.L., Fumagalli, G., Arce, V.M., De Gennaro Colonna, V., Wehrenberg, W.B. and Muller, E.E. (1990) *Endocrinology*, **127**, 1625-1634

Chihara, K., Arimura, A. and Schally, A.V. (1979) *Endocrinology*, **104**, 1656-1662

Chihara, K., Minamitani, N., Kaji, H., Arimura, A. and Fujita, T. (1981) *Endocrinology*, **109**, 2279-2281

Ching, J.L.C., Christofides, N.D., Anand, P., Gibson, S.J., Allen, Y.S., Su, H.C. and Tatemoto, K. (1985) *Neuroscience*, **16**, 343-347

Ciofi, P., Croix, D. and Tramu, G. (1987) *Neuroendocrinology*, **45**, 425-428

Clark, R.G., Corlsson, L.M.S. and Robinson, I.C.A.F. (1988) *Journal of Endocrinology*, **117**, 201-209

Clark, R.G. and Robinson, I.C.A.F. (1985) *Nature*, **314**, 281-283

Collu, R., Fraschini, F., Visconti, P. and Martini, L. (1972) *Endocrinology*, **90**, 1231-1233

Conway, S., McCann, S.M. and Krulich, L. (1985) *Endocrinology*, **117**, 2284-2292

Davis, S.L. and Borgen, M.L. (1973) *Endocrinology*, **92**, 303-309

De Gennaro Colonna, V., Cattaneo, E., Cocchi, D., Muller, E.E. and Maggi, A. (1988) *Peptides*, **9**, 985-988

De Zegher, F., Bettendorf, M., Kappan, S.L. and Grumbach, M.M. (1988) *Endocrinology*, **123**, 658-659

Della-Fera, M.A., Buonomo, F.C. and Baile, C.A. (1986) *Domestic Animal Endocrinology*, **3**, 165-176

Driver, P.M. and Forbes, J.M. (1978) *Journal of Physiology*, **317**, 413-419

Dubreuil, P., Lapierre, H., Pelletier, G., Petitclerc, D., Couture, Y., Gaudreau, P., Monisset, J. and Brazeau, P. (1988) *Domestic Animal Endocrinology*, **5**, 157-164

Dubreuil, P., Petitclerc, D., Pelletier, G., Gaudreau, P., Farmer, C., Mowles, T.F. and Brazeau, P. (1990) *Journal of Animal Science*, **68**, 1254-1268

Dubuis, J.M., Dayer, J.M., Siegrist-Kaiser, C.A. and Burger, A.G. (1988) *Endocrinology*, **123**, 2175-2180

Durand, D., Martin, J.B. and Brazeau, P. (1977) *Endocrinology*, **110**, 722-728

Eden, S., Bolle, P. and Modigh, K. (1979) *Endocrinology*, **105**, 523-528

Eisemann, J.H., Hammond, A.C., Bauman, D.E., Reynolds, P.J., McCutcheon, S.M., Tyrrell, H.F. and Haaland, G.L. (1979) *Journal of Nutrition*, **116**, 157-165

Elsasser, T.H., Caperna, T.J., Kenison, D.C. and Fayer, R. (1989) *Endocrinology Abstracts*, **788**

Enright, W.J., Chaplin, L.T., Moseley, W.M. and Tucker, H.A. (1985) *Federation Proceedings*, **44**, 1358

Estienne, M.J., Schillo, K.K., Green, M.A. and Boling, J.A. (1989) *Endocrinology*, **125**, 85-91

Estienne, M.J., Schillo, K.K., Hileman, S.M., Green, M.A., Hayes, S.H. and Boling, J.A. (1990) *Endocrinology*, **126**, 1934-1940

Etherton, T.D., Wiggins, J.P, Chungs, C.S., Evock, C.M., Rebun, J.F. and Walton, P.E. (1986) *Journal of Animal Science*, **63**, 1389-1397

Foster, D.L., Ebling, F.J.P., Micka, A.F., Vannerson, L.A., Bucholtz, D.C., Wood, R.I., Suttie, J.M. and Fenner, D.E. (1989) *Endocrinology*, **125**, 343-351

Frohman, L.A., Down, T.R., Clarke, I.J. and Thomas, G.B. (1990) *Journal of Clinical Investigation*, **86**, 17-24

Gaillard, R.C., Turnill, D., Sappino, P. and Muller, A.F. (1990) *Endocrinology*, **127**, 101-106

Ganzetti, I., Petraglia, F., Capuano, I., Rosi, F., Wehrenberg, W.B., Muller, E.E. and Cocchi, D. (1987) *Journal of Endocrinology Investigation*, **10**, 241-246

Gluckman, P.D., Marti-Henneberg, C., Kaplan, S.L., Li, C.H. and Grumbach, M.M. (1980) *Endocrinology*, **107**, 76-80

Goodyer, C.G., De Stephano, L., Guyda, H.J. and Posner, B.I. (1984) *Endocrinology*, **115**, 1568-1576

Goodyer, C.G., De Stephano, L., Lai, W.H., Guyda, H.J. and Posner, B.I. (1984) *Endocrinology*, **114**, 1187-1195

Groesbeck, M.D., Parlow, A.F. and Daughaday, W.H. (1987) *Endocrinology*, **102**, 1963-1975

Hart, I.C., Chadwick, P.M.E., Coert, A., James, S. and Simmonds, A.D. (1986) *Journal of Dairy Science*, **61**, 281-289

Hart, I.C., Chadwick, P.M.E., James, S. and Simmonds, A.D. (1985) *Endocrinology*, **105**, 189-196

Hashizume, T. and Kanematsu, S. (1990) *Domestic Animal Endocrinology*, **7**, 451-456

Hertelendy, F. and Kipnis, D.M. (1973) *Endocrinology*, **92**, 402-407

Hokfelt, T., Elde, R., Fuxe, K., Johansson, O., Ljungda, A., Goldstein, M., Luft, R., Efendic, S., Nilsson, G., Terenius, L., Ganten, S., Jeffcoate, L., Rehfeld, F., Said, S., Perez, M., Possain, L., Taia, R., Teran, L. and Palacios, R. (1978) In *The Hypothalamus*, (eds S. Reichlin, R.J. Baldessarinin and J.B. Martin), New York, Raven Press, pp 69-103

Hokfelt, T., Johansson, O. and Goldstein, M. (1984) *Science*, **225**, 1326-1334

Imaki, T., Shibaski, T., Masuda, A., Hotta, M., Yamauchi, N., Demura, H., Shizume, K., Wakabayashi, I. and Ling, N. (1986) *Endocrinology*, **118**, 2390-2399

Imaki, T., Shibasaki, T., Shizume, K., Masuda, A., Hotta, M., Kiuosawa, Y., Jibiki, K., Demura, H., Tsushima, T. and Ling, N. (1985) *Clinics in Endocrinology and Metabolism*, **60**, 290-298

Jacobowitz, D.M., O'Donohue, T.L., Chey, W.Y. and Chang, T.M. (1981) *Peptides*, **2**, 479-487

Kanatsuka, A., Makin, H., Matsushina, Y., Osegawa, M., Yamamoto, M.R. and Kumagai, A. (1983) *Neuroendocrinology*, **29**, 186-190
Katakami, H., Arimura, A. and Frohman, L.A. (1986) *Endocrinology*, **118**, 1872-1877
Kato, Y., Dupre, J. and Beck, J.C. (1973) *Endocrinology*, **93**, 135-140
Katsuura, K.T. (1989) *Biochemical and Biophysical Research Communications*, **156**, 61-62
Klindt, J. and Stone, R.T. (1984) *Growth*, **48**, 1-15
Koenig, J.I., Snow, K., Clark, B.D., Toni, R., Cannon, J.G., Shaw, A.R., Dinarello, C.A., Reichlin, S., Lee, S.L. and Lechan, R.M. (1990) *Endocrinology*, **126**, 3053-3058
Kohler, P.O., O'Malley, B.W., Rayford, P.L., Lipsett, M.B. and Odell, W.D. (1967) *Journal of Clinical Endocrinology Metabolism*, **27**, 219-225
Krieg, R.J., Perkins, S.N., Johnson, J.H., Rogers, J.P., Arimura, A. and Cronin, M.J. (1988) *Endocrinology*, **122**, 531-537
Krulich, L. and McCann, S.M. (1966) *Proceedings of the Society for Experimental Biological and Medicine*, **121**, 1114-1117
Kukucska, I. and Makara, G.B. (1983) *Endocrinology*, **113**, 318-323.
Lacasse, P., Petitclerc, D., Pelletier, G., Delorme, L., Morisset, J., Gaudreau, P. and Brazeau, P. (1991) *Domestic Animal Endocrinology*, **8**, 99-108
Landefeld, T.D., Ebling, F.J.P., Suttie, J.M., Vannerson, L.A., Padmanabhan, V., Beitins, I.Z. and Foster, D.L. (1989) *Endocrinology*, **125**, 351-356
Lapierre, H., Pelletier, G. and Petitclerc, D. (1988) *Journal of Dairy Science*, **71**, 92-98
Lapierre, H., Pelletier, G., Petitclerc, D., Dubreuil, P., Morisset, J., Gaudreau, P., Couture, Y. and Brazeau, P. (1991) *Journal of Animal Science*, **69**, 587-598
Lauterio, T.J. and Scanes, C.G. (1987) *Journal of Nutrition*, **117**, 758-765
Lauterio, T.J. and Scanes, C.G. (1988) *Journal of Endocrinology*, **117**, 223-228
Lea, R.W. and Harvey, S. (1990) *Journal of Endocrinology*, **125**, 409-415
Lechan, R.M., Toni, R., Clark, B.A., Cannon, J.G., Shaw, A.R., Dinarello, C.A. and Reichlin, S. (1990) *Brain Research*, **514**, 135-140
Leshin, L.S., Rund, L.A., Thompson, F.N., Mahaffey, M.B., Chang, W.J., Byerley, D.J. and Kiser, T.E. (1990) *Journal of Animal Science*, **68**, 1656-1665
Lumpkin, M.D. and Hartmann, D.P. (1989) *Endocrinology Abstracts*, **789**
Lumpkin, M.D., Samson, W.K. and McCann, S.M. (1985) *Endocrinology*, **116**, 2070-2074
Machlin, L.J. (1972) *Journal of Animal Science*, **35**, 794-800
Machlin, L.J. (1973) *Journal of Dairy Science*, **56**, 545-551
Martin, J.B., Renaud, L.P. and Brazeau, P. (1974) *Science*, **186**, 538-541
McCann, S.M. (1982) In *Neuroendocrine Perspectives*, (eds E.E. Muller and R.M. MacLeod), Amsterdam, Elsevier, vol 1, pp 1-22
McIntyne, H.B. and Odel, W.D. (1974) *Neuroendocrinology*, **16**, 8-14
Miki, N., Ono, M., Miyoshi, H., Tsushima, T. and Shizume, K. (1989) *Life Sciences*, **44**, 469-476

Minamitani, N., Chihara, K., Kaji, H., Kodama, H., Kita, T. and Fujita, T. (1989) *Endocrinology*, **125**, 2839-2845

Mitsugi, N., Arita, J. and Kimura, F. (1990) *Neuroendocrinology*, **51**, 93-96

Moseley, W.M., Alaniz, G.R., Claflin, W.H. and Krabill, L.F. (1988) *Journal of Endocrinology*, **117**, 253-259

Mueller, G.P., Simpkins, J., Meites, J. and Moore, K.E. (1976) *Neuroendocrinology*, **20**, 121-126

Muller, E.E. (1987) *Physiological Reviews*, **67**, 962-1054

Murakami, Y., Kato, Y., Kabayama, Y., Tojo, K., Inque, T. and Imura, H. (1986) *Endocrinology*, **119**, 1089-1092

Murakami, Y., Kato, Y., Koshiyama, H., Inoue, T., Yanaihara, N. and Imura, H. (1987) *European of Journal Pharmacology*, **136**, 415-419

Murakami, Y., Kato, Y., Shimatsu, A., Koshiyama, H., Hattori, N., Yanaihara, N. and Imura, H. (1989) *Endocrinology*, **124**, 1224-1229

Niswender, G.D., Midgley, A.R., Monroe, S.E. and Reichert, L.E. (1968) *Proceedings of the Society for Experimental Biology and Medicine*, **128**, 807-811

Nordstron, O., Melander, T., Hokfelt, T., Bartfai, T. and Goldstein, M. (1986) *Neuroscience Letters*, **73**, 21-22

O'Donohue, T.L., Beinfeld, M.C., Chey, W.Y., Chang, T.M., Nilaver, G., Zimmerman, E.A., Yajima, H., Adachi, H., Poth, M., McDevitt, R.P. and Jacobowitz, D.M. (1981) *Peptides*, **2**, 467-477

Ogwuegbu, S.O., Hashizume, T. and Kanematsu, S. (1991) *Domestic Animal Endocrinology*, **8**, 29-35

Padmanabhan, V., Enright, W.J., Zinn, S.A., Convey, E.M. and Tucker, H.A. (1987) *Domestic Animal Endocrinology*, **5**, 253-249

Painson, J.C. and Tannenbaum, G.S. (1985) *Endocrinology*, **117**, 1132-1139

Pang, X.P., Hershman, J.M., Mirell, C.J. and Pekary, A.E. (1989) *Endocrinology*, **125**, 76-84

Pastoureau, P., Charrier, J., Blanchard, M.M., Boivin, G., Dulor, J.P., Theriez, M. and Barenton, B. (1989) *Domestic Animal Endocrinology*, **6**, 321-329

Perez, F.M., Malamed, S. and Scanes, C.G. (1985) *IRCS Medical Science*, **13**, 871-872

Rettori, V., Milenkovic, L., Aguila, M.C. and McCann, S.M. (1990) *Endocrinology*, **126**, 2296-2301

Reynart, R., De Paepe, M., Marcus, S. and Peeters, G. (1975) *Journal of Endocrinology*, **66**, 213-218

Richardson, S.B., Naguyen, T. and Hollander, C.S. (1983) *American Journal of Physiology*, **244**, E560-E566

Richman, R.A., Weiss, J.P., Hochberg, Z. and Florini, J.R. (1981) *Endocrinology*, **108**, 2287-2292

Rivier, C., Vale, W., Ling, N., Brown, M. and Guillemin, R. (1977) *Endocrinology*, **100**, 238-241

Robbins, R.J. and Landon, R.M. (1983) *Brain Research*, **273**, 377-386

Robbins, R.J., Leidy, J.W. and Landon, R.M. (1985) *Endocrinology*, **108**, 538-543
Rogers, K.V., Vician, L., Steiner, R.A. and Clifton, D.K. (1988) *Endocrinology*, **122**, 586-591
Rokaeus, A. (1987) *Trends in Neuroscience*, **10**, 158-164
Samson, W.K., Koenig, J.I. and McCann, S.M. (1982) *Brain Research Bulletin*, **8**, 117-121
Samson, W.K., Lumpkin, M.D., Nilaver, G. and McCann, S.M. (1984) *Brain Research Bulletin*, **12**, 57-62
Sartin, J.L., Bartol, F.F., Kemppainen, R.J., Dieberg, G., Buxton, D. and Soyoola, E. (1988) *Neuroendocrinology*, **48**, 627-633
Sartin, J.L., Kemppainen, R.J., Marple, D.N., Carnes, R., Dieberg, G. and Oliver, E.H. (1987) *Domestic Animal Endocrinology*, **4**, 33-41
Sartin, J.L., Pierce, A.C., Kemppainen, R.J., Bartol, F.F., Cummins, K.A., Marple, D.N. and Williams, J.C. (1989) *Acta Endocrinologica*, **120**, 319-325
Scanes, C.G., Carsia, R.V., Lauterio, T.J., Huybrecht, L., Rivier, J. and Vale, W. (1984) *Life Sciences*, **34**, 1127-1134
Scanes, C.G., Griminger, P. and Buonomo, F.C. (1981) *Proceedings of the Society for Experimental Biology and Medicine*, **168**, 344-347
Scarborough, D.E. and Dinarello, C.A. (1989) *Endocrinology*, **124**, 549-551
Sharp, B.M., Matta, S.G., Peterson, P.K., Newton, R., Chao, C. and McAllen, K. (1989) *Endocrinology*, **124**, 3131-3133
Smith, P.J., Brook, G.D.C., Rivier, J., Vale, W. and Thorner, M. (1986) *Clinical Endocrinology*, **25**, 35-44
Smith, R.M., Howe, P.R.C., Oliver, J.R. and Willoughby, J.O. (1984) *Neuropeptides*, **4**, 109-113
Spangelo, B.L., Judd, A.M., Isakson, P.C. and MacLeod, R.M. (1989) *Endocrinology*, **125**, 575-577
Spangelo, B.L., MacLeod, R.M. and Isakson, P.C. (1990) *Endocrinology*, **126**, 582-586
Spencer, G.S.G., Bass, J.J., Hodgkinson, S.C., Edgley, W.H.R. and Moore, L.G. (1991) *Domestic Animal Endocrinology*, **8**, 155-160
Spencer, G.S.G., Macdonald, A.A., Buttle, H. and Carlyle, S.S. (1989) In *Endocrinology of Farm Animals*, (ed K. Boda), Slovak Academy of Sciences, pp 25-30
Stubbs, W.A., Jones, A., Edwards, C.R.W., Delitata, G., Jeffcoate, W.J., Ratler, S.J., Besser, G.M., Bloom, S.R., Alberti, K.G. (1978) *Lancet*, **2**, 2325-2327
Tannenbaum, G.S. (1980) *Endocrinology*, **107**, 2117-2220
Tannenbaum, G.S. (1984) *Science*, **226**, 464-466
Tannenbaum, G.S. (1985) In *Advances in Experimental Medicine and Biology: Somatostatin*, (eds Y.C. Patel and G.S. Tannenbaum), New York, Plenum Press, vol 188, pp 229-259
Tannenbaum, G.S., Guyda, H.J. and Posner, B.I. (1983) *Science*, **220**, 77-79
Tatemoto, K., Rokaeus, A., Joruvall, H., McDonald, T.J. and Mutt, V. (1983) *FEBS Letters*, **164**, 124-126

Terry, L.C. and Martin, J.B. (1981) *Endocrinology*, **109**, 622-627
Thomas, G.B., Mercer, J.E., Karalis, T., Rao, A., Cummins, J.T. and Clarke, I.J. (1990) *Endocrinology*, **126**, 1361-1367
Thomas, G.B., Cummins, J.T., Francis, H., Sudbury, A.W., McCloud, P.I. and Clarke, I.J. (1991) *Endocrinology*, **128**, 1151-1158
Trenkle, A. (1978) *Journal of Dairy Science*, **61**, 281-289
Trout, W.E. and Schanbacher, B.D. (1990) *Journal of Endocrinology*, **125**, 123-129
Trudeau, V.L., Meijer, J.C., van de Wiel, D.F.M. and Erkens, J.H.F. (1988) *Journal of Endocrinology*, **119**, 501-508
Vankelecom, H., Carmeliet, P., Damme, J.V., Billiau, A. and Denef, C. (1989) *Neuroendocrinology*, **49**, 102-106
Varner, M.A., Davis, S.L. and Reeves, J.J. (1980) *Endocrinology*, **106**, 1027-1032
Wakabayshi, I., Myazawa, Y., Kondon, M., Miki, N., Demura, R., Demura, H. and Shizume, K. (1977) *Endocrinology Japan*, **24**, 601-604
Walton, P.E. and Cronin, M.J. (1989) *Endocrinology*, **125**, 925-929
Wehrenberg, W.B., Brazeau, P., Luben, R., Bohlen, P. and Guillemin, R. (1982) *Endocrinology*, **111**, 2147-2148
Wehrenberg, W.B., Ling, N., Bohlen, P., Esch, F., Brazeau, P. and Guillemin, R. (1982) *Biochemical and Biophysical Research Communications*, **109**, 562-567
Wehrenberg, W.B., Bloch, B. and Ling, N. (1985) *Neuroendocrinology*, **41**, 13-16
Wehrenberg, W.B., Bloch, B. and Phillips, B.J. (1984) *Endocrinology*, **115**, 1218-1219
Willoughby, J.O., Menadue, M., Zeegers, P., Wise, P. and Oliver, J.R. (1980) *Journal of Endocrinology*, **86**, 165-169
Yanaihara, C., Sato, H., Yanaihara, N., Naruse, S., Foresmann, W.G., Helmstaedter, V., Fujita, T., Yamaguchi, K. and Abe, K. (1978) *Advances in Experimental Medicine and Biology*, **106**, 269-283
Zainur, A.S., Tassell, R., Kellaway, R.C. and Dodemaide, W.R. (1989) *Australian Journal of Agriculture Research*, **40**, 195-206

7

MOLECULAR CONTROL OF IGF GENE EXPRESSION

R.S. GILMOUR, J.C. SAUNDERS, M.C. DICKSON*, J. LI, J.M. PELL
Department of Molecular and Cellular Physiology, Institute of Animal Physiology, Babraham, Cambridge CB2 4AT, UK
**Present address: Department of Medical Genetics, Duncan Guthrie Institute, Yorkhill Hospital, Glasgow 3, UK*

A.L.F. OWDEN and M. SILVER
Department of Physiology, University of Cambridge, Cambridge CB2 3EG, UK

Introduction

The insulin-like growth factors (IGF-I and IGF-II) are polypeptides with marked structural similarities to insulin and possess both anabolic and mitogenic properties both *in vivo* and *in vitro* (Blundell, Bedarker and Humbel, 1983; Rinderknecht and Humbel, 1978a; b). IGF-I mediates many of the growth promoting effects of growth hormone (GH) in the neonate (Van Wyk, 1984). Because circulating IGF-II levels are higher in the foetus and IGF-I levels are higher after birth it has been conjectured that IGF-II regulates prenatal while IGF-I regulates postnatal growth and development (Daughaday, Parker, Borowsky, Trivedi and Kapadia, 1982; Moses, Nissley, Short, Rechler, White, Knight and Higa, 1980). Direct evidence for a physiological role of IGF-II in embryonic growth has come from experiments in which transgenic mice, heterozygous for a disrupted IGF-II gene, show a growth-deficient phenotype (DeChiara, Efstratiadis and Robertson, 1990). In post-natal animals the liver was originally considered to be the only source of IGF-I. It is now known, however, that virtually all tissues of the body produce IGF-I and IGF-II to some degree (D'Ercole, Stiles and Underwood, 1984). There is no doubt that the liver is the major source of circulating IGF-I and that its synthesis in this tissue is highly GH-sensitive (Froesch, Schmid, Schwander and Zapf, 1985). Much lower levels of production by non-hepatic tissues has led to the idea that the liver acts in an endocrine manner to promote skeletal growth while IGF-I produced from extra-hepatic tissues may sustain various functions in a paracrine or autocrine fashion. Thus the synthesis of IGFs may be regulated in a tissue specific manner by a number of stimuli including GH to control diverse cellular processes. IGF-I, and almost certainly IGF-II, do not cross the placenta (D'Ercole and Underwood, 1981) and, as will be discussed later, in the foetus both endocrine and paracrine IGF-I and II synthesis also occur. However, there is considerable evidence that GH does not play a significant role in the regulation of foetal growth (Gluckman, 1986). The question of what does regulate foetal IGF synthesis is largely unresolved. Thus the mechanisms regulating foetal growth are quite distinct from those directing postnatal growth although the same IGF genes are active in both situations. Most of the early information about regulation and sites of synthesis was based on *in vivo* plasma concentrations. Immunological localisation at the cellular level was complicated by the fact that plasma IGF concentrations are higher relative to those of tissues (D'Ercole *et al*, 1984). The advent of recombinant DNA technology provided an alternative strategy for analysis of IGF biosynthesis that could overcome these difficulties. Complementary DNAs (cDNAs) encoding human IGF-I and II were isolated and sequenced (Jansen, Van Schaik, Ricker, Bullock, Woods, Gabby, Nussbaum, Sussenbach and Van Den Brande, 1983; Bell, Merryweather, Sachez-Pascador, Sempien, Priestly and Rall, 1984; Dull, Gray, Hayfield and Ullrich, 1984). This and the subsequent characterisation of the IGF genes themselves have increased our understanding of the molecular mechanisms that underlie IGF synthesis. The complexity of the IGF-I and II genes and their messenger RNA (mRNA) structure and expression is much greater than expected. A consequence of this complexity is the potential for multifactorial regulation compatible with the broad metabolic actions of IGFs. A study of the molecular biology of IGF genes offers an alternative and complementary approach to understanding IGF action *in vivo*; the challenge is to extend the physiology to a molecular dimension without compromising whole animal relevance. The aim of this article is to review current knowledge of IGF regulation at this "molecular physiology" interface and to provide a strategy for future investigations.

The molecular biology of the IGFs

An exhaustive account of IGF gene structure is beyond the scope of this article and the reader is referred to a recent review for more detailed information (Daughaday and Rotwein, 1989). The following discussion summarises the principal features of the IGF genes drawn mainly from work on rat and man.

Both IGF-I and II peptides, like pro-insulin consist of B, C and A domains. However, unlike insulin, the mature IGF molecule retains the C peptide region. In addition the mature circulating IGFs differ from insulin in that they contain an additional D domain at the carboxyterminus. When IGF-I and II cDNAs were cloned and their mRNA sequences deduced it was predicted that in both IGFs the core B, C, A and D domains were flanked by precursor sequence at both the amino terminus, *ie* the 5′ end of the cDNA (putative signal peptide region) and carboxyterminus or 3′ end of the cDNA (E domain). The possible relevance of these additional peptide sequences will be considered later. The cloned cDNA sequences were also found to be heterogeneous. They revealed that for both IGF-I and II there is a choice of at least 3 or 4 alternative 5′ sequences and for IGF-I only, two alternative 3′ (E domain) sequences. At the DNA level these features imply a structure for the IGF gene which is much more complex than might have been predicted from the mature protein structure. Figure 1 shows a schematic for a hypothetical IGF gene which incorporates all of these features and explains the origins of the mRNA heterogeneity. The regions of the gene which encode the B, C, A and D domains are contained in two exons (filled boxes in

Figure 1) which are present in every RNA transcript. Transcription may initiate at one of three pre-pro sequences (exons 1-3) which are spliced separately to the mature domains (exons 4 and 5). In addition, IGF-I transcripts contain one of two alterative E domains (exons 6 and 7); for IGF-II there is a single E domain. The length of sequence that encodes the E peptide domain is relatively short compared to the total length of the exon; there is a long tail of non-coding of 3′ untranslated (UT) sequence which in the case of IGF-I can contribute as much as 6 kilobases (kb) to the size of the mRNA (Lund, Hoyt and Van Wyk, 1989). After translation, the precursor IGF peptide is processed further by proteolytic cleavage of the signal and E peptide regions to give the mature hormone.

The IGF genes of all species so far characterised have many if not all of the features described in Figure 1. There are at least four aspects which are relevant to the control of IGF gene expression.

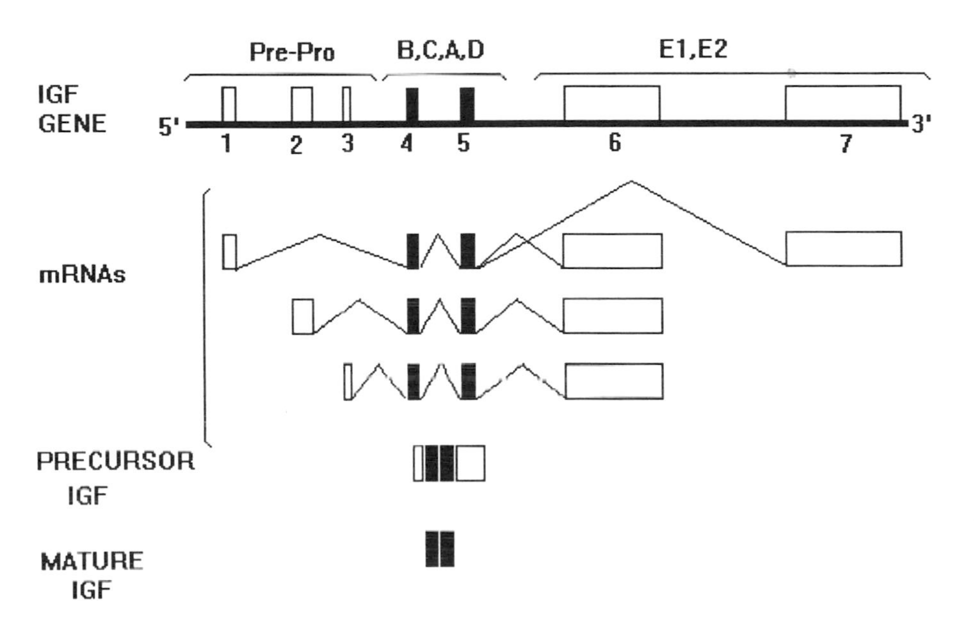

Figure 1 Schematic to illustrate the relationship between a hypothetical mature IGF peptide and its gene structure. Many IGF-I and II genes from a number of species have all or most of the features shown. The mature peptide is encoded in exons 4 and 5 of the gene. The open boxes are exons which are alternately spliced to exons 4 and 5. Separate transcriptional initiation from the 5′ exons, 1, 2 and 3 give rise variant pre-pro peptide regions. In IGF-I each of these forms contains one of two alternate E domains (E1 or E2, exons 6 and 7). For the sake of clarity only the first variant shows both alternate E domain spliced forms. In IGF-II there is only one E domain exon. Only a relatively small part of the terminal exon 6 and 7 encode the E domain peptide; the remainder is 3′ untranslated (UT) sequence which is processed to give multiple sizes depending on choice of multiple polyadenylation sites. Both the pre-pro signal region and the E peptide is removed by proteolytic cleavage to give mature circulating peptide.

MULTIPLE 5′ EXONS

The occurrence of multiple 5′ sequences in mRNA implies the existence of multiple start points for RNA synthesis on the gene. The initiation and control of transcription is normally associated with a region of DNA, called the promoter, situated near to the start. The promoter DNA contains a variety of sequences involved in establishing the transcriptional initiation complex and for regulating the transcription process in response to external signals. The existence of multiple promoters suggests that the synthesis of IGF mRNAs are controlled by diverse factors.

VARIANT PRE-PRO SIGNAL PEPTIDES

The predominant 5′ exon for IGF-I encodes a substantial portion of the prepeptide sequence which together with an additional common segment of prehormone sequence at the beginning of the B domain contributes to an unusually large precursor signal peptide. There is good evidence to show that this sequence is co-translationally removed by microsomal membranes and therefore functions during the secretory process as a conventional signal peptide despite its length (Rotwein, Folz and Gordon, 1987). There is also coding potential in the other alternative 5′ exons which could give rise to quite different distal segments of the pre-pro IGF peptide. It is known that the 5′ variant IGF mRNA forms are expressed in a cell and tissue specific fashion (Lowe, Roberts, Lasky and Le Roith, 1987; Hoyt, Van Wyk and Lund, 1988) and therefore the amino terminal precursor peptides could play a role in regulating precursor processing or secretion.

E DOMAINS

All known IGF sequences encode carboxyterminal E peptide extensions. Differential splicing in the E coding region produces mRNAs encoding two different forms of IGF-I, namely IGF-Ia and IGF-Ib (Rotwein, 1986). IGF-II mRNAs have a single invariant E domain encoded in the gene. At present the biological significance of these domains is not known but their evolutionary conservation across species suggests a biological role. Little is known about the pathway of post-translational processing and secretion of IGF precursors. Antisera which recognise the Ia type E domain have detected pro-IGF-Ia peptide in plasma (Powell, Lee, Chang, Lin and Hintz, 1987). In the rat IGF-I mRNAs encoding Ib type E domain are expressed primarily in liver (Hoyt *et al*, 1988; Lowe, Lasky, Le Roith and Roberts, 1988) which appears to be a primary source of circulating IGF-I. In contrast, rat IGF-I mRNAs encoding Ia type E domains are expressed in both liver and non-hepatic tissues. This might suggest a local, paracrine function for this form. Further analysis of biosynthesis, processing and secretion are required to define the roles of precursor E domain peptides; however, the possibility exists that they could be important in IGF-I action at a point downstream from secretion.

mRNA SIZE

Both IGF-I and II mRNAs appear as multiple size forms, some of which are greatly in excess of that expected to encode for the pre-pro mRNA molecule. For example, in rat liver, IGF-I mRNA exists as two predominant forms of estimated size 7.0-7.5 kb and 0.9-1.2 kb as well as minor forms of intermediate size (Lund, Moats-Staats, Hynes, Simmons, Jansen, D'Ercole and Van Wyk, 1986; Lowe *et al*, 1987). The additional size is not derived from the variable 5′ exons but from unusually long 3′ UT, in some cases greater that 6 kb in length, which are subsequently processed at various alternate polyadenylation sites along its length. The possible involvement of 3′ UTs in the regulation of IGF turnover is discussed in a later section which deals with nutritional regulation.

It is clear from this description that behind the relatively uncomplicated mature IGF peptide structure lies a complex pattern of biosynthetic variation and the potential for a high degree of specific regulation in response to metabolic or developmental signals. Not all of this potential regulation need operate at the DNA level and hence require the synthesis of new transcripts; effective controls could operate on pre-existing RNA and protein precursors through post-translational and post-transcriptional processing events.

Regulation of IGF mRNAs

Most existing knowledge on the regulation of IGF expression has come from studies performed on the rat. Relatively little has been done with large animals. From the point of view of understanding animal growth and productivity it is undesirable to extrapolate entirely from the rodent model. Not only are there major differences in metabolic and growth rates but, in the case of the ruminant, areas of intermediary metabolism are quite different. In this discussion we will emphasise recent data on the regulation of IGF in the sheep with reference to the existing rodent data. A prerequisite for studying mRNA regulation is a quantitative and sensitive assay of mRNA levels. This is particularly important since IGF mRNA levels in large animals are generally around one-tenth of those found in the rat. We have adopted solution hybridisation/RNase protection assays to quantitate IGF mRNA (Dickson, Saunders and Gilmour, 1991). Briefly, a short segment of IGF-coding DNA is cloned into a vector from which a highly radiolabelled complementary (or anti-sense) RNA probe is transcribed *in vitro*. The probe is hybridised to purified tissue RNA and then digested thoroughly with ribonuclease (RNase). Only a perfect hybrid survives this step. Gel separation of the RNase resistant material to give a discrete species of predicted size is diagnostic for the mRNA and is quantifiable. This method is up to two orders of magnitude more sensitive than conventional Northern blot analysis of RNA.

DEVELOPMENT

Figure 2 shows RNase protection analyses of IGF-I and II mRNAs in samples of total RNA isolated from tissues of the sheep foetus and neonate.

IGF-I

IGF-I mRNAs are detected in sheep liver during foetal development. Substantial synthesis occurs in late gestation and continues to rise after birth with the onset of GH-dependent growth. This pattern is similar to that reported for the rat (Lund *et al*, 1986; Hoyt *et al*, 1988; Norstedt, Levinovitz, Moller, Eriksson and Anderson, 1988). Foetal GH concentrations are markedly elevated compared to postnatal concentrations (Gluckman, Grumbach and Kaplan, 1981) and it is

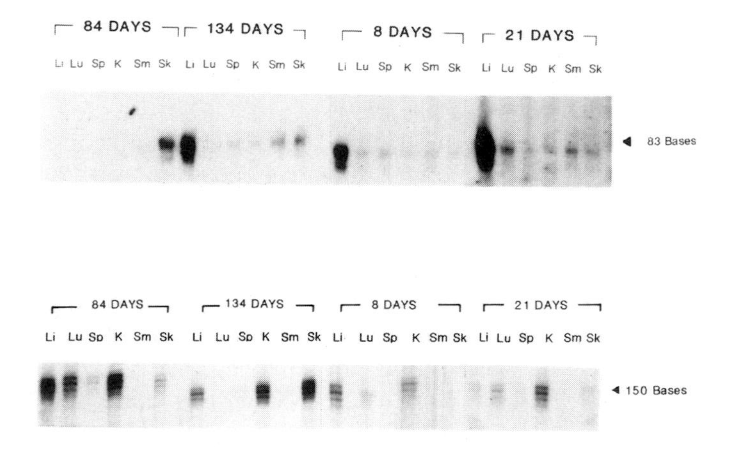

Figure 2 Analysis of the ontogeny of IGF-I and IGF-II in sheep tissues by solution hybridisation/RNase assay. Total RNA was prepared from liver (Li), lung (Lu), kidney (K), spleen (Sp), smooth muscle (Sm) and skeletal muscle (Sk) obtained from foetuses at 84 and 134 days of gestation and from 8- and 12-day-old neonates. A protection assay was carried out on 100 μg of RNA with an exon 3 ovine IGF-I probe containing an 83 base complementary region (top panel) or on 50 μg of RNA with an exon 1 ovine IGF-II probe containing a 150 base complementary region (bottom panel). Methodology is described in Dickson, Saunders and Gilmour (1991). Panels show autoradiograms exposed for either 4 days (IGF-I) or 1 day (IGF-II).

tempting to speculate that this synthesis represents the initiation of GH-responsive hepatic IGF-I production. Gluckman, Butler and Elliot (1983) reported that somatogenic receptors are essentially absent from ovine liver until after birth however their data show a small but significant specific binding of GH to hepatic membranes of 133 and 137 day foetuses. GH receptor mRNA transcripts have also been detected in foetal livers just prior to parturition (Adams, Baker, Fiddes and Brandon, 1990). It is therefore possible that the induction of GH receptors late in gestation could signal the onset of adult-type IGF-I synthesis, however, previous to this time there is considerable evidence to show that the synthesis of IGF-I and II in the ovine foetus is not dependent on GH and that GH does not play a significant role in foetal growth (Liggins, 1974; Gluckman and Butler, 1983). The occurrence of IGF-I in non-hepatic foetal tissues is also seen. Particularly striking is the burst of synthesis in skeletal muscle around 84 day gestation at a time when there is none in the liver. Soon after birth most non-hepatic tissues establish a low and variable level of IGF-I synthesis in line with the proposal that IGF-I has a paracrine action in addition to an endocrine function. The ontogeny of IGF-I expression is therefore complex. The regulation of IGF-I synthesis appears to be different between foetus and neonate.

IGF-II

The pattern of expression of IGF-II in the developing ovine foetus also shows characteristic developmental variation in each positive tissue suggesting that they are subject to different controls. In the rat, IGF-II mRNA abundance is highest in the foetal liver and rapidly declines postnatally to levels that are barely detectable (Soares, Ishii and Efstratiadis, 1985; Gray, Tam, Hayflick, Pintar, Carense, Koutos and Ullrich, 1987). While this is generally true for the sheep, IGF-II synthesis continues at appreciable levels in most tissues of the neonate. There has been considerable speculation that placental lactogen (PL) which is structurally similar to GH may regulate foetal IGF-II levels. Adams, Nissley, Handwerger and Rechler (1983) showed that PL, but not GH, stimulated IGF-II but not IGF-I secretion in primary foetal rat fibroblasts. However, a direct effect of PL has yet to be shown *in vivo* and seems unlikely since PL levels are higher in the foetal lamb up to 90 days gestation and fall off thereafter during which time IGF-II levels remain high (Gluckman, Kaplan, Rudolph and Grumbach, 1979; Gluckman and Butler, 1983). The widespread occurrence of mRNA in foetal tissues (Figure 2) might suggest a predominantly paracrine rather than endocrine mode of action for IGF-II and the lack of parallel ontogeny of expression between tissues could indicate multiple controls.

Hormonal regulation of IGF expression

GROWTH HORMONE

Background

Previous studies, carried out mainly with growing rats, have shown that following hypophysectomy circulating IGF-I levels fall dramatically in parallel with a decrease in liver IGF-I production. Non-hepatic tissue IGF-I levels also fall, though to a lesser extent (D'Ercole *et al*, 1984). After the administration of GH, tissue-extractable IGF increased even before the rise in serum IGF-I (Daughaday and Rotwein, 1989). Thus, it was concluded that GH regulates IGF-I production in non-hepatic tissues in addition to its effect on liver IGF-I synthesis. The rat IGF-I gene has two alternative 5′ exons and two alternative E domains. These features have been identified in separate cDNA clones. The expression of IGF-I mRNAs with the different 5′ exons was differentially regulated by GH in a tissue specific fashion. GH administered to hypophysectomised rats caused about a 2-fold increase in abundance of one of the alternative 5′ exons across all tissues while the other 5′ exon showed an 8-fold increase only in the liver (Lowe *et al*, 1987). Thus, in rat, GH regulates the multiple IGF-I mRNAs with variant 5′ exons differently in different tissues. With regard to alternative usage of E domains, GH stimulated the levels of both IGF-Ia and Ib forms in all tissues. In liver the IGF-Ib mRNA increased 8-fold compared with a 2.5-fold increase in IGF-Ia; in all other tissues both forms increased 1 to 3-fold (Lowe *et al*, 1988).

Ovine IGF-I gene

Three IGF-I cDNAs with three different 5′ exons have been documented (Wong, Ohlsen, Godfredson, Dean and Wheaton, 1989) and their relationship to exon 2, the first coding exon for mature IGF-I, is shown in Figure 3. Exons 1, 1W, and 1A are individually spliced to exon 2 to give three variant pre-pro IGF-I mRNAs. Exon 1A is homologous to the more GH-responsive of the two alternative 5′ rat IGF-I exons described above. In order to verify this for the sheep, RNA from the livers of lambs treated with or without GH was analysed by an RNase protection assay. An antisense RNA probe was made which was complementary to exon 1A and about 400 bp of 5′ sequence. On carrying out the assay this yielded a series of protected fragments between 46-61 nucleotides in length (Figure 4B) which were much more pronounced in RNA from GH-treated animals (lanes 6 and 8) than from untreated animals (lanes 5 and 7). Figure 5 shows the DNA sequence in the region of exon 1A and the predicted 5′ termini of the protected fragments. The protection assays only measure continuous sequence homologies. Therefore it is not possible to distinguish a genuine start site of RNA synthesis from a break in homology due to a spliced sequence transcribed from a more distant part of the DNA. This latter possibility seems unlikely since examination of the DNA sequence in the vicinity of the multiple starts does not show a prerequisite splice

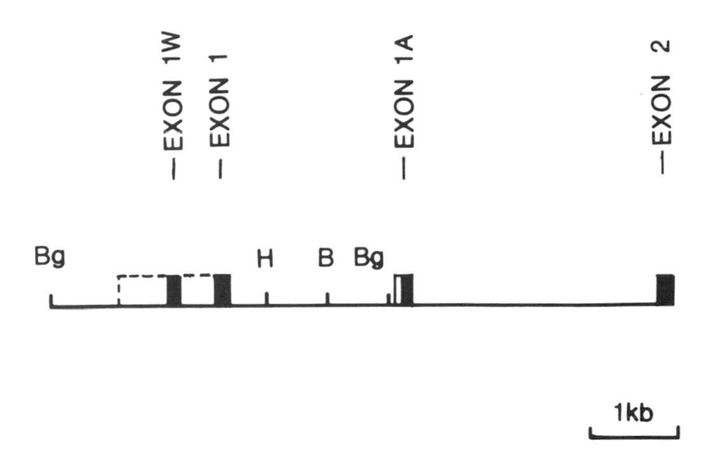

Figure 3 Genomic map of exons 1W, 1 and 1A and 2 for the ovine IGF-I gene. Isolation and characterisation have been described previously (Dickson, Saunders and Gilmour, 1991). Exons are shown as boxes; coding regions are solid, untranslated regions are open. The exact extents of exons 1W and 1 untranslated regions are not known and are denoted with dotted lines. Restriction enzyme sites are shown: Bg, Bgl II; H, Hind III, B, Bam HI.

acceptor site. More formal evidence comes from primer extension analysis. Here, reverse transcriptase is used to copy mRNA starting from an internal primer and extending it to the 5′ end of the RNA; the length of the extended primer indicates the start of the transcript. Figure 4A shows that a 17 base primer complementary to exon 1A was extended to 46 bases by this procedure; this point coincides with the most proximal of the start sites shown in Figure 5. Taken together these two pieces of data indicate that the transcription start for exon 1A is between 46-61 nucleotides 5′ to the point at which it splices to exon 2; the spread of sites predicted by these two techniques could indicate genuine heterogeneity in the RNA or more likely, could result from operational differences between the two methods.

The data of Figure 4B indicate that in animals not receiving GH the levels of exon 1A transcripts are low; some variation between the two animals is seen which may reflect differences in background metabolic status. Unlike the rat data previously cited, the sheep in these experiments were not hypophysectomised to reduce background noise. From densitometric measurements it was estimated that GH treatment caused a 5 to 10-fold enhancement in exon 1A transcripts. A parallel analysis with an exon 3 probe, common to all IGF-I mRNAs, was also carried out to give an estimate of total IGF-I mRNA levels. The results show (Figure 4A) that there are high levels of total IGF-I mRNA in untreated liver

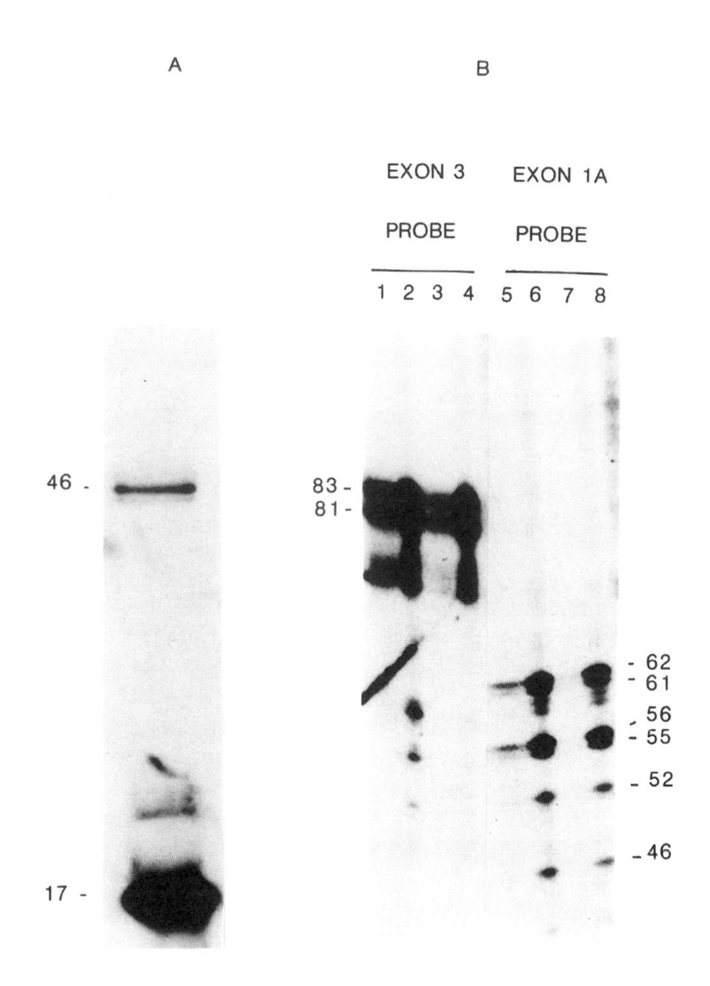

Figure 4 (A) Primer extension of ovine liver mRNA. Primer extension was carried out with poly A^+ liver mRNA (3 μg) using a primer (5′TGTAGGTGTAAC CATTT3′) complementary to exon 1A (see Figure 5). The autoradiogram shows the original primer and the 46 nucleotide extension product.

(B) Solution hybridisation/RNase protection analysis of total ovine liver RNA probes specific for exons 3 and 1A. Protection assays were carried out with total liver RNA (100 μg) from control and GH treated lambs with antisense probe to exon 3 (as for Figure 2) and to exon 1A. The 1A probe extended 600 bp from the Bam H1 site to the 3′ end of exon 1A (Figure 3). Lanes 1, 3, 5 and 7 are control animals; Lanes 2, 4, 6 and 8 are GH treated. Protected fragments were sized by comparison against labelled DNA markers (not shown).

(lanes 1 and 3) and that GH treatment causes an approximate doubling in levels (lane 2 and 4). It is estimated that about 50% of this increase is due to enhancement of exon 1A transcripts.

From mRNA analysis it is concluded that one of the multiple 5′ alternative exons (exon 1A) is extremely sensitive to GH. It is possible that not all of the increase in hepatic IGF-I mRNA is accounted for by exon 1A transcripts, however, we have not been able to show a GH effect on either exons 1 and 1W (data not shown). This would imply that the untranscribed DNA sequence 5′ to exon 1A (Figure 5) contains the promoter and signals for GH-regulated transcription.

Exon 1A promoter

Transient cell transfection assays were performed to test for the existence of a promoter for exon 1A. A 1.3 kb Hind III-Bgl II fragment 5′ to exon 1A (Figure 1) was inserted in both 5′-3′ and 3′-5′ orientations into a vector carrying the bacterial chloramphenicol acetyltransferase (CAT) gene to give plasmids pCAT-1A and pCAT-1AR respectively.

The ability of the inserted fragment to act as a promoter for CAT gene expression was determined by transfecting plasmid DNA into a pre-adipose mouse cell line (OB 1771) which on stimulation with GH has been shown previously to differentiate into adipoctyes. One of the early responses to GH treatment is the *de novo* synthesis of IGF-I mRNA (Doglio, Dani, Fredrikson, Grimaldi and Aiehaud, 1987). Table 1 gives the levels of CAT activity obtained with the plasmids compared with pRSV CAT in which the CAT gene is driven by the Rous Sarcoma virus (RSV) promoter. In the absence of GH the exon 1A promoter is about 20% as effective as the RSV promoter when inserted in the same orientation as the CAT gene (pCAT-1A). In the presence of GH a 2-fold increase is obtained in levels of activity expressed from this plasmid. In contrast no effect is observed with pRSV CAT. These transfection assays demonstrate that 1.3 kb of 5′ DNA can act as a promoter in an orientation-specific manner to give expression levels about one-fifth as high as those from a strong viral (RSV) promoter. The dependence on GH for CAT expression in transfected OB1771 cells however is not as clear as it is in liver *in vivo*. About half of the response in GH-stimulated cells appears to be permissive, whereas in OB1771 cells the levels of endogenous IGF expression in the absence of GH is negligible (Doglio *et al*, 1987).

In view of the elevated activity of this promoter in rat liver compared with other tissues it is possible that mouse pre-adipocytes lack liver specific factors which contribute to the stringency; alternatively the promoter fragment might not encompass all of the necessary control elements. In future studies attempts will be made to use primary sheep hepatocytes for these studies. DNA transfection of cultured cells provides a powerful tool for dissecting the molecular control of gene expression. The present study has established the identity of a GH responsive promoter for the ovine IGF-I gene and partially demonstrated

Table 1 DETERMINATION OF CHLORAMPHENICOL ACETYLTRANSFERASE (CAT) ACTIVITY IN TRANSFECTED OB1771 CELLS

Plasmid	CAT activity pmoles chloramphenicol acetylated/hr/100 μg protein	
	-GH	+GH
pRSVCAT	45.3	43.6
pCAT-1A	10.1	21.6
pCAT-1AR	1.3	0.84
None	0.98	0.91

Each result is the average of duplicate determinations for 3 separate transfection experiments

characteristics seen *in vivo*. Further analysis of the promoter may reveal additional information on the molecular signalling pathways and the synergistic stimuli involved in GH action. For example, within 400 bp of the exon 1A start region there are potential binding sites for transcription factors AP1, AP2 and SP1 in addition to a direct and an inverted octonucleotide repeat sequence (Figure 5). The role of these features in the maintenance of promoter function and GH responsiveness is under investigation. Interestingly, the DNA sequence within 100 bp 5′ to the start sites does not contain the TATA or CCAAT elements normally seen in conventional eukaryotic promoters. However the sequence surrounding the most proximal start has some of the features of a recently described "initiator" sequence found in viral promoters; this sequence can direct specific initiation of transcription in the absence of a TATA box (Smale and Baltimore, 1989). Also the sequence ATCAGTCT which includes the most distal transcriptional start has been recognised by Smale, Schmidt, Bert and Baltimore (1990) as an initiation sequence which acts in conjunction with SP1 to initiate transcription in the absence of a TATA consensus.

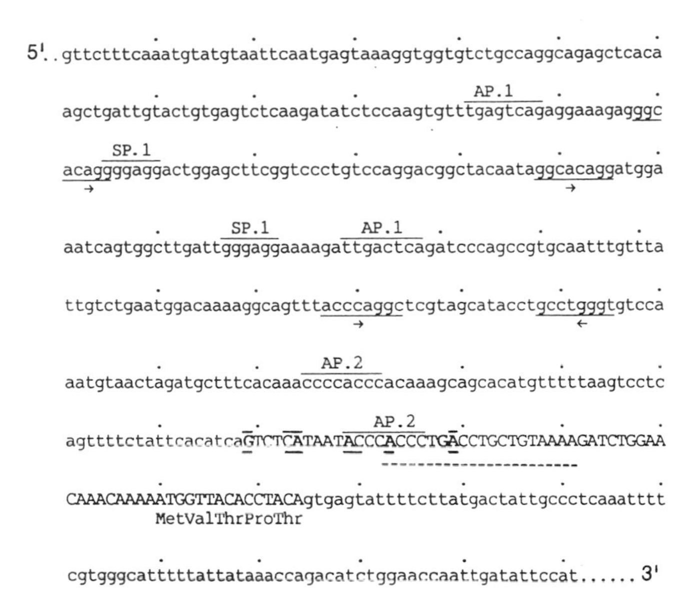

Figure 5 DNA sequence of the ovine IGF-I exon 1A and 5'-flanking region. The upper case represents exon 1A and lower case intron. Codon assignments are given for a potential pre-propeptide sequence that starts from an in-frame Met codon. Lines above and below individual nucleotides in the exon sequence denote the transcriptional starts determined from primer extension and RNase protection assays. Also shown are possible DNA binding motifs for transcriptional factors AP1, AP2 and SP1, directed and inverted repeat sequences (indicated by arrows) and the putative "initiator" described in the text (dashed line).

Distribution of exon 1A transcripts

Evidence has already been given for GH induced IGF-I synthesis in non-hepatic tissues of the hypophysectomised rat. Also in the neonatal sheep establishment of the GH-hepatic axis is accompanied by lower levels of IGF synthesis in other tissues (Figure 2). These findings prompted us to look for exon 1A transcripts in tissues other than liver. Since levels were likely to be low this was done by polymerase chain reaction (PCR) in order to maximise sensitivity. The strategy was to carry out PCR between either exon 1 or 1A and exon 5 (E domain) using appropriate primers and to look for the diagnostic products of 530 bp (1-5) and 390 bp (1A-5) with specific radiolabelled probes. Figure 6 (panels 1-3) shows the results of an analysis of RNA from liver, spleen, lung, smooth and skeletal muscle of 11-day-old lambs for the presence of either exon 1 or exon 1A in IGF-I transcripts. Gel electrophoresis of the PCR products gave the predicted sizes of 530 and 390 bp and these were verified by hybridisation to probes specific for exons 1, 1A and 3. The results show that both exon 1 and 1A transcripts exist as separately identifiable species using their exon specific probes (panels 1 and 2) and that both also hybridise to a shared exon 3 sequence (panel 3). An interesting and consistent finding is that similar experiments with material from older lambs (28 days onward) showed only exon 1 in RNA from tissues other than liver (panel 4). It is concluded that exon 1A transcripts are confined principally to liver during the life of the sheep except for a short period immediately after birth when they also occur in a number of peripheral tissues and that the low levels of IGF-I mRNA found in these tissues at later stages are of the exon 1 type. This would suggest a fundamental difference in the GH response between liver and peripheral tissues at these stages. At the mechanistic level, a comparison of GH regulation of the IGF-I gene in liver and peripheral tissues may help to understand the relative physiological significance of endocrine and autocrine/paracrine IGF-I synthesis.

GLUCOCORTICOIDS

Background

In children glucocorticoids can inhibit growth (Loeb, 1976) and development (Devenport and Devenport, 1985) but the mechanism by which this occurs is unclear. More recent data suggest that growth inhibitory effects of glucocorticoids involve IGF-I. In pituitary intact rats, dexamethasone decreases liver mRNA levels and in hypophysectomised animals, dexamethasone attenuates the increase in liver IGF-I mRNA induced by GH (Murphy and Luo, 1989).

Rat liver IGF-II mRNA abundance is also down-regulated by glucocorticoids. The effect is specific to dexamethasone and is not seen with oestrogen or testosterone (Levinovitz and Norstedt, 1989). In the rat there is an abrupt decrease in hepatic IGF-II expression between 18 and 20 days after birth which coincides with a surge in plasma cortisol levels. This effect can be induced earlier in the post-natal period by injection of cortisone acetate (Carlile and Beck, 1983) and subsequent work demonstrated that this treatment rapidly extinguished IGF-II mRNA expression in the neonatal liver, the major site of endocrine synthesis, but had less effect on mRNA levels in skeletal muscle (Beck, Samani, Senior, Byrne, Morgan, Gebhard and Brammar, 1988). In normal development, skeletal muscle IGF-II mRNA levels fall with the onset of differentiation and the appearance of heavy-chain isozymes of myosin which takes place in the first three weeks of life. In the choroid plexus of the rat brain IGF-II mRNA synthesis continues throughout foetal life and persists at that site in the adult (Beck, Samani,

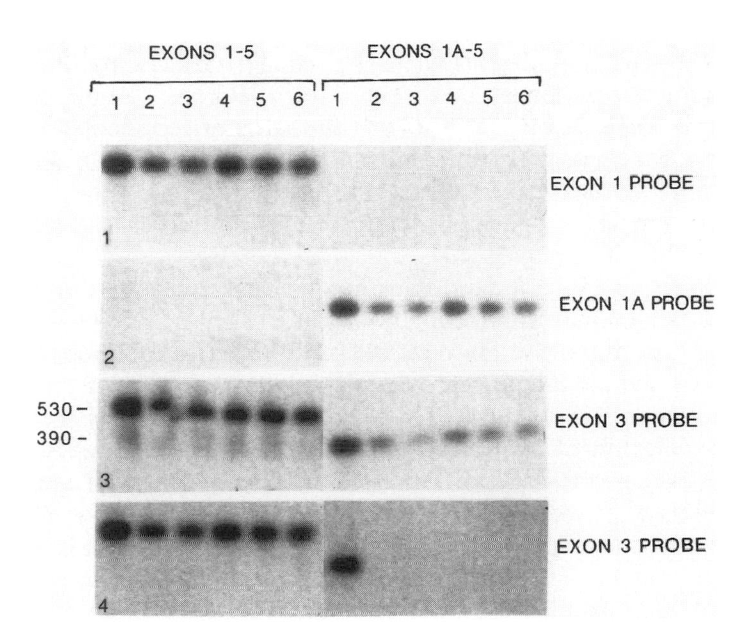

Figure 6 Exon 1A expression in liver and peripheral tissues. cDNAs were made from 20 μg total RNA from liver, spleen, smooth muscle, lung and skeletal muscle (lanes 1-6 respectively) of 11-day-old lambs and 40 cycles of PCR amplification between exon 1 and 5 and 1A and 5 performed for each sample. The exon 1 primer, (5′,GACATTGCTCTCAACAT3′) begins 34 bases 5′ to the exon 1 coding region and the exon 5 primer (5′CACAAGCAGAGGGAGTG3′) is within the E domain coding region. Gel analysis gave single ethidium staining bands of 530 bp for exons 1 5 and 390 bp for exons 1A-5 (not shown). Gels were blotted and probed with ^{32}P-labelled oligonucleotides derived from the internal regions of exon 1 (5′TTATTTAAGTGCTGCTTTTGTGATTT3′) and exon 3 (5′ACAAGCCCACGGGTA3′). The same sequence was used for the exon PCR primer and oligonucleotide probe (5′AAATGGTTACACCTACA3′).

Penshow, Thorley, Tregear and Coughlan, 1987; Ichimiya, Emson, Northrop and Gilmour, 1988). It has been proposed that IGF-II secretion by the choroid plexus has a neuromodulatory role as a maintenance hormone in the adult nervous system (Sara, Hall, von Holz, Humbel, Sjorgren and Wetterberg, 1982). Cortisone treatment has no effect on IGF-II mRNA synthesis in this tissue (Beck *et al*, 1988). This suggests that cortisone inhibition is less marked or absent at sites where IGF-II might act in an autocrine or paracrine fashion and where it does not appear to be directly involved with growth and cell division.

Cortisol and the late gestation sheep foetus

The importance of cortisol for prepartum maturation of the foetus and for parturition itself has been well established. Birth does not occur in the sheep in the absence of the foetal pituitary (Liggins, Fairclough, Grieves, Kendal and Knox, 1973) and gestation is prolonged by adrenalectomy (Barnes, Comline and Silver, 1977). Furthermore, infusion of cortisol into the catheterised sheep foetus several weeks prior to term elicits premature delivery. A number of important tissue and organ maturation changes occur during the pre- or perinatal period in which corticosteroids have been implicated and the reader is referred to a recent review for further details (Silver, 1990).

In the sheep (term 146 ± 2 days) there is a gradual rise in cortisol levels from about 3 weeks before parturition culminating in a more rapid rise in the final 3 to 5 days (Silver, Comline and Fowden, 1983). In order to examine the status of IGF-II synthesis during this period total RNA was prepared from tissues of foetuses in the ranges 127-130 days, 138-142 days and 143-145 days gestation and mRNA levels determined by RNase protection assays. Figure 7A shows the relationship between relative IGF-II mRNA levels and gestational age. Over the period 127-144 days gestation there is about a 50% drop in IGF-II mRNA in liver however, no statistically significant effect was seen in the mRNA levels in kidney, lung or skeletal muscle. The liver levels also correlate with the increase in foetal plasma cortisol levels over the same period (Figure 7B). Foetuses were also taken at 130-140 days when cortisol levels were <20 ng/ml and these levels raised to around 60 ng/ml by maternal fasting or direct infusion of cortisol into the foetus. These treatments caused a similar drop in IGF-II mRNA which correlated directly with serum cortisol. These results indicate that the changes in circulating foetal IGF-II that occur around parturition are probably due in large measure to down-regulation of IGF mRNA synthesis in the liver. This could be seen as part of the more general role of cortisol in tissue maturation in preparation for birth.

It will be of interest to determine whether there is a concomitant up-regulation of GH receptor mRNA which might explain the prepartum surge of IGF-I mRNA synthesis that occurs in the 134 day foetal liver (Figure 2). The mechanism of regulation of IGF-II synthesis in non-hepatic tissues is not known and it seems unlikely from these data that cortisol is involved.

IGF-II gene regulation

Information on the promoters for IGF-II is confined to work done on man and rat. The overall structural homology of the genes from the two species is considerable; a major difference is that the human gene has four promoters while the rat has three (PI-P3). In the rat IGF-II transcripts have characteristic development profiles in various tissues and the major species in the neonatal liver is clearly derived from promoter P2 (Soares, Turken, Ishii, Mills, Episkou, Cotter, Zeitlin and Efstratiadis, 1986). This promoter has been extensively characterised (Evans, DeChiara and Efstratiadis, 1988) and shown to contain the basic elements

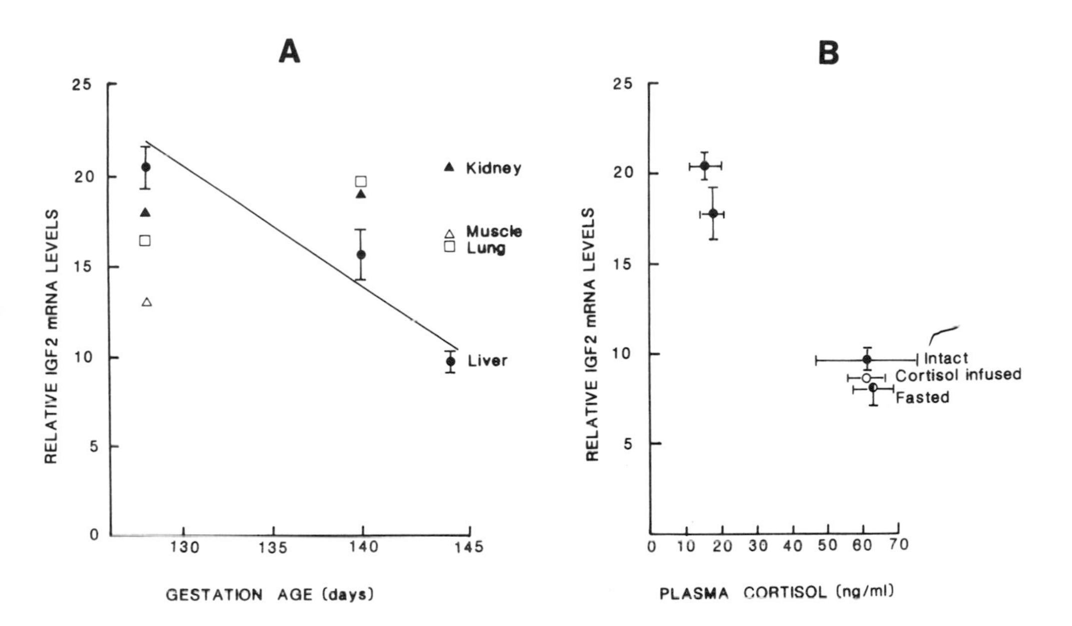

Figure 7 Relationship of foetal tissue IGF-II mRNA levels to gestation age and foetal plasma cortisol levels. (A) Total RNA was prepared from liver, skeletal muscle, kidney and lung from groups of 4 foetal lambs taken between 127-130 days and 143-145 days gestation. RNase protection assays were carried out as described for Figure 2 and the protected bands quantitated by densitometry. A positive correlation with gestational age was obtained for liver IGF-II mRNA but not with mRNAs from other tissues.
(B) Groups of 4 foetuses around 130-140 days gestation were taken when cortisol levels were less than 20 ng/ml and directly infused with cortisol or subjected to nutritional stress by maternal fasting. In both cases cortisol levels rose to around 60 ng/ml. IGF-II mRNA levels were determined in total liver RNA as described in A. An inverse linear relationship was found between IGF-II mRNA levels and foetal plasma cortisol levels.

necessary to support transcription in a rat liver cell line.

The glucocorticoid receptor controls gene expression by hormone-mediated interaction with specific DNA sequences found in close proximity to the regulated promoter. Although in most cases, glucocorticoids act to induce gene expression there are several examples of repression by glucocorticoids (Eberwine and

Roberts, 1984; Siminoski, Murphy, Rennert and Heinrich, 1987). In the case of glucocorticoid regulation of phosphoenolpyruvate carboxykinase its gene is repressed in adipose tissue and induced in liver and kidney (Nechushtan, Benvenisty, Brandeis and Reshef, 1987). Glucocorticoid-repressible genes appear to carry copies of the consensus glucocorticoid-responsive element (GRE) 5′ to the coding region (Camper, Yao and Rottman, 1985; Israel and Cohen, 1985; Charron and Druin, 1986) but it has not been established how the hormone receptor complex interacts with these sequences to cause down-regulation.

Examination of the sequence of the rat IGF-II gene (Soares *et al*, 1986) shows the presence of several sequences with strong homology to the GRE (Scheidereit, Westphal, Carlson, Boshard, and Beato, 1986; Jantzen, Strahle, Gloss, Stewart, Schmid, Boshard, Miksicek and Schuitz, 1987) located between promoters P1 and P2. There are two pairs of putative GREs one about 1.2 kb and the other about 0.5-0.6 kb from the start of transcription from the P2 promoter. Pairs of closely positioned GREs have shown to act synergistically in other systems. However, given that these pairs of elements are roughly equidistant between P1 and P2 and that many enhancer elements can act in either orientation (Banerji, Rusconi and Schaffner, 1981) the GREs could control transcription from either promoter.

Experiments are in progress to isolate and characterise the IGF-II promoters for sheep and to confirm the presence of glucocorticoid regulatory elements. Using the approach already described for GH induction of IGF-I we hope to shed light on the mechanism whereby cortisol represses IGF-II gene activity in foetal liver prior to parturition and possibly directs the switch to IGF-I synthesis in neonatal liver.

NUTRITION

Nutrition is a modulator of IGF-I synthesis and consequently this regulation may be an important means by which nutrition influences growth. In rats, fasting results in a decrease in cartilage growth and in immunoreactive IGF (Philips and Young, 1976; Furlanetto, Underwood, Van Wyk and D'Ercole, 1979). Although GH levels are low in fasting rats (Tannenbaum, Rorstad and Brazeau, 1979), administration of GH is ineffective in restoring IGF levels (Philips and Young, 1976), suggesting a state of GH resistance. In young rats serum IGF-I is correlated with growth rate and appears to be influenced primarily by dietary protein (Prewitt, D'Ercole, Switzer and Van Wyk, 1982). In man, both energy and protein are important determinants of IGF-I synthesis, with energy possibly being more important than protein (Underwood, Clemmons, Maes, D'Ercole and Ketelslegers, 1986). The biosynthesis of foetal IGFs is also regulated by nutrition. In late-gestation foetal rats, maternal fasting causes growth retardation with a concommitant decline in total bioassayable plasma IGF (Hill, Fekete, Milner, DePrins and Van Assche, 1982). Very little is known about the molecular mechanisms by which nutrition controls IGF synthesis. There is considerable evidence to suggest that for post-natal animals malnutrition reduces the number of GH binding sites and that this correlates with the decrease in IGF-I production.

Since insulin secretion is also depressed in malnutrition it has been suggested that this may be a regulator of GH receptor levels (Baxter, Bryson and Turtle, 1980; Postel-Vinay, Cohen-Tanugi and Charrier, 1982; Maes, Underwood and Ketelslegers, 1986; Elmer and Schalch, 1987; Bornfeldt, Arnquist, Enberg, Mathews and Norstedt, 1989). However this is clearly an oversimplification. In protein-restricted diabetic rats which had been treated with insulin, circulating IGF-I levels remain depressed (Maes *et al*, 1986; Maiter, Fliesen, Underwood, Maes, Gerard, Davenport and Ketelslegers, 1989). This conclusion is also supported by the observation that GH treatment of protein restricted rats induced hepatic GH binding but did not restore serum IGF-I levels (Thissen, Triest, Underwood, Maes and Ketelsleger, 1990). It is therefore concluded that nutritional regulation of IGF-I involves post-receptor as well as receptor-mediated processes. Attempts to measure IGF-I gene transcription levels in isolated nuclei from fasting and fed animals do not reveal differences sufficiently large to account for the much larger decreases in cellular IGF-I mRNA (Straus and Takemoto, 1990). These results suggest that the decrease in IGF-I mRNA that accompanies fasting is caused at least partly by a post-transcriptional regulatory mechanism.

The distribution of alternatively processed IGF-I forms that exist in the fasted rat have been carried out using solution hybridisation/RNase protection assays. In all tissues studied fasting reduced the level of 5′ exon variants to the same extent within individual tissues (Lowe, Adams, Werner, Roberts and LeRoith, 1989). With respect to the E peptide variants both fasting and diabetes reduced steady state levels of IGF-Ia and IGF-Ib mRNA to the same extent in all tissues, except in liver where the effect on IGF-Ib levels was much greater (Fagin, Roberts, LeRoith and Brown, 1989; Lowe *et al*, 1989). These findings also suggest that the effects of fasting on IGF-I gene expression may be partially related to the induction of GH resistance but also argue a strong case for a non-GH effect. The co-ordinate decrement in the 5′ exon IGF-I forms is evidence that other specific factors play an important role in the regulation of IGF-I gene expression in fasting. Evidence for post-transcriptional regulation of mRNA comes from the observation that the largest (3.5 kb) rat IGF-II mRNA which differs from the smaller (1 kb) IGF-II mRNA only in the length of 3′ UT did not translate in a cell free protein synthesising system (Graham, Rechler, Brown, Frunzio, Romanus, Bruni, Whitfield, Nissley, Seeling and Berry, 1986). It has also been shown that deprivation of cultured BRL-3A rat liver cells for a single amino acid caused a decrease in the levels of the 3.6 kb IGF-II mRNA while the 1 kb species was unaffected (Straus and Takemoto, 1988). Significantly, IGF-II gene transcription was not decreased in amino acid deprivation neither was the level of mRNA for α-tubulin affected. Recent studies have shown that the IGF-I 7.0-7.5 kb mRNAs have shorter half-lives than the smaller forms *in vitro* and *in vivo* (Hepler, Van Wyk and Lund, 1990). Furthermore, the long 3′ UT region of these large mRNAs contain structural features implicated in destabilisation of mRNAs (Steenbergh, Kooner-Reemst, Cleutjens and Sussenbach, 1991). This suggests the existence of specific regulatory sequences in the long 3′ UT region of both IGF-I and IGF-II mRNA that regulate their stability in response to changes in metabolic status. In support of this as a general mechanism it is already known that the 3′ UTs of a number of other growth factor and oncogene mRNAs contain sequences that influence mRNA stability (Brawerman, 1987). Future studies on the structure and the molecular mechanisms of turnover of the 3′ UT regions of IGF mRNAs may provide an insight into one of the possible post-receptor events in the nutritional regulation of IGF synthesis.

INSULIN

Insulin is measurable in foetal sheep plasma from about 40 days gestation (Alexander, Britton, Cohen, Nixon and Palmer, 1968). The presence of a correlation between foetal size and circulating insulin concentrations for a number of species including the sheep (Gluckman, Butler and Barry, 1985) is indirect evidence for a physiological role for insulin in regulating foetal growth.

Experiments on malnourished and pancreatectomised lambs suggest that IGF-II activity is not dependent on insulin but rather is regulated by glucose availability. Pancreatectomy late in gestation leads to a significant reduction in immunoreactive IGF-I and an elevation in IGF-II activity (Gluckman, Fowden, Butler and Comline, 1985). Elevated foetal IGF-II however does not directly support foetal growth in the presence of low IGF-I and insulin levels. Insulin may have a direct effect on foetal growth via actions on foetal glucose uptake, however other actions may well be mediated by the IGFs. IGF-I appears likely to be a regulatory factor for foetal somatic growth, but it is difficult to separate the effects of insulin and IGF-I experimentally.

In the adult rat, insulin is implicated in the regulation of serum IGF-I concentrations and IGF-I mRNAs in liver. In experimentally induced diabetes both are reduced significantly (Goldstein, Asertich, Levan and Phillips, 1988; Murphy, 1988). The effect of insulin does not appear to be due to indirect effects on GH since administration of exogenous GH does not normalise serum IGF-I levels and GH binding sites appear to be normal in the diabetic rat (Maes *et al*, 1986). This implies a post-receptor effect of insulin deficiency. On the basis of liver GH binding sites the mechanisms responsible for GH resistance in diabetes and malnutrition would therefore seem to be different, although both conditions involve decreases in insulin levels (Baxter *et al*, 1981; Postal-Vinay *et al*, 1982; Maes *et al*, 1984). One possible interpretational complication is the observation that insulin deficiency impairs protein synthesis and in particular that of secretory proteins (Roy *et al*, 1980).

At the molecular level, a number of genes have been described to be regulated by insulin at the level of transcription (Magnuson, Quin and Granner, 1987; Meisler, Keller, Howard, Ting, Samuelion and Rosenberg, 1989; Nasrin, Ercolani, Denaro, Kong, Kang and Alexander, 1990). In the case of the promoter for the glyceraldehyde-3-phosphate dehydrogenase gene, an insulin responsive element (IRE) has been identified in the nucleotide sequence which confers insulin responsiveness on CAT gene plasmids when introduced into liver derived cell lines. Gene activation correlated with the appearance of a nuclear protein which complexed with the IRE. A similar complex was obtained with nuclear extracts prepared from rats shifted from a low-insulin to high-insulin state *in vivo* by nutritional manipulation, indicating that the phenomenon is a potentially relevant component of insulin action in the whole animal. The identification and localisation of molecular mechanisms associated with insulin action *in vivo* offers a degree of analysis which is difficult to achieve in whole animal studies. The application of this knowledge to the structure of potential targets (*eg* GH receptor gene) should provide independent and confirmatory evidence in support of the physiological data.

Future perspectives

The physiology of normal and manipulated animals has defined many of the parameters which influence the levels of circulatory IGFs. However the finding that most tissues exhibit paracrine IGFs synthesis which may be subject to local regulation has also led to investigations at the cellular level. It is at this important interface that molecular studies can contribute to whole animal physiology. The actions of pleiotropic factors like GH are clearly more amenable to molecular analysis than for example the responses to nutritional change which involve hierarchies of interdependent controls. The study of IGF gene structure, although complex, can yield information on the various metabolic cues which affect its activity directly. The same strategy applied to the regulation of metabolically linked genes (*eg* GH receptor genes) could also explain indirect mechanisms of IGF-I gene control. Although not all metabolic controls act at the gene level those that do elicit their effects by interacting with control sequences in the gene promoter. As demonstrated for GH, introduction of responsive control elements into an appropriate cell type provides a "proof of principle" model for individual regulatory signals. A particularly intriguing possibility is that some post-receptor mechanisms may not act on the nucleus but rather on IGF mRNA pools maintained in the cytoplasm. The hitherto neglected long 3′ UTs of these and other mRNAs could become important targets for transient metabolic control signals. The appropriate application of molecular biology to these areas holds considerable promise for extending our present understanding of the physiology of growth.

References

Adams, S.O., Nissley, S.P., Handwerger, S. and Rechler, M.M. (1983) *Nature*, **302**, 150-153

Adams, T.E., Baker, L., Fiddes, R.J. and Brandon, M.R. (1990) *Molecular and Cellular Endocrinology*, **73**, 135-145

Alexander, D.P., Britton, H.G., Cohen, N.M., Nixon, D.A. and Palmer, R.A. (1968) *Journal of Endocrinology*, **40**, 389-390

Banerji, J., Rusconi, S. and Schaffner, W. (1981) *Cell*, **27**, 299-308
Barnes, R.J., Comline, R.S. and Silver, M. (1977) *Journal of Physiology*, **264**, 429-447
Baxter, R.C., Bryson, J.M. and Turtle, J.R. (1980) *Endocrinology*, **107**, 1176-1181
Baxter, R.C., Bryson, J.M. and Turtle, J.R. (1981) *Metabolism*, **30**, 1086-1090
Beck, F., Samani, N.J., Penshow, J.D., Thorley, B., Tregear, G.W. and Coughlan, J.P. (1987) *Development*, **101**, 175-184
Beck, F., Samani, N.J., Senior, P., Byrne, S., Morgan, K., Gebhard, R. and Brammar, W.J. (1988) *Journal of Endocrinology*, **1**, R5-R8
Bell, G., Merryweather, J., Sachez-Pascador, R., Stempien, M., Priestly, L.S. and Rall, L. (1984) *Nature*, **310**, 775-777
Blundell, T.L., Bedarker, S. and Humbel, R.E. (1983) *Federation Proceedings*, **42**, 2592-2596
Bornfeldt, K.E., Arnqvist, H.J., Enberg, B., Mathews, L.S. and Norstedt, G. (1989) *Journal of Endocrinology*, **122**, 651-656
Brawerman, G. (1987) *Cell*, **48**, 5-6
Camper, S.A., Yao, Y.A.S. and Rottman, F.M. (1985) *Journal of Biological Chemistry*, **260**, 12246-12251
Carlile, A.E. and Beck, F. (1983) *Journal of Anatomy*, **137**, 357-369
Charron, C.T. and Druin, J. (1986) *Proceedings of the National Academy of Sciences, USA*, **83**, 8903-8907
Daughaday, W.H. and Rotwein, P. (1989) *Endocrine Reviews*, **10**, 68-91
Daughaday, W.H., Parker, K.A., Borowsky, S., Trivedi, B. and Kapadia, M. (1982) *Endocrinology*, **110**, 575-580
DeChiara, T.M., Efstratiadis, A. and Robertson, E.J. (1990) *Nature*, **345**, 78-80
D'Ercole, A.J. and Underwood, L.E. (1981) In *Fetal Endocrinology* (eds M.J. Nory and J.A. Resko), Academic Press, New York, pp 155-182
D'Ercole, A.F., Stiles, A.D. and Underwood, L.E. (1984) *Proceedings of the National Academy of Sciences, USA*, **81**, 935-939
Devenport, L. and Devenport, J. (1985) *Experimental Neurology*, **90**, 44-52
Dickson, M.C., Saunders, J.C. and Gilmour, R.S. (1991) *Molecular Endocrinology*, **6**, 17-31
Doglio, A., Dani, C., Fredrikson, G., Grimaldi, P. and Aiehaud, G. (1987) *EMBO* **6**, 4011-4016
Dull, T.J., Gray, S., Hayfield, J.S. and Ullrich, A. (1984) *Nature*, **310**, 777-781
Eberwine, J.H. and Roberts, J.L. (1984) *Journal of Biological Chemistry*, **259**, 2166-2170
Elmer, C.A. and Schalch, D.S. (1987) *Endocrinology*, **120**, 832-834
Evans, T., DeChiara, T. and Efstratiadis, A. (1988) *Journal of Molecular Biology*, **199**, 61-81
Fagin, J.A., Roberts, C.T., LeRoith, D. and Brown, A.T. (1989) *Diabetes*, **38**, 428-434
Froesch, E.R., Schmid, C., Schwander, J. and Zapf, J. (1985) *Annual Review of Physiology*, **47**, 443-460

Furlanetto, R.W., Underwood, L.E., Van Wyk, J.J. and D'Ercole, A.J. (1979) In *International Symposium on Somatomedins and Growth, Proceedings of the Serrano Symposium*, London, Academic Press, vol 23, pp 123-135

Gluckman, P.D. (1986) In *Oxford Reviews of Reproductive Biology*, (ed J.R. Clarke), Oxford, Clarendon Press, vol 8, pp 1-60

Gluckman, P.D. and Butler, J.H. (1983) *Journal of Endocrinology*, **99**, 223-232

Gluckman, P.D., Kaplan, S.L., Rudolph, A.M. and Grumbach, M.M. (1979) *Endocrinology*, **104**, 1828-1833

Gluckman, P.D., Grumbach, M.M. and Kaplan, S.L. (1981) *Endocrine Reviews*, **2**, 363-395

Gluckman, P.D., Butler, J.H. and Elliott, T.B. (1983) *Endocrinology*, **112**, 1607-1612

Gluckman, P.D., Butler, J.H. and Barry, T.N. (1985) *Pediatric Research*, **19**, 620

Gluckman, P.D., Fowden, A., Butler, J. and Comline, R. (1985) *Journal of Developmental Physics*, **9**, 79-88

Goldstein, S., Asertich, G.L., Levan, K.R. and Phillips, L.S. (1988) *Molecular Endocrinology*, **2**, 1093-1099

Graham, D.E., Rechler, M.M., Brown, A.L., Frunzio, R., Romanus, J.A., Bruni, C.B., Whitfield, H.J., Nissley, S.P., Seeling, S. and Berry, S. (1986) *Proceedings of the National Academy of Sciences, USA*, **83**, 4519-4523

Gray, A., Tam, T.J., Hayflick, J., Pintar, J., Carense, W.K., Koufos, A. and Ullrich, A. (1987) *DNA*, **6**, 283-295

Hepler, J.E., Van Wyk, J.J. and Lund, P.K. (1990) *Endocrinology*, **127**, 1550-1552

Hill, D., Fekete, M., Milner, D., DePrins, F. and Van Assche, A. (1982) In *Insulin-like growth factors/Somatomedins Basic Chemistry, Biology, Clinical Importance*, (ed E.M. Spencer), Berlin, Walter de Gruytcs, pp 345-352

Hoyt, E., Van Wyk, J. and Lund, P.K. (1988) *Molecular Endocrinology*, **2**, 1077-1086

Ichimiya, Y., Emson, P.C., Northrop, A.J. and Gilmour, R.S. (1988) *Brain*, **464**, 167-170

Israel, A. and Cohen, S.N. (1985) *Molecular and Cellular Biology*, **5**, 2443-2453

Jansen, M., Van Schaik, F., Ricker, A., Bullock, B., Woods, D., Gabby, K., Nussbaum, A., Sussenbach, J. and Van Den Brande, J. (1983) *Nature*, **306**, 609-611

Jantzen, H.M., Strahle, U., Gloss, B., Stewart, F., Schmid, W., Boshard, M., Miksicek, R. and Schutz, G. (1987) *Cell*, **49**, 29-38

Levinovitz, A. and Nortstedt, G. (1989) *Molecular Endocrinology*, **3**, 797-804

Liggins, G.C. (1974) In *Size at Birth, Ciba Foundation Symposium*, Amsterdam, Elsevier, no 27, pp 165-183

Liggins, G.C., Fairclough, R.J., Grieves, S.A., Kendal, J.Z. and Knox, B.S. (1973) *Recent Progess in Hormone Research*, **29**, 111-150

Loeb, J. (1976) *New England Journal of Medicine*, **295**, 547-552

Lowe, W., Roberts, C., Lasky, S. and LeRoith, D. (1987) *Proceedings of the National Academy of Sciences, USA*, **84**, 8946-8950

Lowe, W., Lasky, S., LeRoith, D. and Roberts, C. (1988) *Molecular Endocrinology*, **2**, 528-535

Lowe, W., Adams, M., Werner, H., Roberts, C.T. and LeRoith, D. (1989) *Journal of Clinical Investigation*, **84**, 619-626

Lund, P., Hoyt, E. and Van Wyk, J. (1989) *Molecular Endocrinology*, **3**, 2054-2061

Lund, P.K., Moats-Staats, B.M., Hynes, M.A., Simmons, J.G., Jansen, M., D'Ercole, A.J. and Van Wyk, J.J. (1986) *Journal of Biological Chemistry*, **261**, 14539-14544

Maes, M., Underwood, L.E. and Ketelslegers, J.M. (1986) *Endocrinology*, **118**, 377-382

Magnuson, M.A., Quin, P.G. and Granner, D.K. (1987) *Journal of Biological Chemistry*, **262**, 14917-14920

Maiter, D., Fliesen, T., Underwood, L.E., Maes, M., Gerard, G., Davenport, M.L. and Ketelslegers, J.M. (1989) *Endocrinology*, **124**, 2604-2611

Meisler, M., Keller, S., Howard, G., Ting, C.N., Samuelson, L. and Rosenberg, M. (1989) *Diabetes, Supplement 21A-249A,* 63a

Moses, A.C., Nissley, S.P., Short, P.A., Rechler, M.M., White, R.M., Knight, A.B. and Higa, O.Z. (1980) *Proceedings of National Academy of Sciences, USA*, **77**, 3649-3654

Murphy, L. (1988) *Diabetologia*, **31**, 842-847

Murphy, L. and Luo, J. (1989) *Endocrinology*, **125**, 165-171

Nechushtan, H., Benvenisty, N., Brandeis, R. and Reshef, L. (1987) *Nucleic Acids Research*, **15**, 6405-6417

Nasrin, N., Ercolani, L., Denaro, M., Kong, X.F., Kang, I. and Alexander, M. (1990) *Proceedings of National Academy of Science USA*, **87**, 5273-5277

Norstedt, G., Levinovitz, A., Moller, C., Eriksson, L. and Anderson, G. (1988) *Carcinogenesis*, **9**, 209-213

Phillips, L.S. and Young, H.S. (1976) *Endocrinology*, **99**, 304-314

Postel-Vinay, M.C., Cohen-Tanugi, E. and Charrier, J. (1982) *Molecular and Cellular Endocrinology*, **18**, 657-662

Powell, D., Lee, P., Chang, D., Lin, F. and Hintz, R. (1987) *Journal of Clinical Endocrinology and Metabolism*, **65**, 868-875

Prewitt, T.E., D'Ercole, A.J., Switzer, B.R. and Van Wyk, J.J. (1982). *Journal of Nutrition*, **112**, 144-150

Rinderknecht, E. and Humbel, R.E. (1978a) *Journal of Biological Chemistry*, **253**, 2769-2774

Rinderknecht, E. and Humbel, R.E. (1978b) *FEBS Letters*, **89**, 283-286

Rotwein, P. (1986) *Proceedings of the National Academy of Sciences, USA*, **83**, 77-81

Rotwein, P., Folz, J. and Gordon, J.I. (1987) *Journal of Biological Chemistry*, **262**, 11807-11812

Roy, A.K., Chatterjee, B., Prasad, M.S.K. and Unakar, M.J. (1980) *Journal of Biological Chemistry*, **255**, 11614

Sara, V.R., Hall, K., von Holz, H., Humbel, R., Sjogren, B. and Wetterberg, L. (1982) *Neuroscience Letters*, **34**, 39-44

Scheidereit, C., Westphal, H.M., Carlson, C., Boshard, H. and Beato, M. (1986) *DNA*, **5**, 383-391

Shaw, G. and Kamen, R. (1986) *Cell*, **48**, 659-667

Silver, M., Comline, R.S. and Fowden, A.L. (1983) *Journal of Developmental Physiology*, **5**, 307-321

Silver, M. (1990) *Experimental Physiology*, **75**, 285-307

Siminoski, K., Murphy, R.A., Rennert, P. and Heinrich, G. (1987) *Endocrinology*, **121**, 1432-1437

Smale, S.T., Schmidt, M.C., Berk, A.J. and Baltimore, D. (1990) *Proceedings of the National Academy of Sciences, USA*, **87**, 4509-4513

Smale, S.T. and Baltimore, D. (1989) *Cell*, **59**, 103-112

Soares, M., Ishii, D. and Efstratiadis, A. (1985) *Nucleic Acids Research*, **13**, 1119-1134

Soares, M.B., Turken, A., Ishii, D.N., Mills, L., Episkopou, V., Cotter, S., Zeitlin, S. and Efstratiadis, A. (1986) *Journal of Molecular Biology*, **192**, 737-752

Steenbergh, P.H., Koonen-Reemst, A.M.C.B., Cleutjens, C.B.J.M. and Sussenbach, J.S. (1991) *Biochemical and Biophysical Research Communications*, **175**, 507-514

Straus, D.S. and Takemoto, C.D. (1990) *Molecular Endocrinology*, **4**, 91-100

Straus, D.S. and Takemoto, C.D. (1988) *Journal of Biological Chemistry*, **263**, 18404-18410

Tannenbaum, G.S., Rorstad, O. and Brazeau, P. (1979) *Endocrinology*, **104**, 1733-1738

Thissen, J.P., Triest, S., Underwood, L.E., Maes, M. and Ketelslegers, J.M. (1990) *Endocrinology*, **126**, 908-913

Turcotte, B., Guertin, M., Chevrette, M. and Bellanger, L. (1985) *Nucleic Acids Research*, **13**, 2387-2398

Underwood, L.E., Clemmons, D.R., Maes, M., D'Ercole, A.J. and Ketelslegers, J.M. (1986) *Hormone Research*, **24**, 166-176

Van Wyk, J.J. (1984) In *Hormonal Proteins and Peptides*,(ed C.H. Li), New York, Academic Press, pp 22-125

Wong, E.A., Ohlsen, S.M., Godfredson, J.A., Dean, D.M. and Wheaton, J.E. (1989) *DNA*, **8**, 649-657

8

NUTRITIONAL CONTROL OF THE GROWTH HORMONE AXIS

J.J. BASS, G.S.G. SPENCER and S.C. HODGKINSON
Growth and Fibre Physiology, Ruakura Agricultural Centre, Private Bag, Hamilton, New Zealand

Introduction

Growth of animals is under nutritional control, throughout development, from the foetus until maturity. Even when skeletal growth has ceased nutritional changes have a profound effect on body composition. Foetal growth appears to be pituitary independent, whereas post-natal growth is pituitary dependent, with a transition period occurring immediately after birth and before the GH axis is fully operational (Gluckman, 1984; 1986).

Post-natal pituitary dependent growth is considered to be mainly driven by the GH axis, which develops after birth as the GH receptors increase in the liver (Gluckman, Butler and Elliott, 1983). The post-natal biological actions of GH are considered to be mediated by insulin-like growth factor-I (IGF-I) and this axis is sensitive to nutritional changes. Although foetal growth is pituitary independent, both GH and IGF-I are present in foetal tissues and plasma (Clemmons and VanWyk, 1981) and both hormones are sensitive to availability of nutrients to the foetus (Gluckman, Douglas, Ambler, Breier, Hodgkinson, Koea and Shaw, 1991). Plasma IGF-I is also sensitive to nutritional changes during the fast growing transitional neonatal period, when the GH axis is not fully operational (Breier, Gluckman and Bass, 1988c). Therefore during foetal and neonatal development plasma GH can be manipulated by altering nutritional status at a time when the GH appears not to control growth. This possibly indicates that some of the changes seen in post-natal growth and IGF-I levels, when the GH axis is operational, may still be modified by mechanisms more fundamental than GH alone.

Growth hormone axis

GROWTH HORMONE (GH)

GH is a major factor controlling post-natal growth. Slow-growing, GH-deficient animals show an increase in growth when treated with GH (Tindal and

Yokoyama, 1964). GH treatment of normal animals stimulates growth in well-fed animals (Wagner and Veenhuizen, 1978). GH does not directly stimulate cartilage or muscle growth *in vitro*, yet plasma from GH-treated hypophysectomised rats has been shown to stimulate growth *in vitro* (Daughaday and Reeder, 1966). Thus it was hypothesised (Daughaday, Hall, Raben, Salmon, Van Den Brande and Van Wyk, 1972) that GH anabolic actions in promoting skeletal and muscle growth are mediated by an intermediary factor or factors. These factors have been identified as insulin-like growth factors which have previously been named somatomedins, non-suppressable, insulin-like activity (NSILA) and sulphation factors. Although GH is generally regarded as an anabolic hormone it also has catabolic actions by promoting lipolysis in fat depots, which may result from GH directly antagonising the anabolic actions of insulin on adipocytes (Vernon, 1982) and/or potentiating the effects of lipolytic hormones such as epinephrine (Boyd and Bauman, 1989).

This review will concentrate on the direct and indirect effects of nutrition on the GH/IGF axis (Figure 1) in domestic animals. In post-natal animals, when the

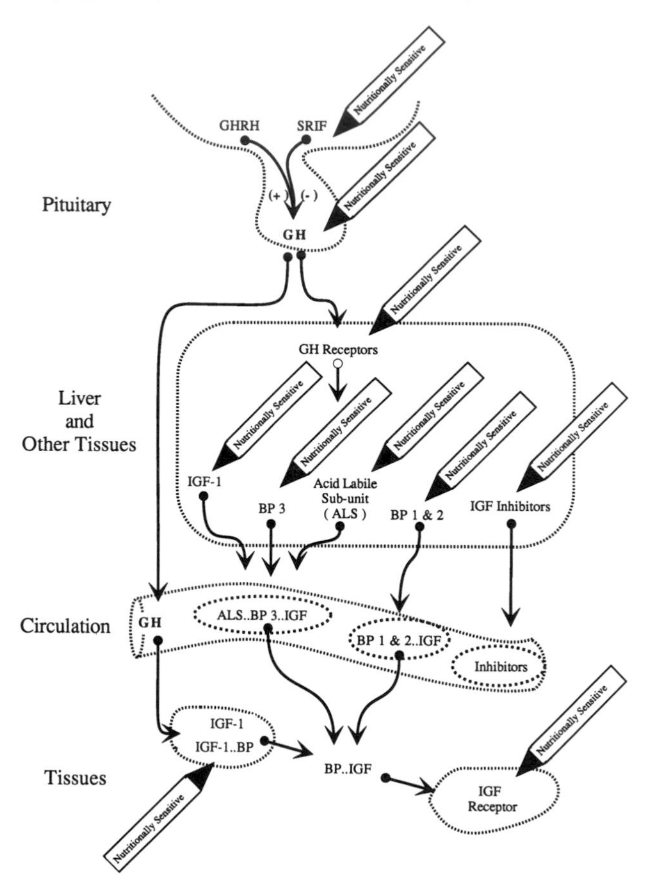

Figure 1 Nutritional sensitivity (Nutr. Sens.) of the growth hormone axis and the relationship of binding proteins (BP), insulin-like growth factor (IGF) and the acid labile sub-units (ALS).

GH axis is operational, nutritional deprivation increases plasma GH in pigs, sheep, steers and man (see review, Buonomo and Baile, 1990) but not in rats. In steers, reduced levels of nutrition increase mean plasma GH and the amplitude of GH pulses but do not affect baseline GH (Figure 2). In sheep GH plasma concentrations also increase when feed is reduced (Bassett, Weston and Hogan, 1971). Nutritional status not only alters plasma GH but also levels of GH and GH mRNA in the pituitary, which are increased following restricted feeding and decreased after *ad libitum* feeding in sheep (Thomas, Mercer, Karalis, Rao, Cummins and Clarke, 1990). Nutritional status also affects the half-life of GH with fasting increasing the half-life of GH in sheep and in calves (Trenkle, 1976), possibly by decreasing the number of GH binding sites.

GH BINDING PROTEINS (GHBP)

GH circulates in both a free and protein bound form. The level of binding varies between species, with it being reported that pigs have high levels of GHBP whereas cattle have low levels of GHBP (Clarke, personal communication). The serum GHBP has a close structural relationship to the extracellular portion of the GH receptor (Barnard and Waters, 1986). In humans GHBPs have been shown to complex approximately 50% of circulating hGH. This binding affects the bioactivity of GH, because the binding protein increases GH half life, modifies its distribution (Baumann, Shaw and Winter, 1987) and reduces the biological effect of GH activity on preadipocytes (Lim and Waters, 1989). Lim and Waters (1989)

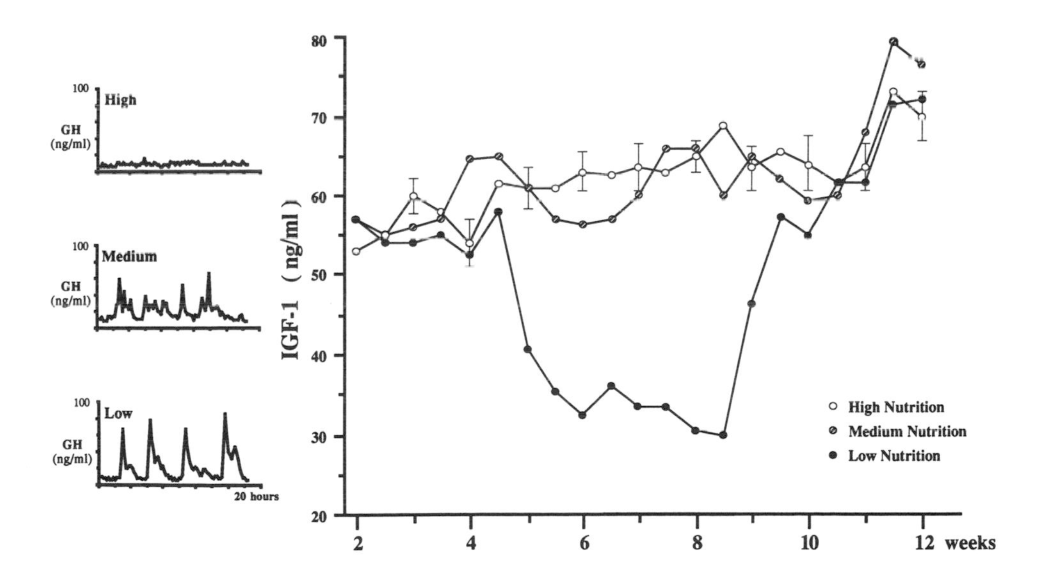

Figure 2 Effect of nutrition on GH & IGF-1 of steers

suggest that another function of GHBP is to dampen the biological effects of pulsatile GH secretion by reducing free GH during secretory pulses. As the GHBP appears to be produced by the proteolytic cleavage of the GH receptor (Leung, Spencer, Cachianes, Hammonds, Collins, Henzel, Barnard, Waters and Wood, 1987; Spencer, Hammonds, Henzel, Rodriguez, Waters and Wood, 1988) and hepatic GH receptors are nutritionally sensitive, it is likely that the GHBP is also sensitive to nutritional changes.

GH RECEPTORS

The first step in GH action is the binding of the hormone to a specific cell receptor. Receptors for GH have been identified in a number of different tissues, including liver, muscle, bone and adipose tissue (Barnard, Haynes, Werther and Waters, 1988). Hepatic GH receptors develop post-natally in cattle (Gluckman *et al*, 1983). Malnutrition induces resistance to GH and this is due in part to changes in GH receptor numbers (Maes, Underwood and Ketelslegers, 1983) and possibly also post-receptor mechanisms (Maes, Underwood and Ketelslegers, 1986). Specific GH binding to hepatic membrane preparations is decreased after chronic underfeeding (Baxter, Bryson and Turtle, 1981; Breier, Gluckman and Bass, 1988a). The GH receptor, although controlled by a single gene has been found to have different affinity states in a number of species (rats, Baxter *et al*, 1981; bovine, Breier *et al*, 1988a; ovine, Sauerwein, Breier, Bass and Gluckman, 1991). The high affinity hepatic GH receptor is particularly sensitive to nutritional manipulation (Breier *et al*, 1988a). The capacity of the high affinity GH receptor decreases with malnutrition and correlates positively with growth rate and plasma IGF-I. The capacity of the low affinity GH receptor is not correlated with growth or IGF-I. So it has been postulated that the high affinity GH receptor may mediate the GH control of IGF-I and explain the paradoxical negative correlation between GH and IGF-I/growth found in many animals, when nutritional status is altered (Figure 3).

GH has been shown to control its own receptor capacity, with GH administration increasing specific GH binding to the liver (Chung and Etherton, 1986). At a medium plane of nutrition GH secretion in cattle was shown to increase sufficiently to maintain plasma IGF-I, whereas the degree of GH resistance at a low level of nutrition was such that the resulting increase of plasma GH was insufficient to maintain plasma IGF-I levels (Figure 2).

The increase in plasma GH at low planes of nutrition has been postulated as a homeostatic mechanism which maintains basal metabolism during times of low nutrition by mobilising energy from adipose tissue (Bauman and Currie, 1980; Bines and Hart, 1982).

Nutrition Level	Plasma GH	GH Specific binding	GH Receptor Affinity		Plasma IGF-1
			high	low	
High (positive energy balance)	↓	↑	↑	↔	↑
Low (negative energy balance)	↑	↓	↓	↔	↓

Figure 3 GH and nutrition interactions

Insulin-like growth factor-I (IGF-I)

Somatomedin (IGF-I) has been identified as the intermediary of GH action on sulphate uptake by cartilage. It was postulated that IGF-I mediates many of the known *in vivo* effects of GH (Daughaday *et al*, 1972).

The liver is considered the major source of GH-controlled circulating IGF-I, contributing about 55% of the total plasma IGF-I (D'Ercole, Stiles and Underwood, 1984). A number of other tissues also produce IGF-I (D'Ercole *et al*, 1984) under GH control. These findings have led Green, Morikawa and Nixon (1985) to postulate "the dual effector theory" in which GH has two separate actions on tissue growth. GH not only stimulates IGF-I production but also stimulates cells to differentiate, which can then respond to IGF-I. IGF-I also seems to mediate a broad range of intracellular effects which include manipulating metabolic changes as well as growth and differentiation (see review DePablo, Scott and Roth, 1990).

D'Ercole *et al* (1984) found that IGF-I tissue levels of liver, kidney, lung, muscle and testes of the rat are strongly dependent on GH although IGF-I dependence on GH varies among tissues. The local production of IGF-I in ovine kidney and muscle tissue is partly under nutritional control, with tissue IGF-I levels being lower in lambs after 5 days starvation. When the total organ or muscle weight was taken into consideration there was less IGF-I present in the tissues of liver, kidney and skeletal muscle from starved sheep than those on high level of nutrition, although no significant long-term effect of GH on tissue IGF-I was seen (Figure 4) (Hua, Bass, Hodgkinson, Spencer and Ord, in press). These results indicate that in lambs the local production of IGF-I in liver, kidney and muscle are sensitive to starvation because tissue levels are derived from local production and plasma IGF-I.

In a study on the infusion of recombinant N-Met IGF-I from plasma into tissues and lymph (Hodgkinson, Bass, Davis, Spencer and Gluckman, 1990) it was found that muscle connective tissue, not muscle fibres, had the highest percentage of locally produced IGF-I. The muscle fibres contained the highest percentage of recombinant N-Met IGF-I establishing that the IGF-I in muscle fibres had originated from blood, and thus indicating a possible endocrine role for IGF-I in muscle fibres, but not necessarily for connective tissue. This may indicate that specific tissues within muscle, particularly connective tissue rather than muscle fibres, may be actively producing IGF-I at certain stages of maturation (Hodgkinson, Bass and Gluckman, 1991). It is also the IGF-I receptors in connective tissue which appear sensitive to nutritional changes, possibly indicating that local control of both production and reception of IGF-I in tissues are sensitive to changes in nutritional status.

CIRCULATING IGF-I

Although tissues produce IGF-I locally, which appears to be sensitive to changes in nutrition and GH status, there is still evidence that circulating IGF-I is biologically active and acts as a hormone. IGF-I when administered to rats stimulates growth of particular organs (Clark and Cronin, 1991). In sheep short-term infusion studies have shown IGF-I decreases protein degradation rates at a

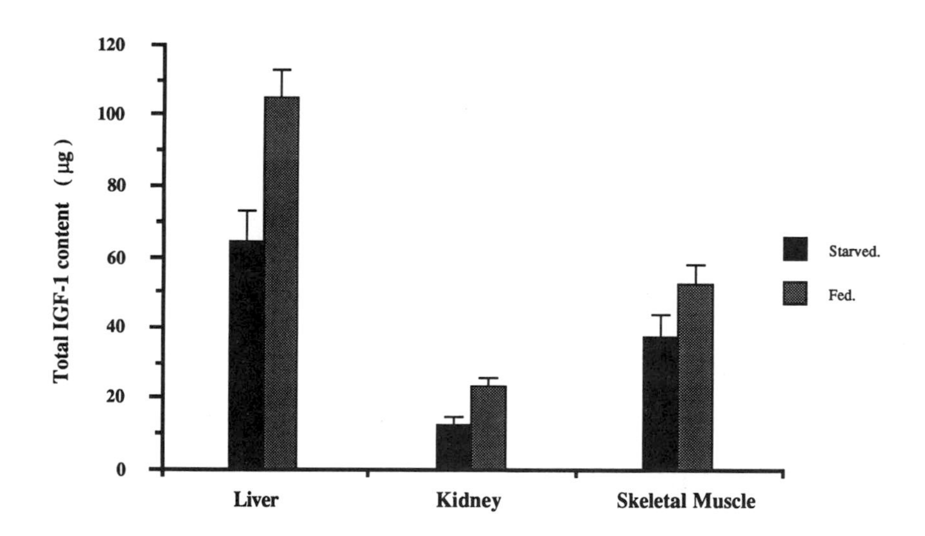

Figure 4 Total organ/muscle IGF-1 (μg) from starved and fed sheep (Hua *et al*, 1991)

greater rate than protein synthesis in infused hind limb studies (Oddy, Warren, Moyse and Owens, 1991). Short-term whole body IGF-I infusions in sheep have been shown to decrease net protein loss, which could not be explained by insulin-like actions and at a dose of IGF-I which has been shown not to induce hypoglycaemia (Douglas, Breier, Ball, Shaw and Gluckman, 1990). The administration of IGF-I does appear to stimulate growth but is not as effective as GH *in vivo*.

These studies indicate that circulating IGF-I can increase protein deposition in muscles in sheep and stimulate the growth of rats, supporting the hypothesis that IGF-I has an endocrine function. This endocrine control of body growth by circulating IGF-I has recently been challenged by studies on the growth effects after passively immunising GH-treated dwarf rats against IGF-I. Immunisation had no effect on the GH-stimulated growth of the GH-deficient rats (Spencer, Hodgkinson and Bass, 1991). However short-term IGF-I passive immunisation studies in sheep have shown that initially net protein deposition is decreased, indicating an endocrine role for IGF-I in sheep (Gluckman *et al*, 1991). An explanation for these apparently contradictory results is that the natural homeostatic mechanisms can compensate for the loss of circulating IGF within a relatively short period of time, with one possible mechanism being a compensatory increase in local production of IGF-I or binding proteins.

Plasma concentrations of IGF-I are clearly under GH and nutritional control. The ability of GH to stimulate plasma IGF-I is impaired under reduced nutrition (Bass, Oldham, Hodgkinson, Fowke, Sauerwein, Molan, Breier and Gluckman, 1991). The GH/IGF-I axis in post-natal sheep (Bass, Davis, Peterson, Gluckman and Gray, 1984) and cattle (Breier, Bass, Butler and Gluckman, 1986; Elsasser, Rumsey and Hammond, 1989) is sensitive to changes in nutrition. The IGF-I response to GH in lambs on low nutrition showed a declining response with time on limited feed (Figure 5). This decline may be because the nutritional status had not stabilised or alternatively, the initial presence of excess GH maintained the receptor binding potential of the liver for GH, so allowing a declining but positive IGF-I response to GH in the lambs on low nutrition (Bass *et al*, 1991).

Although nutrition has a major effect on growth rate and plasma IGF-I of farm animals, in steers IGF-I levels rise from 6-18months and then plateau regardless of nutritional status (Bass, Carter, Duganzich, Kirton, Breier and Gluckman, 1989). This rise and plateauing of IGF-I plasma concentrations of steers on two different nutritional levels occurs at the same time and does not appear to be affected by nutrition or growth rate.

IGF binding proteins (IGFBP)

IGFs circulate in blood bound to specific binding proteins (Table 1), which may modulate IGF action by restricting the bound hormone to the circulatory system and inhibiting IGF-I cellular action (Baxter and Martin, 1989). However in some studies the reverse has been reported with BP stimulating IGF-I action (Blum,

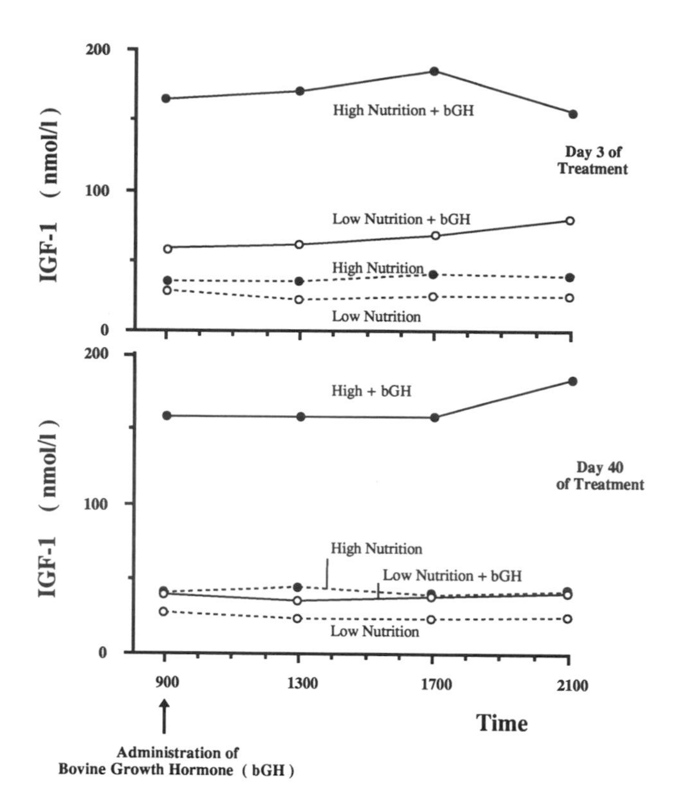

Figure 5 Effect of GH treatment on lambs on high or low nutrition (Bass *et al*, 1991)

Jenne, Keitzmann, Bierich and Ranke, 1987). An explanation of this is that the biological actions of the binding proteins may be altered by their post-translational modification, particularly phosphorylation and the inhibitory/stimulatory actions of IGFBPs may relate to changes in IGFBP phosphorylation (Clemmons, Busby and McCusker, 1991). At present it is unknown whether these changes can be modified by nutritional status. The IGF binding proteins do not merely play a passive role of transporting IGF-I to their receptor binding and it has been suggested that IGFBP bind to specific cell surfaces and extracellular matrices, thus possibly targeting IGFs to specific cells.

Three major families of IGFBP have been identified and preliminary purification and NH_2 sequence data for the ovine IGFBP have been reported (Walton, Grant, Seamark, Francis, Ross, Wallace and Ballard, 1991).

IGFs circulate predominantly in a large complex of about 150 kDa (Table 1). This large unit is composed of two subunits, an acid-labile non-IGF-binding

Table 1 IGF BINDING PROTEINS

		Molecular Weight (k DA)	Binds	Found	Relative Affinity	Biological Effect
Type 2 receptor circulating (Mannose 6 phosphate)		220 kDA	IGF-2	Foetal Plasma	IGF-2 > IGF-1	
→		150 kDA		Blood, Lymph	IGF-1 = IGF-2	Retains IGF in Plasma, Decreases IGF Biological activity
Acid labile Sub-unit (ALS)		85 kDA				
	IGF-1 & 2					
BP 3 (Acid stable)		53 kDA	IGF-1 & 2		IGF-1 = IGF-2	
	BP 1	25 kDA	IGF-1 & 2	Blood, Amniotic Fluid, Milk	IGF-1 = IGF-2	Inhibits IGF action
	BP 2	32 kDA	IGF-2 & 1	Foetal Tissue, Amniotic Fluid, Uterus	IGF-2 > IGF-1	Inhibits IGF action
	Other BPs	20 - 30 kDA	IGF-1 & 2	Blood, Lymph	IGF-1 = IGF-2	Unknown

subunit (85 kDa) and an acid-stabile subunit (BP3 53 kDa) which joins with IGF to form the 150 kDa complex (Baxter and Martin, 1989). The 150 kDa complex carries most of the endogenous circulating IGF (Figure 6).

The half life of the 150 kDa complex in sheep has been shown to be approximately 500-800 min and this was not altered by 48 h of starvation (Hodgkinson, Davis, Burleigh, Henderson and Gluckman, 1987). However the 150 kDa bound IGF-I was shown to increase in sheep when on a high plane of nutrition and was further increased by GH treatment (Hodgkinson *et al*, 1991). IGFBP3, a component of the 150 kDa complex, is IGF-I-dependent and is increased by IGF-I in hypophysectomised and diabetic rats (Zapf, Hauri, Waldvogel, Futo, Husler, Binz, Guler, Schmid and Froesch, 1989). Production of the acid-labile subunit which forms part of the 150 kDa IGFBP complex is under the control of GH which is nutritionally sensitive. The 150 kDa complex level does not vary acutely but is chronically related to GH status (Baxter and Martin,

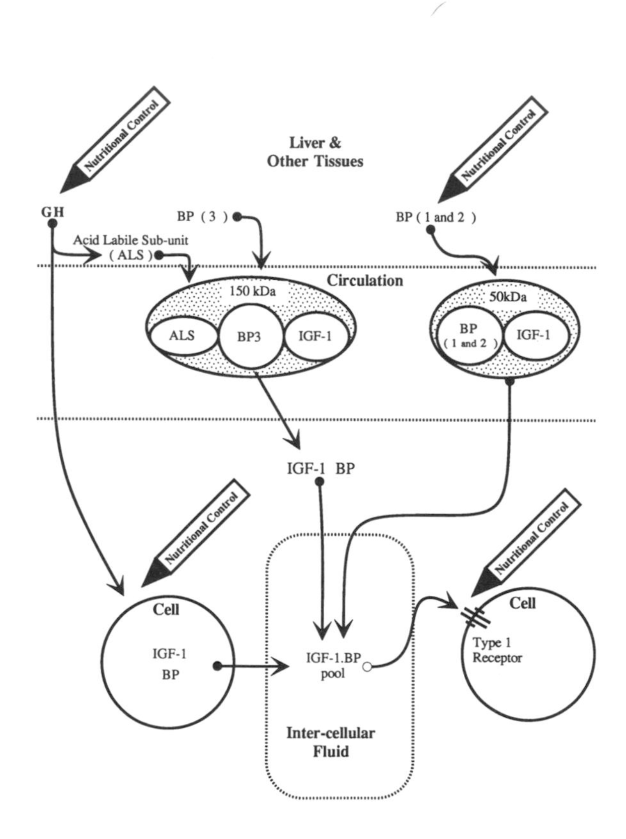

Figure 6 Nutritional control (Nut cont) of insulin-like growth factor (IGF) binding protein (BP)

1989). The 150 kDa BP complex therefore appears to be modulated by the metabolic status of animals through GH. The post-natal appearance of the 150 kDa complex in sheep has been related to the post-natal appearance of the hepatic GH receptor (Butler and Gluckman, 1986). As the 150 kDa complex is mainly maintained within the circulation (Binoux, Hossenlop, Lassarre and Segovia, 1991), its function appears to be to act as a large reservoir of IGF-I. Unlike many other hormones IGF-I is not stored, but immediately secreted from tissues as produced. Although the 150 kDa complex is predominantly retained in the circulation, it has been detected in extravascular fluids such as lymph (Hodgkinson *et al*, 1990) and milk (Baxter and Martin, 1989). IGF-I activity has been shown to be decreased when bound in the complex, with DNA synthesis in liver and skin fibroblasts being reduced and also an inhibition of lipogenesis and glucose oxidation in porcine adipose tissues (Walton, Gopinath and Etherton, 1989). 150 kDa IGF-I complex, which is under metabolic control, has direct effects on IGF-I action as well as maintaining a pool of circulating IGF-I.

BP1 AND 2

The low molecular weight BPs are found in the circulation and are also widely distributed in tissues. The low molecular weight 30-40 kDa BP (BP1/2) (Table 1) usually carry a minor proportion of the circulating IGF (Binoux and Hossenlop, 1988). The metabolic clearance of IGF-I bound to the low molecular weight BPs is only 52 min in sheep and no long-term nutritional effect was seen on their clearance rate in sheep (Hodgkinson *et al*, 1987). Saturation analysis of the 30-40 kDa IGFBP revealed, that despite no obvious changes in bound IGF-I occurring at different planes of nutrition, the available low molecular weight BPs in the 30-40 kDa fraction were elevated in the low nutrition group (Hodgkinson *et al*, 1991). This elevation may be related to low insulin levels in poorly-fed animals as BP1 and 2 have been shown to be inversely related to insulin (Holly and Wass, 1989). In humans, IGFBP undergoes a marked circadian variation (Baxter and Cowell, 1987) which has also been shown to be inversely related to insulin (Holly and Wass, 1989). These observations have led to the suggestion that BP1 is regulated by metabolic (Baxter and Martin, 1989) and/or dietary factors (Busby, Snyder, and Clemmons, 1988). IGFBP1 and 2 are considered to inhibit the cellular actions of IGF-I (Holly and Wass, 1989) although there have been reports of BP1 enhancing IGF-I action (Clemmons *et al*, 1991).

The metabolic control of IGF-I action through specific binding proteins may occur in two physiologically distinct stages. The insulin sensitive BP1/2 can respond rapidly to changes in nutrition, whereas the long-term changes in IGF-I and BP3 production occur via the nutritional down-regulation of GH receptors.

Other IGF-I BPs have been identified from the culture media of different cell types, and these are considered to be different from BP1, 2 and 3. Their biological actions are unknown and whether they are under nutritional control has not been established.

Serum protease

IGFBP3 can be regulated by a specific serum protease (Binoux *et al*, 1991), which has been found in pregnant animals (Rosenfeld, Oh, Beukers, Pham, Cohen, Deal, Fielder, Donovan, Neely, Giudice, Ling and Lamson, 1991) and critically ill patients (Holly, Davies, Cotterill, Coulson, Ross, Miell, Abdulla and Wass, 1991). The IGFBP3 specific protease in critically ill patients has been shown to be dependent on nutritional state with the specific protease activity increasing with time of fasting and decreasing with parenteral feeding (Holly *et al*, 1991). However protease activity has not been demonstrated in ruminants.

Although IGFBP-3 is still able to bind IGF-I after specific proteolysis there is a reduction in binding affinity allowing the easy removal of the IGF-I (Rosenfeld *et al*, 1991). The marked increase in circulating specific protease plus a decrease of IGFBP3 in critically ill patients could increase IGF-I availability from the circulation to tissues from patients in a catabolic state (Holly *et al*, 1991).

Somatomedin inhibitors

The somatomedin inhibitors are peptides between 20-30 kDa which are produced mainly by the liver and are under nutritional control. The inhibitors can blunt the action of IGFs and insulin on cartilage, fat and muscle (Phillips, 1986). Although the IGF inhibitors have similar biological inhibitory actions to the small IGFBPs, purified preparations of the IGF inhibitor do not react with binding protein antisera and do not bind IGF-I in ligand blots, suggesting that they are distinct from IGFBP1/2. Starvation and streptozotocin-induced diabetes have been shown to induce plasma somatomedin inhibitors which would have the effect of reducing the availability of bioactive IGF-I. The effect of diabetes and starvation on the IGF inhibitors indicates that they are modulated by nutrition.

IGF receptors

The insulin and Type 1 IGF receptors are very similar in structure, consisting of two extra-cellular subunits with a molecular weight of 130 kDa and two transmembrane α-subunits of 95 kDa (Pessin and Treadway, 1989). Despite the structural similarity between the insulin and Type 1 receptor the binding affinity for the hormone is very different (Figure 7), with Type 1 receptors preferentially binding IGF-I followed closely by IGF2 and then insulin.

IGF receptors are widely distributed throughout the body, with most tissues containing Type 1 and 2 IGF and insulin receptors (Sara and Hall, 1990). The capacity and affinity of the IGF receptors change during development. In rats Type 2 receptors are dominant in foetal muscle and they decline rapidly early in post-natal life. Type 1 receptors are present in the foetus at term and initially remain high post-natally, declining to adult levels from 4 weeks of age. The

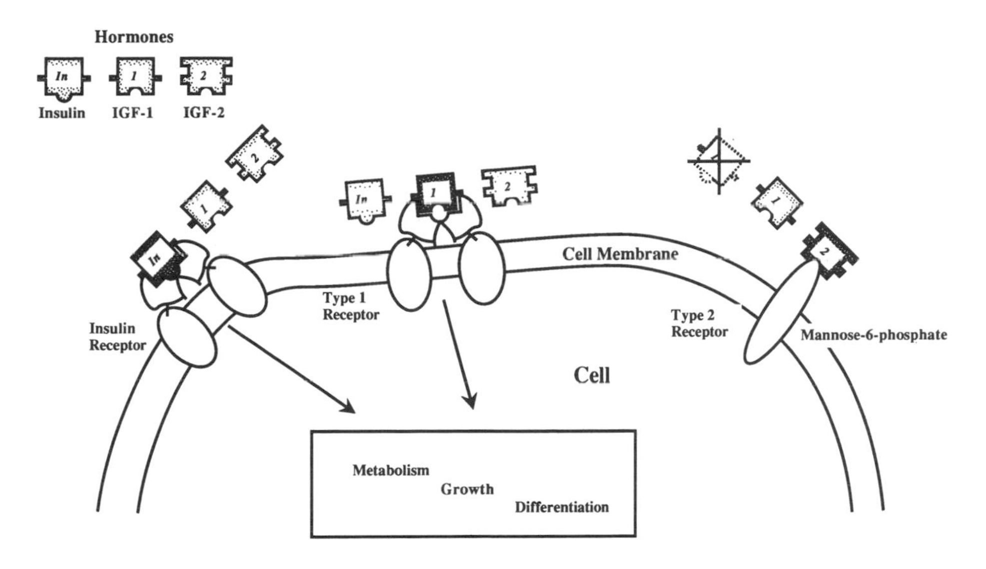

Figure 7 Receptors for insulin and insulin-like growth factors 1 and 2 (IGF-1 & IGF-2) (After DePablo *et al*, 1990)

insulin receptor rises post-natally and reaches a peak at 4 weeks of age (Alexandrides, Moses and Smith, 1989). Preliminary studies using histological autoradiography (Oldham, Bass, Hodgkinson, Hodges, Duganzich and Molan, 1989) indicate that specific binding of IGF-I to skeletal muscle fibres is high in 2-week-old lambs and then decreases, so by 9 months of age the binding of IGF-I is very low in skeletal muscle fibres. However, connective tissue of muscle does not follow the same developmental pattern as muscle fibre, with the ability of connective tissue to bind IGF-I being maintained until 9 months of age.

Evidence for the nutritional control of IGF receptors is limited. Lowe, Adams, Werner, Roberts and LeRoith (1989) demonstrated an increase in IGF-I binding and Type 1 receptor mRNA when rats were fasted. Oldham *et al* (1989) have also reported an increase in IGF-I specific binding (Type 1 receptor) to muscle connective tissue after starvation of 6-month-old lambs (Figure 8). However by nine months, although still able to bind IGF-I to the Type 1 receptor, the connective tissue of ovine skeletal muscle was no longer responsive to changes in nutritional status (Oldham *et al*, 1989). Type 1 IGF receptor capacity and sensitivity to nutrition is probably dependent on the tissue, its degree of maturation, plus its overall biological environment. It would appear that the Type 1 IGF receptors' responses to nutritional changes are not synchronised between muscle fibres and muscle connective tissue (Figure 8). The sensitivity to GH treatment not only varies between tissues within muscles but possibly between muscles. Studies on 5-month-old lambs have shown that GH stimulated collagen

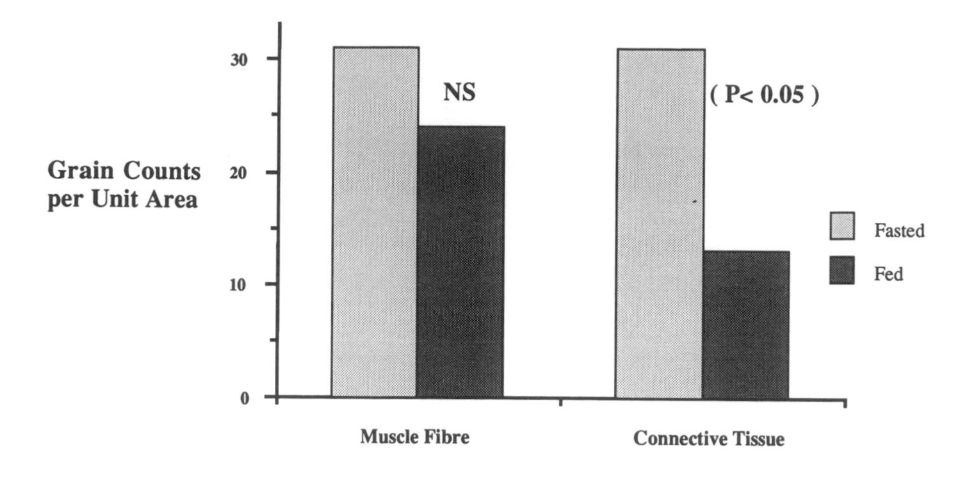

Figure 8 Specific Binding of ^{125}I-IGF-1 to ovine muscle tissues from 6 month old lambs (Oldham *et al*, 1990)

and non-collagen protein synthesis in the *M.bicepsfemoris*, but failed in the *M.semitendinosis* (Pell and Bates, 1987). This indicates that there appears to be a differential response by Type 1 receptors to GH and nutritional status by different muscles and tissues within muscles and this differential response may vary with age. This is in agreement with the findings of Glasscock, Gelber, Lamson, McGee-Tekula and Rosenfeld (1990) who found that at least during the neonatal period not all growing tissues were necessarily pituitary dependent.

The sensitivity of tissues to nutritional status obviously varies, *ie* adipose tissue is very sensitive, as opposed to the relatively insensitive nutritionally-independent growth of neural tissues. It has been proposed that at certain developmental phases of growth, the growth of specific tissues is programmed to be maintained regardless of nutritional state and the growth is possibly fuelled at the expense of other body tisues. An example is skeletal tissue just after birth which continues to grow rapidly even though initially there is a loss in body weight (Bishop, King and Lucas, 1990). Another example of programmed growth may be the seasonal growth of antlers in deer, which are very fast growing. Type 1 and 2 receptors have been localised in different histological zones in the antler tip (Elliott, Oldham, Suttie, Bass, Hodgkinson and Spencer, 1990). Antler growth and binding capacity of the IGF Type 1 and 2 receptors to the antler do not appear to be affected by a significant weight loss induced by underfeeding, indicating that, like neonatal skeletal growth, antler growth continues regardless of short-term changes in nutrient availability.

Neuropeptides controlling GH

Hypophysial stalk transection in pigs decreases episodic secretion of GH and elevates circulating GH concentrations (Klindt, Ford, Berardinelli and Anderson, 1983). The hypothalamus was thus considered to be necessary to control the episodic release and tonic inhibition of basal GH secretion from the pituitary. Two major neurohormones have been identified in the hypothalamus which control GH secretion, growth hormone releasing hormone (GHRH) which stimulates GH and somatostatin (SRIF) which inhibits GH secretion (Buonomo and Baile, 1990). The inhibiting actions of SRIF at physiological concentrations have been difficult to demonstrate in sheep when SRIF was administered IV or ICV (Spencer, Bass, Hodgkinson and Dobbie, 1991), or when SRIF was manipulated by treatment with cysteamine (Spencer, Bass and Hodgkinson, 1989). However, Frohman and Jansson (1986) found that in sheep, in which hypoglycaemia had been induced by insulin, there was a rapid and brief increase of SRIF secretion into the hypophysial portal system, followed by a decline in circulating GH. The decrease found in GH may be because SRIF decreases the GH response to GHRH (Frohman and Jansson, 1986).

Infused metabolites

When the major metabolites are infused or manipulated in animals they have profound effects on the GH and the GH axis. The infusion of glucose into sheep stimulates GH. In some studies hypoglycaemia induced in sheep by infusing insulin results in a rapid but brief stimulation of SRIF, followed by a decline in GH. However, Spencer, Johnston, Berry, Hodgkinson and Bass (1990) found the reverse, with an insulin tolerance test resulting in marked hypoglycaemia and an elevation ($P<0.01$) in plasma GH. An infusion of free fatty acids suppresses GH secretion, which appears to act by stimulating SRIF. Imaki, Shibasaki, Masuda, Hotta, Yamauchi, Demura, Shizume, Wakabayushi and Ling (1986) have shown that anti-SRIF serum abolishes the inhibitory effects of free fatty acids on GH secretion. The stimulatory effect on GH of glucose infusion and the inhibitory effect of FFA infusions provides serum profiles which are at variance with the profiles seen in starved animals, when plasma glucose is low and FFA are high. This difference may be related to the effect of infusion which are short term, whereas the starvation effects do not usually appear for 10-24 h in ruminants and are a result of changes in the GH axis.

Feeding

Plasma GH of lambs (Driver and Forbes, 1981) and sheep (Bassett, 1974b) is decreased immediately after feeding or in anticipation of feeding. This effect is so rapid it is unlikely to be mediated by hormone or metabolic changes and so it has been postulated that the actions are through a neural reflex pathway (Bassett, 1974a). Mosley, Alaniz, Claflin and Krabill (1988) reported that GHRH-induced GH secretion was reduced immediately after feeding in cattle, indicating that SRIF might also be involved in the GH response to feeding.

Feedback

The two main hormones within the GH axis which feedback on the GH axis, are GH and IGF-I. GH exerts negative feedback inhibition on further GH release. This effect is probably mediated through increased release of hypothalamic SRIF (Tannenbaum, 1981). There is also the possibility of a direct effect of GH on the pituitary somatotrophs although this appears unlikely from *in vitro* studies (Buonomo and Baile, 1990).

IGF-I has been shown to exert negative feedback inhibition on GH release, (Abe, Molitch, Van Wyk and Underwood, 1983). The effect in sheep does not appear to be via the hypothalamus (Spencer, Bass, Hodgkinson, Edgley and Moore, 1991) and so appears to be a direct effect on the pituitary. However, Clarke, Corlsson and Robinson (1988) considered that the slow response of GH to changes in IGF-I would preclude it from a major physiological role in the control of GH release.

GH treatment

GH treatment increases milk production and growth of veal calves (Karg, 1987), sheep (Wagner and Veenhuizen, 1978) and pigs (Etherton, Wiggins, Evock, Chung, Rebhun, Walton and Steele, 1987). The increase in growth involves an increase in carcass protein and a decrease in carcass lipid. In cattle, short-term (Breier *et al*, 1988a) and long-term (Elsasser *et al*, 1989) dietary changes alter the IGF-I response to GH treatment. Elsasser *et al* (1989) suggest that plasma IGF-I response to GH is mainly determined by the crude protein in the diet, although this response can be affected by availability of metabolisable energy.

Peters (1986) found that GH treatment decreased average daily gain (ADG) increased lean and decreased fat in steers fed *ad libitum*, but increased ADG with small effects on carcass composition on steers on a restricted diet. These results contrast with those for sheep (Bass *et al*, 1991) where the lambs on the higher plane of nutrition showed the greater weight response and the smaller decrease in carcass fat from GH treatment in comparison with lambs on a low plane of nutrition. The effect of GH on growth at different dietary levels needs to be established for the major farm species, before the maximum effect of GH treatment on lean growth can be determined.

Conclusion

Nutritional regulation is only one of the many interacting systems which control the growth hormone axis and growth in farm animals. In general, farm animals being fed on high energy and high protein diets, have low GH and high IGF-I plasma concentrations.

The main sites for the direct action of nutrition on the GH axis appear to be at the hypothalamus and liver. The effect of glucose and FFA on GH secretion is thought to be mediated by direct or indirect stimulation of the GH inhibitor SRIF in the hypothalamus.

The action of GH is mediated by the effect of nutrition on the GH receptors in the liver and explains the paradox that plasma GH is low in well-fed farm animals when plasma IGF-I is high. The direct effect of nutrition on GH receptors in other tissues is not known, although it presumably interacts with IGF-I production from the liver and possibly other tissues, because the autocrine/paracrine functions of IGF-I must relate to the endocrine role of IGF-I to achieve co-ordinated growth, if IGF-I does play a major role in GH-controlled growth. The main research areas which need extending to further our understanding of the nutritional control of the GH axis are the nutritional interaction with the hypothalamic/pituitary factors and how the IGFBPs are involved in controlling the distribution and action of IGF-I at the cellular level.

The authors are indebted to P.Dobbie for the figures, J.Martyn and M.Bates for the preparation of the manuscript and S.Davis and C.Prosser for useful criticism.

References

Abe, H., Molitch, M.E., Van Wyk, J.J. and Underwood, L.E. (1983) *Endocrinology*, **113**, 1319-1324

Alexandrides, T., Moses, A.C. and Smith, R.J. (1989) *Endocrinology*, **124**, 1064-1076

Barnard, R., Haynes, K.M., Werther, G.A. and Waters, M.J. (1988) *Endocrinology*, **122**, 2562-2569

Barnard, R. and Waters, M.J. (1986) *Biochemical Journal*, **237**, 885-892

Bass, J.J., Carter, W.D., Duganzich, D.M., Kirton, A.H., Breier, B.H. and Gluckman, P.D. (1989) *Livestock Production Science*, **21**, 303-308

Bass, J.J., Davis, S.R., Peterson, A.J., Gluckman, P.D. and Gray, M. (1984) *Proceedings of Endocrine Society of Australia*, **27**, (Suppl), 15

Bass, J.J., Oldham, J.M., Hodgkinson, S.C., Fowke, P.J., Sauerwein, H., Molan, P., Breier, B.H. and Gluckman, P.D. (1991) *Journal of Endocrinology*, **128**, 181-186

Bassett, J.M. (1974a) *Australian Journal of Biological Science*, **27**, 157-166
Bassett, J.M. (1974b) *Australian Journal of Biological Science*, **27**, 167-181
Bassett, J.M., Weston, R.H. and Hogan, J.P. (1971) *Australian Journal of Biological Science*, **24**, 321-330
Bauman, D.E. and Currie, W.B. (1980) *Journal of Dairy Science*, **63**, 1514-1529
Baumann, G., Shaw, M.A. and Winter, R.J. (1987) *Journal of Clinical Endocrinology and Metabolism*, **65**, 814-816
Baxter, R.C, Bryson, J.M. and Turtle, J.R. (1981) *Metabolism*, **30**, 1087-1090
Baxter, R.C. and Cowell, C.T. (1987) *Journal of Clinical Endocrinology and Metabolism*, **65**, 432-440
Baxter, R.C. and Martin, J.L. (1989) *Progress in Growth Factor Research*, **1**, 49-68
Bines, J.A. and Hart, I.C. (1982) *Journal of Dairy Science*, **65**, 1375-1389
Binoux, M., Hossenlopp, P., Lassarre, C. and Segovia, B. (1991) In *Proceedings of Second International Symposium on Insulin-like Growth Factors/Somatomedins*, (chairman E.M. Spencer), San Francisco, p 20
Binoux, M. and Hossenlop, P. (1988) *Journal of Clinical Endocrinology and Metabolism*, **67**, 509-514
Bishop, N.J., King, F.J and Lucas, A. (1990) *Archives of Diseases in Children*, **65**, 707-708
Blum, W.F., Jenne, E., Keitzmann, K., Bierich, J.R. and Ranke, K.M.B. (1987) *Journal of Endocrinology Investigations*, **10**, (Suppl 4), 25 (Abstract)
Boyd, R.D. and Bauman, D.E. (1989) In *Current Concepts of Animal Growth Regulation* (eds D.R. Campion, G.J. Hausman and R.J. Martin), New York, Plenum Publishing Corp, pp 257-293
Breier, B.H., Bass, J.J., Butler, J.H. and Gluckman, P.D. (1986) *Journal of Endocrinology*, **111**, 209-215
Breier, B.H., Gluckman, P.D. and Bass, J.J. (1988a) *Journal of Endocrinology*, **116**, 169-177
Breier, B.W., Gluckman, P.D. and Bass, J.J. (1988b) *Journal of Endocrinology*, **118**, 243-250
Breier, B.H., Gluckman, P.D. and Bass, J.J. (1988c) *Journal of Endocrinology*, **119**, 43-50
Buonomo, F.C. and Baile, C.A. (1990) *Domestic Animal Endocrinology*, **7**, 435-450
Busby, W.H., Snyder, D.K. and Clemmons, D.R. (1988) *Journal of Clinical Endocrinology and Metabolism*, **67**, 1225-1230
Butler, J.H. and Gluckman, P.D. (1986) *Journal of Endocrinology*, **109**, 333-338
Chung, C.S. and Etherton, T.D. (1986) *Endocrinology*, **119**, 780-786
Clark, R.G., Corlsson, L.M.S. and Robinson, I.C.A.F. (1988) *Journal of Endocrinology*, **119**, 201-209
Clark, R.G. and Cronin, M.J. (1991) In *Proceedings of the Second International Symposium on Insulin-like Growth Factors/Somatomedins*, (chairman E.M. Spencer), San Francisco, p 276
Clemmons, D.R. and Van Wyk, J.J. (1981) In *Handbook of experimental pharmacology*, (ed R. Baseiga), New York: Springer, Vol **157**, pp 161-208

Clemmons, D.R., Busby, W.H. and McCusker, R.H. (1991) In *Proceedings Second International Symposium on Insulin-like Growth Factor/Somatomedins*, (chairman E.M. Spencer), San Francisco, p 28

D'Ercole, A.J., Stiles, A.D. and Underwood, L.E. (1984) *Proceedings National Academy Sciences USA*, **81**, 935-939

Daughaday, W.H. and Reeder, C. (1966) *Journal of Laboratory and Clinical Medicine*, **68**, 367-368

Daughaday, W.H., Hall, K., Raben, M.S., Salmon, W.D. Jr., Van Den Brande, J.L. and Van Wyk, J.J. (1972) *Nature*, **235**, 107

De Pablo, F., Scott, L.A. and Roth, J. (1990) *Endocrine Reviews*, **11**, 558-577

Douglas, R.G., Breier, B.H., Ball, K.T., Shaw, J.H.F. and Gluckman, P.D. (1990) *Proceedings New Zealand Society of Endocrinology*, abstract 1

Driver, P.M. and Forbes, J.M. (1981) *Journal of Physiology*, **317**, 413-424

Elliott, J., Oldham, J., Suttie, J., Bass, J.J., Hodgkinson, S.C. and Spencer, G.S.G. (1990) *Proceedings of New Zealand Society of Endocrinology*, **33** (Supplement), abstract 5

Elsasser, T.H., Rumsey, I.S. and Hammond, A.C. (1989) *Journal of Animal Science*, **67**, 128-141

Etherton, T.D., Wiggins, J.P., Evock,C.M., Chung, C.S., Rebhun, J.F., Walton, P.E. and Steele, N.C. (1987) *Journal of Animal Science*, **64**, 433-443

Frohman, L.A. and Jansson, J.O. (1986) *Endocrine Reviews*, **7**, 223-253

Glasscock, G.F., Gelber, S.E., Lamson, G., McGee-Tekula, R. and Rosenfeld,R.G. (1990) *Endocrinology*, **127**, 1792-1803

Gluckman, P.D. (1984) *Journal of Developmental Physiology*, **6**, 301-312

Gluckman, P.D. (1984) In *Oxford Reviews of Reproductive Biology*, (ed I.R. Clarke), Oxford, Canadian Press, **8**, 1-59

Gluckman, P.D., Butler, J.H. and Elliott, T.B. (1983) *Endocrinology*, **112**, 1607-1612

Gluckman, P.D., Douglas, R.G., Ambler G., Breier, B.H., Hodgkinson, S.C., Koea, J. and Shaw, J.H.F. (1991) *Second International Symposium on Growth and Growth Disorders*, Madrid, pp 1-4

Green, H., Morikawa, M. and Nixon T. (1985) *Differentiation*, **29**, 195-198

Hodgkinson, S.C., Bass, J.J., Davis, S.R., Spencer, G.S.G. and Gluckman, P.D. (1990) *Proceedings of 72nd Annual Meeting of the Endocrine Society*, abstract 1138

Hodgkinson, S.C., Bass, J.J. and Gluckman, P.D. (1991) *Domestic Animal Endocrinology* (in press)

Hodgkinson, S.C., Davis, S.R., Burleigh, B.D., Henderson, H.V. and Gluckman, P.D. (1987) *Journal of Endocrinology*, **115**, 233-240

Holly, J.M.P. and Wass, J.A.H. (1989) *Journal of Endocrinology*, **122**, 611-618

Holly, J.M.P., Davies, S.C., Cotterill, A.M., Coulson, V.J., Ross, R.J.M., Miell, J.P., Abdulla, A.F. and Wass, J.A.H. (1991) In *Proceedings of the Second International Symposium on Insulin-like Growth Factors/Somatomedins*, (chairman, E.M. Spencer), San Francisco, p 219

Hua, K.M., Bass, J.J., Hodgkinson, S.C., Spencer, G.S.G. and Ord, R. (1991) *Endocrine Society* (in press)

Imaki, T., Shibasaki, T., Masuda, A., Hotta, M., Yamauchi, N., Demura, H., Shizume, K., Wakabayushi, K. and Ling, N. (1986) *Endocrinology*, **118**, 2390-2399

Karg, H. (1987) *Ubers Trevevnahrung*, **15**, 1-28

Klindt, J., Ford, J.J., Berardinelli, J.G. and Anderson, L.L. (1983) In *Proceedings of the Society for Experimental Biology and Medicine*, **172**, 508-513

Leung, D.W., Spencer, S.A., Cachianes, G., Hammonds, R.G., Collins, C., Henzel, W.J., Barnard, R., Waters, M.J. and Wood, W.I. (1987) *Nature*, **330**, 537-543

Lim, L. and Waters, M. (1989) *Proceedings of the Combined Meeting of the Endocrine Society of Australia and New Zealand Society of Endocrinology*, **32**, abstract 204

Lowe, W.L. Jnr, Adams, M., Werner, H., Roberts, C.T. Jnr, and Le Roith, D. (1989) *Journal of Clinical Investigation*, **84**, 619-626

Maes, M., Underwood, L.E. and Ketelslegers, J.M. (1983) *Journal of Endocrinology*, **97**, 243-252

Maes, M., Underwood, L.E. and Ketelslegers, J.M. (1986) *Endocrinology*, **118**, 377-382

Mosley, W.M., Alaniz, G.R., Claflin, W.H. and Krabill, L.F. (1988) *Journal of Endocrinology*, **117**, 253-259

Oddy, V.H., Warren, H.M., Moyse, K.J. and Owens, P.C. (1991) In *Second International Symposium on Insulin-like Growth Factors/Somatomedins*, (chairman, E.M. Spencer), San Francisco, p 28

Oldham, J., Bass, J.J., Hodgkinson, S.C., Hodges, A., Duganzich, D. and Molan, P.C. (1989) *Proceedings of the Combined Meeting of the Endocrine Society of Australia and New Zealand Society of Endocrinology*, **32**, abstract 212

Pell, J.M. and Bates, P.C. (1987) *Journal of Endocrinolgy*, **115**, R1-R4

Pessin, J.E. and Treadway, J.L. (1989) In *Molecular and Cellular Biology of Insulin-like Growth Factors and their Receptors*, (eds D. LeRoith and M.K. Raizada), New York, Plenum Press, pp 261-284

Peters, J.P. (1986) *Journal of Nutrition*, **116**, 2490-2503

Phillips, L.S. (1986) *Metabolism*, **35**, 78-87

Rosenfeld, R.G., Oh, Y., Beukers, M.W., Pham, H., Cohen, P., Deal, C., Fielder, P., Donovan, S., Neely, E.K., Giudice, L.C., Ling, N. and Lamson, G. (1991) In *Proceedings of the Second International Symposium on Insulin-like Growth Factors/Somatomedins*, (chairman E.M.Spencer), San Francisco, p 27

Sara, V.K. and Hall, K. (1990) *Physiological Reviews*, **70**, 591-614

Sauerwein, H., Breier, B.H., Bass, J.J. and Gluckman, P.D. (1991) *Acta Endocrinologica* (in press)

Spencer, G.S.G., Hodgkinson, S.C. and Bass, J.J. (1991) *Endocrinology*, **128**, 2103-2109

Spencer, G.S.G., Bass, J.J. and Hodgkinson, S.C. (1989) *Proceedings of the Combined Meeting of the Endocrine Society of Australia and the New Zealand Endocrine Society*, 32, abstract A198

Spencer, G.S.G., Bass, J.J., Hodgkinson, S.C. and Dobbie, P.M. (1991) *Domestic Animal Endocrinology*, **8**, 377-383

Spencer, G.S.G., Bass, J.J., Hodgkinson, S.C., Edgley, W.H.R. and Moore, L.G. (1991) *Domestic Animal Endocrinology*, **8**, 155-160

Spencer, G.S.G., Johnston, S.J., Berry, C., Hodgkinson, S.C. and Bass, J.J. (1990) *Proceedings of New Zealand Society of Endocrinology*, **33** (Supplement), abstract 19

Spencer, S.A., Hammonds, R.G., Henzel, W.J, Rodriguez, H., Waters, M.J. and Wood, W.I. (1988) *Journal of Biological Chemistry*, **263**, 7862-7867

Tannenbaum, G.S. (1981) *Endocrinology*, 108, 76-82

Thomas, G.B., Mercer, J.B., Karalis, T., Rao, A., Cummins, J.T. and Clarke, I.J. (1990) *Endocrinology*, **126**, 1361-1367

Tindal, J.S. and Yokoyama, A. (1964) *Journal of Endocrinology*, **31**, 45-51

Trenkle, A. (1976) *Journal of Animal Science*, **43**, 1035-1043

Vernon, R.G. (1982) *International Journal of Biochemistry*, **14**, 255-258

Wagner, J.F. and Veenhuizen, E.L. (1978) *Journal of Animal Science*, **47** (Suppl), 397-403

Walton, P.E., Grant, P., Seamark, R.F., Francis, G.L., Ross, M., Wallace, J.G. and Ballard, F.J. (1991) In *Second International Symposium on Insulin-like Growth Factors/Somatomedins*, (chairman E.M. Spencer), San Francisco, Abstract C57 p 257

Walton, P.E., Gopinath, R. and Etherton, T.D. (1989) *Proceedings of the Society of Experimental Biology and Medicine*, **190**, 315-319

Zapf, J., Hauri, C., Waldvogel, M., Futo, E., Husler, H., Binz, K., Guler, H.P., Schmid, C. and Froesch, E.R. (1989) *Proceedings of the National Academy of Sciences of the USA*, **86**, 3813-3817

9
METABOLIC EFFECTS OF GUT PEPTIDES

L.M. MORGAN, J. OBEN, V. MARKS
School of Biological Sciences, University of Surrey, Guildford, Surrey GU2 5XH, UK

and J. FLETCHER
Unilever Research, Colworth House, Sharnbrook, MK44 1LQ, UK

Introduction

The gut could fairly be described as the body's largest endocrine organ. The very term "hormone" was first used by Bayliss and Starling in 1902 to describe the action of the gastrointestinal hormone secretin, although the firm establishment of the gut's endocrine status is comparatively recent. Most of our knowledge has accumulated over the last two decades, when the advent of new investigative technologies has led to an explosion of information about individual gut peptides. In particular, immunological techniques have led to the precise localisation of peptides in endocrine or paracrine cells and in neurones. Advances in molecular biology have provided a better understanding of the transcription, translation and tissue expression of gut peptides. In contrast, the physiological role of many gut peptides remains elusive, particularly in the case of the neuronal gut peptides, whose functions are only just beginning to be elucidated. Many of the biological functions of the gut hormones are confined to the gastrointestinal tract and are concerned with the absorption and digestion of food. However, there is increasing evidence that the endocrine gut hormones, at least, play a metabolic role outside the gastrointestinal tract. The following discussion focuses on this area, in particular the role of gastrointestinal hormones in stimulating insulin secretion, their effects on adipose tissue and their possible role in appetite regulation.

Characteristics of gut hormones

The gastrointestinal hormones are low molecular weight polypeptides with a chain length of generally less than 50 amino acids. They were first identified within gut endocrine cells, scattered diffusely throughout the epithelium lining the gastrointestinal tract (Feyrter, 1938). Gut hormones have, however, subsequently been found to have a much wider distribution (Table 1). In addition to endocrine mucosal cells, peptides have been isolated from neural tissue both within the gut and throughout the entire central nervous system, including brain tissue (Pearse, 1984; Keast, Furness and Costa, 1985). Many of the gut peptides can thus act

Table 1 DISTRIBUTION AND FUNCTIONAL CATEGORIES OF GUT PEPTIDES

Peptide	Distribution			Functional Category
	Brain	Gastro-Intestinal Tract		
		Neural	Endocrine	
Secretin			+	Endocrine
Glucagon peptides			+	Endocrine
GIP			+	Endocrine
Motilin			+	Endocrine/ Paracrine
Gastrin	?+		+	Endocrine
PP Family	+	+	+	Endocrine/ Neurocrine
CCK	+	+	+	Endocrine/ Neurocrine
Somatostatin	+	+	+	Endocrine/ Paracrine/ Neurocrine
Opioid peptides	+	+		Neurocrine
Tachykinins	+	+		Neurocrine
VIP	+	+		Neurocrine

as paracrine or neurocrine transmitters as well as in an endocrine manner. Studies of their distribution have supported the hypothesis that many of the secretory, absorptive and motor functions of the digestive tract are regulated by the interaction between the endocrine and nervous systems. For example, gastrin is a major endocrine stimulator of parietal gastric acid secretion. However gastrin secretion is itself modulated by somatostatin, acting from adjacent cells in a paracrine manner to inhibit gastrin release. In contrast, gastrin secretion is increased by cholinergic stimulation, which inhibits somatostatin secretion and stimulates the release of the neurotransmitter peptide bombesin, or gastrin releasing peptide, which in turn stimulates gastrin secretion (Makhlouf and Grider, 1989). Endocrine gut peptides are involved in biological actions outside the gut, and in addition several gut/brain neuropeptides may be involved in the regulation of such centrally mediated functions as food intake.

Many of the gastrointestinal hormones show a large degree of structural homology and can be divided up into distinct families on the basis of similarities in their amino acid sequence. There is also considerable overlap in the biological activities of hormones in the same group. The secretin family form one such group, consisting of secretin, gastric inhibitory polypeptide (GIP), vasoactive intestinal peptide (VIP), peptide histidine methionine (PHM), and the glucagon-like peptides. The glucagon-like peptides were initially characterised in terms of their ability to cross-react with antisera raised against pancreatic glucagon (Samols, Tyler, Marri and Marks, 1965). The mammalian glucagon precursor is processed differently in the pancreas and small intestine (Mojsov, Heinrich, Wilson, Ravazzola, Orci and Habener, 1986). In the pancreas the major biologically active hormone is glucagon. In the intestine the main products are glicentin (Thim and Moody, 1981), oxyntomodulin (Bataille, Coudray, Carlqvist, Rosselin and Mutt, 1982) and two glucagon-like peptides GLP-1 and GLP-2 (Ørskov, Holst, Knuhtsen, Baldiserra, Poulsen and Vagn Nielsen, 1986). GLP-1 is further cleaved and the major natural form found in porcine intestine is GLP-1 (7-36) amide from which the N-terminal hexapeptide of the larger form has been deleted (Holst, Ørskov, Nielson and Schwartz, 1987). It is this truncated form of GLP-1 which is the major circulating form following ingestion of food (Ørskov, Holst, Poulsen and Kirkegaard, 1987). It has recently attracted attention because of its potency in stimulating insulin secretion.

Nutrient ingestion is a major stimulus to gut hormone secretion and the secretory response can be very selective. For example, GIP secretion is stimulated in response to fat and carbohydrate ingestion. However, it is stimulated by carbohydrates only after their conversion to actively-transported monosaccharides, such as glucose and galactose, and not by passively-absorbed sugars such as fructose, or polyol sweeteners such as xylitol (Sykes, Morgan, English and Marks, 1980; Salminen, Salminen and Marks, 1982). Fat-induced GIP secretion is stimulated by long-chain fatty acids and triglycerides but not by short or medium chain fatty acids (Kwasowski, Flatt, Bailey and Marks, 1985). There is also some evidence that chronic dietary change can affect gut hormone secretion. Increasing the fat content of the diet, for example, has been shown to increase nutrient stimulated GIP secretion in both experimental animals (Hampton, Kwasowski, Tan, Morgan and Marks, 1983; Ponter, Salter, Morgan and Flatt, 1990) and man (Morgan, Hampton, Tredger, Cramb and Marks, 1988).

The entero-insular axis

A major, if indirect, means by which gut hormones can exert their influence beyond the gastrointestinal tract is via the potentiation of insulin secretion. Oral glucose is much more effective in raising circulating insulin levels than intravenous glucose given in amounts sufficient to produce similar degrees of arterial hyperglycaemia (McIntyre, Holdsworth and Turner, 1964; Elrick, Stimmler, Hlad and Arai, 1964). This ability of the gut to stimulate insulin secretion has been termed the entero-insular axis (Unger and Eisentraut, 1969). Gut peptides potentiate insulin secretion via both neural and endocrine pathways, but in man, the endocrine role of the gut hormones (collectively called incretins) is considered to be quantitatively the more important (Clark, Wheatley, Brons, Bloom and Calne, 1989).

All members of the secretin family of gut peptides have the ability, to a greater or lesser extent, to stimulate insulin secretion. However, in order for them to qualify for a physiological role as endocrine mediators of insulin release, the peptides must not only stimulate insulin release at concentrations that circulate *in vivo*, but must also be released in response to the ingestion of carbohydrate. GIP and GLP-1 7-36 amide are the only two members to date which qualify on these grounds as incretins.

GIP is generally recognised as being a major component of the entero-insular axis and until recently has been thought to be the peptide with the most potent insulin-stimulating activity secreted in response to glucose. Originally characterised in terms of its gastric acid inhibitory properties, GIP was subsequently shown to have a powerful insulinotropic effect. Both natural porcine and synthetic human GIP have been shown to potentiate glucose-induced insulin secretion in physiological concentrations in man (Jones, Owens, Moody, Luzio, Morris and Hayes, 1987; Füessl, Yiangou, Ghatei, Goebel and Bloom, 1988). Fat is a more potent stimulus to GIP secretion in man than is carbohydrate (Penman, Wass, Medback, Morgan, Lewis, Besser and Rees, 1981) but in other animals, for example rats and pigs, who customarily consume a diet with a very much lower fat content than western man, the situation is reversed, and carbohydrate provides the greater stimulus to GIP secretion (Hampton *et al*, 1983 ; Ponter *et al*, 1990). Post-prandial circulating GIP levels after oral glucose in man are positively correlated with the size of the glucose load ingested (Hampton, Morgan, Tredger, Cramb and Marks, 1986). Recent measurements of rates of GIP secretion in pigs in which post-prandial glucose absorption rates have also been quantitated, have also shown a close correlation between the rates of glucose absorption and GIP secretion (Morgan, Roberts, Low and Ellis, unpublished). GIP potentiation of insulin secretion may therefore be most marked when large carbohydrate loads are ingested, or when mixed meals of fat and carbohydrate are consumed.

Studies involving immunoneutralisation of endogenous GIP in rodents show that GIP accounts for about 50% of the augmentation of insulin release seen after the administration of intraduodenal compared with intravenous glucose (Ebert and Creutzfeldt, 1982; Ebert, Unger and Creutzfeldt, 1983). Recent infusion studies suggest that the insulin secretory potency of GLP-1 (7-36) amide is more powerful than GIP in molar terms in isolated perfused pancreas preparations (Shima, Hirota and Ohboshi, 1988) or, *in vivo*, in human volunteers, (Kreymann, Williams, Ghatei and Bloom, 1987). Circulating levels of GLP-1 (7 - 36) amide do not however rise as high as GIP post-prandially in man (Kreymann *et al*, 1987; Takahashi, Manak, Katsuyuki Fukase, Tominaga, Sasaki, Kawai, and Ohashi, 1990). The experimental data on GLP-1 (7-36) amide are still very scanty, but present evidence suggests that it is likely to play an important role in the entero-insular axis, as a major incretin of the lower gut, where it is found in the highest concentrations (Kreymann, Yiangou, Kanse, Williams, Ghatei and Bloom, 1988). Studies in rat isolated perfused pancreas suggest a synergistic stimulatory effect of GLP-1 (7-36) amide and GIP on insulin secretion, supporting the concept of several incretin factors in the entero-insular axis with functional interactions at pancreatic B cell level (Fehmann, Göke, Göke, Trautman and Arnold, 1989).

Although the endocrine arm of the entero-insular axis is better characterised, the pancreas is innervated by peptidergic neurones, many of which contain gut peptides that function as neurotransmitters. The carboxyl-terminal peptides of cholecystokinin (CCK), CCK-4 and CCK-8 are potent insulin stimulators *in vitro*, and both CCK and VIP have been implicated as neurotransmitters stimulating insulin secretion (Hermansen, 1984), although their precise role remains speculative. There is evidence that single neurones can contain more than one neurotransmitter (Furness and Costa, 1987) and it is quite likely that the function of these gut peptides is to modulate the action of the classical adrenergic and cholinergic neurotransmitters.

Effects of gut hormones on adipose tissue metabolism

Evidence is beginning to accumulate implicating gut peptides in the regulation of adipose tissue metabolism. Both GIP and GLP-1 (7-36) amide, in addition to their role in stimulating insulin secretion, demonstrate direct anabolic effects on adipose tissue. GIP has been shown to stimulate the synthesis and release of lipoprotein lipase (LPL) by cultured mouse pre-adipocytes (Eckel, Fujimoto and Brunzell, 1978), to stimulate fat incorporation into rat epididymal adipose tissue (Beck and Max, 1983) and to promote the clearance of chylomicron triglycerides from the circulation in dogs (Wasada, McCorkle, Harris, Kawai, Howard and Unger, 1981), as well as inhibiting glucagon stimulated lipolysis in adipose tissue (Dupré, Greenidge, McDonald, Ross and Rubinstein, 1976). We have recently shown (Oben, Morgan, Fletcher and Marks, 1989) that GIP and GLP-1 (7-36) amide, in common with insulin but in contrast to GLP-2 and glucagon, stimulate fatty acid synthesis in explants of rat adipose tissue, as measured by the incorporation of [^{14}C]-acetate into saponifiable fat (See Figure 1). Both GIP and GLP-1 (7 - 36) amide had the ability to stimulate fatty acid synthesis within the physiological ranges of the circulating hormones. At lower (0.1 nM) concentrations of the hormone, GLP-1 (7 - 36) amide was a more potent stimulator of fatty acid synthesis than GIP, consistent with the observation that GLP-1 (7-36) amide circulates in man at lower concentrations than GIP. GLP-1 (7-36) amide and GIP may thus contribute *in vivo* to a more effective uptake of glucose post-prandially by virtue of increased insulin secretion and, in addition, their direct effect on adipose tissue may enhance the effect of insulin on fatty acid synthesis from glucose as precursor.

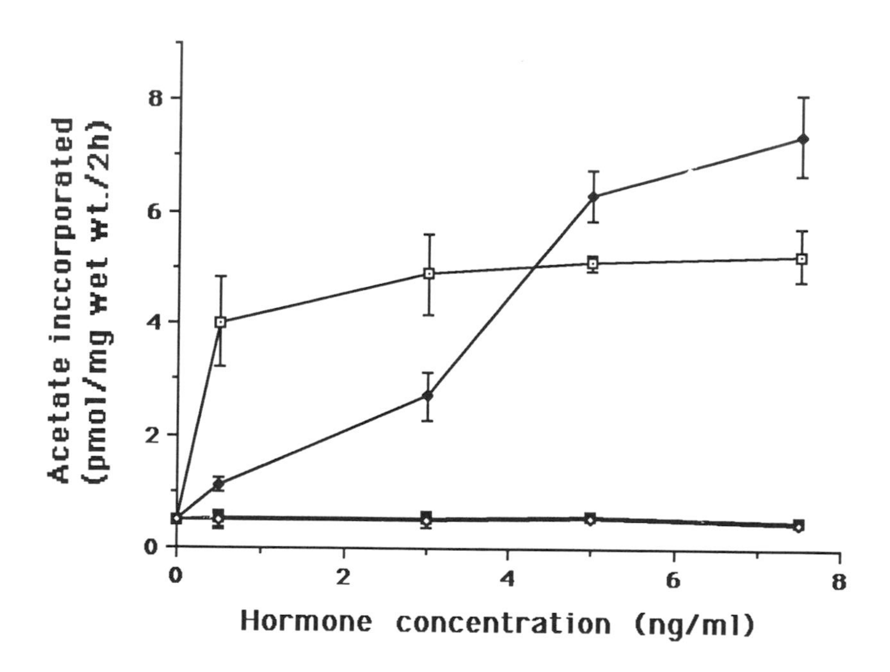

Figure 1 Comparison of the effects of GIP (...♦...), GLP-1 (7-36) amide (..□..), GLP-2 (..■.) and glucagon (.◊) on [14C] acetate incorporation into fatty acids in explants of rat omental adipose tissue.
Values are mean ± SEM (n = 6).

Dietary modification of post-heparin lipoprotein lipase activity has previously been demonstrated in man (Romsos and Leveille, 1975) and adipose tissue LPL activity has been shown to be influenced by both insulin and steroid hormones, in addition to GIP (Rebbuffe-Scrive, 1987). We have shown that GIP, in common with insulin, stimulated LPL activity in explants of rat adipose tissue. However, GLP-1 (7-36) amide over the same concentration range as GIP (0 - 4 nM) was without effect on LPL activity. Rats which had been subjected to a high fat diet for 14 days exhibited significantly elevated basal and hormone stimulated LPL activity compared with control animals (Figure 2). A high fat diet, similar to the

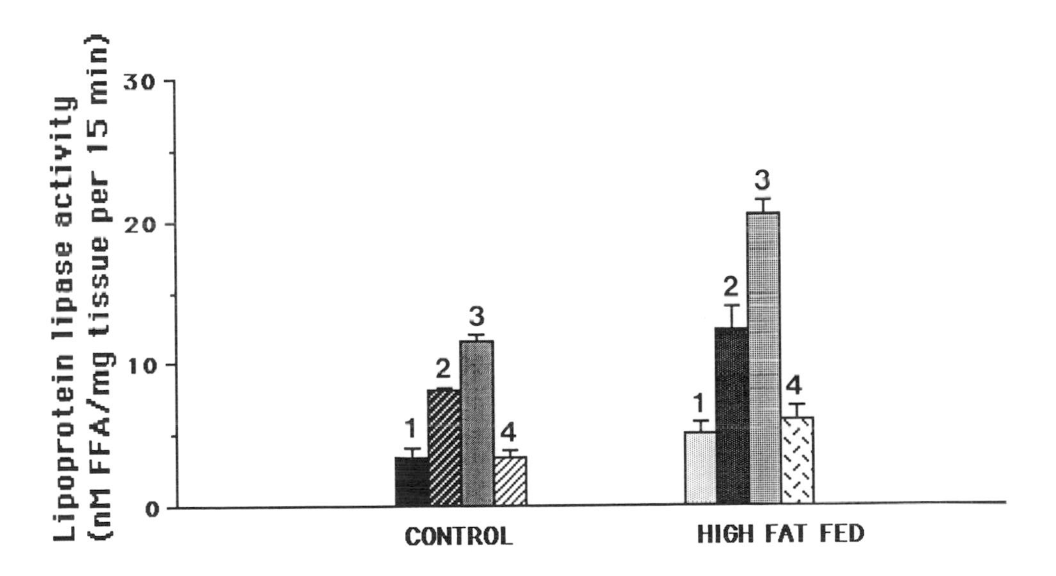

Figure 2 The effects of incubation with insulin, GIP or GLP-1 (7-36) amide on lipoprotein lipase (LPL) activity in explants of rat omental adipose tissue, in rats fed standard rat chow and those whose diets were supplemented with 1ml triolein daily for 14 days.
1 = basal LPL activity, 2 = Insulin (1 nM), 3 = GIP (4 nM), 4 = GLP-1 (7-36) amide (4 nM). Mean ± SEM (n = 6).

one used in this study, has been shown to increase nutrient stimulated GIP and insulin secretion in rats (Hampton *et al*, 1983). Under conditions of high fat feeding, higher circulating levels of GIP and insulin, together with an enhanced sensitivity of LPL to these hormones, may therefore facilitate the uptake of circulating triglycerides and contribute to increased weight gain.

Regional fat distribution has emerged in man as an independent risk factor predisposing towards a variety of metabolic disorders, including maturity onset diabetes (Kissebah, Peiris and Evans, 1989) and cardiovascular disease (Peiris, Sothmann, Hoffmann, Hennes, Wilson, Gustafson and Kissebah,1989). Excessive intra-abdominal fat, in particular, is associated with increased risk. Regional site differences in adipose tissue metabolism have been demonstrated (Smith, 1985). The mechanism by which fat is preferentially deposited in various adipose tissue sites is not known, although androgens and oestrogens play a part in this determination, possibly by altering the activity of LPL (Lithell, 1987). Our studies in rats have indicated that there are regional site differences in adipose tissue sensitivity to gut peptides. *De novo* fatty acid synthesis in omental adipose tissue is more sensitive to the stimulatory action of GLP-1(7-36) amide and GIP than is subcutaneous adipose tissue; similarly, GIP-stimulated LPL activity is greater in omental than subcutaneous adipose tissue. Site differences in the hormonal sensitivity of adipose tissue may be species specific. Preliminary findings in humans suggest that LPL activity in omental adipose tissue may also be more sensitive to GIP than subcutaneous fat. In unweaned piglets, however, we have found that fatty acid synthesis in subcutaneous adipose tissue taken from the belly region is not responsive to stimulation by GIP, in contrast with back fat, where 4 nM GIP causes an approximately six-fold increase in the rate of fatty acid synthesis. Pigs are born with little fat but make rapid fat gains in the first three weeks of life (Wood and Groves, 1965). This site sensitivity to GIP in pigs may be of particular importance in their fat deposition, as the bulk of their fat is deposited as back fat as a result of *de novo* fatty acid synthesis.

Abnormalities of GIP secretion or control have been implicated in the hyperinsulinaemia characteristic of obesity. Considerable evidence has accumulated in some genetically obese rodents, although the evidence in man is less secure (Morgan, Flatt and Marks, 1988). The genetically obese hyperglycaemic (ob/ob) mouse exhibits hyperinsulinaemia, insulin resistance and hyperplasia of intestinal GIP, but not GLP-1 (7-36) amide secreting cells (Bray and York, 1979; Flatt, Bailey, Kwasowski, Swanston-Flatt and Marks, 1983). We have found that adipose tissue from obese (ob/ob) mice, in contrast to their thin litter mates, is unresponsive to the stimulation of fatty acid synthesis and activation of LPL by insulin and GIP. In contrast GLP-1 (7-36) amide retained the ability to stimulate adipose tissue fatty acid synthesis in obese animals although fatty acid synthesis rates were lower in the adipose tissue taken from obese animals (Figure 3). Prolonged exposure to high circulating levels of GIP and insulin may have induced resistance to these hormones in the adipose tissue of obese animals. Other animal models of obesity, for example the Zucker fatty (fa/fa) rat have no obvious change in circulating GIP concentrations (Morgan, 1979; Chan, Pederson,

Buchan, Tubesing and Brown, 1984). These animals, in contrast to ob/ob mice, exhibit increased sensitivity to the insulin-releasing action of GIP (Chan *et al*, 1984) and its stimulatory action on fatty acid incorporation into adipose tissue (Beck and Max, 1987).

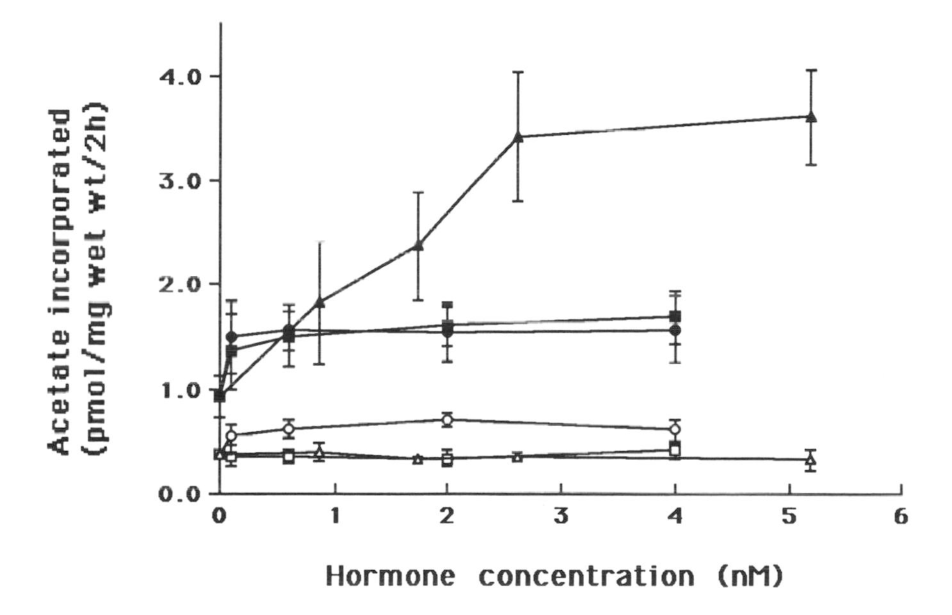

Figure 3
The effects of GIP, GLP-1 and insulin on [14C] acetate incorporation into fatty acids in explants of omental adipose tissue in obese (ob/ob) mice and their thin littermates.
(Obese = ...△.. insulin, .□.GIP, .○..GLP-1 (7-36) amide;
Lean = ...▲.. insulin, ..■.GIP, .●..GLP-1 (7-36) amide)
Mean ± SEM,(n = 6.)

In contrast to the anabolic role played by GIP and GLP-1 (7-36) amide in adipose tissue, some peptides show catabolic effects. The N-terminal portion of CCK has been reported to stimulate lipolysis in human adipose tissue (Richter and Schwandt, 1989). The consequent release of free fatty acids could have a satiating effect which would reinforce the direct action of CCK as a satiety hormone, described in the following section. Both glucagon (Billington, Bartness, Briggs, Levine and Morley, 1987) and peptide YY (Morley and Flood, 1987) induce weight loss when administered chronically, effects which may be secondary to the activation of brown adipose tissue. Peptide YY is present in the blood vessels associated with brown adipose tissue and it may induce weight loss by acting synergistically with insulin to activate brown adipose tissue. However, peptide YY has been reported to inhibit lipolysis in white adipose tissue (Valet, Berlan, Beauville, Crampes, Montastruc and Lafontan, 1990) and its physiological role is obscure. Both glucagon and peptide YY have also been implicated in the control of appetite regulation.

Regulation of food intake by gut peptides

It is generally accepted that food intake is regulated by a central feeding drive which is held in check during feeding by a peripheral satiety system responsible for terminating the meal (Morley, 1980). Studies in the control of food intake initially focused on potential metabolic signals, linking the effects of ingested nutrients and subsequent changes of cellular energy metabolism with satiety. However, with the availability of relatively pure peptides for experimentation, it has become clear that a major component of the peripheral satiety system appears to be the release of gastrointestinal peptides in response to the passage of food through the gut. These two components of the satiety system, peptidergic and metabolic, are not mutually exclusive and a combination of all three factors is responsible for the regulation of food intake in any given individual.

A number of gut peptides released in response to food ingestion have been shown to modulate food intake when adminstered exogenously. Considerable evidence has accumulated to implicate CCK as a physiological satiety peptide. Since the report in 1973 that CCK-8 reduced food intake in rats (Gibbs, Young and Smith, 1973), CCK has been shown to decrease food intake in multiple species, including man (Morley, Levine, Bartness Nizielski, Shaw, and Hughes, 1985; Shaw, Hughes, Morley, Levine, Silvis, and Shafer, 1985). CCK appears to be more potent when administered peripherally than when injected directly into the brain, suggesting its major site of action for the satiety effect is peripheral rather than central. Intraperitoneal injections of CCK in animals produce the greatest satiety, implicating the GI tract and surrounding structures in its site of action. However, there are species differences and in sheep the brain appears to be CCK's predominant site of action (Morley *et al*, 1985). Selective vagotomy attenuates the effect of CCK on food intake (Smith, Jerome, Cushin, Eterno and Simansky, 1981). The vagus nerve contains CCK receptors (Zarbin, Wamsley,

Innis and Kuhar, 1981). CCK may therefore act directly in a paracrine manner to activate the vagus, which in turn sends messages modulating the central feeding drive within the brain. The lowest doses of CCK that induce satiety produce circulating CCK levels similar to those observed after feeding (Smith, Greenberg, Falusco, Gibbs, Liddie and Williams, 1985) and are within the range expected for a paracrine effect. Although CCK remains effective in reducing meal size over relatively long intervals, body weight remained stable when CCK was used alone in a rodent study as the animals compensated by eating more frequently (West, Fey and Woods, 1984).

In addition to CCK, other peptides including gastrin releasing peptide and somatostatin have been reported to reduce food intake in animal studies (Morley, 1987; Lotter, Krinsky, McKay, Treneer, Porter and Woods, 1981). They are less potent on a molar basis than CCK, but it is possible that gut hormones released from various levels of the gastrointestinal tract during feeding can act together in an additive manner to terminate a meal.

Whilst the majority of gut peptides studied exhibit a satiety effect, some actually increase meal size. These include the opioid peptides and peptides of the pancreatic polypeptide family. These peptides must be administered centrally as opposed to peripherally to be effective and their mode of action must therefore be fundamentally different from the majority of the satiety peptides. Interaction between a number of neurotransmitters, including norepinephrine, dynorphin and neuropeptide Y is believed to be responsible for the feeding drive, which is located in a number of anatomical sites within the brain including the paraventricular and ventromedial nuclei of the hypothalamus (Morley, 1987).

Conclusions

Gut peptides have a very wide distribution, as evidenced by their occurrence throughout the central nervous system in addition to endocrine and neural elements within the gastrointestinal tract. Their biological activities within the gut, concerned with regulating the secretory, absorptive and motor functions of the gastrointestinal tract, are the result of complex interactions between endocrine, neurocrine and paracrine systems.

Gut peptides, however, also have a wider role outside the GI tract, concerning the metabolic fate of ingested nutrients. The entero-insular axis makes a considerable contribution to total insulin secretion, modifying the pancreatic insulin response against the "set" determined by the circulating blood glucose concentration. In addition, the two major hormones of the entero-insular axis, GIP and GLP-1 (7-36) amide show direct anabolic effects on adipose tissue in common with insulin, including the stimulation of fatty acid synthesis and incorporation and activation of lipoprotein lipase. The observation that GIP secretion is determined very selectively by nutrient composition and can be modified by chronic dietary changes, provides a mechanism linking adipose tissue metabolism to food intake, and has implications concerning weight gain and obesity. The demonstration of site differences in the sensitivity of adipose tissue to gut hormones may have implications regarding our understanding of the links between diet and cardiovascular disease risk. Gut peptides are also concerned with the very regulation of appetite itself, and evidence is accumulating that their combined influence on metabolic events outside the gut is a very significant one.

References

Bataille, D., Coudray, A., Carlqvist, M., Rosselin, G. and Mutt, V. (1982) *FEBS Letters*, **146**, 73-78

Beck, B. and Max, J. (1983) *Regulatory Peptides*, **7**, 3-8

Beck, B. and Max, J.P. (1987) *Cellular and Molecular Biology*, **33**, 555-562

Billington, C.J., Bartness, T.J., Briggs, J., Levine, A.S. and Morley, J.E. (1987) *American Journal of Physiology*, **252**, R160-165

Bray, G.A. and York, D.A. (1979) *Physiological Reviews*, **59**, 719-809

Chan, C.B., Pederson, R.A., Buchan, A.M.J., Tubesing, K.B. and Brown, J.C. (1984) *Diabetes*, **33**, 536-542

Clark, J.D.A., Wheatley, T., Brons, I.G.M., Bloom, S.R. and Calne, R.Y (1989) *Diabetic Medicine*, **6**, 813-817

Dupré, J., Greenidge, N., McDonald, T.J., Ross, S.A. and Rubinstein, D. (1976) *Metabolism*, **25**, 1197-1199

Ebert, R. and Creutzfeldt, W. (1982) *Endocrinology*, **111**, 1601-1606

Ebert, R., Unger, R.H. and Creutzfeldt, W. (1983) *Diabetologia*, **24**, 449-454

Eckel, R.H., Fujimoto, W.J. and Brunzell, J.D. (1978) *Diabetes*, **28**, 1141-1142

Elrick, H., Stimmler, L., Hlad, C.J. and Arai, Y. (1964) *Journal of Clinical Endocrinology and Metabolism*, **24**, 1076-1082

Fehmann, H-C, Göke, R., Göke, M.E., Trautman, M.E. and Arnold, R. (1989) *FEBS Letters*, **252**, 109-112

Feyrter, F. (1938) *Über diffuse endokrine epitheliale organe*, Leipzig, Barth, pp 6-17

Flatt, P.R., Bailey, C.J., Kwasowski, P., Swanston-Flatt, S.K. and Marks, V. (1983) *Diabetes*, **32**, 433-435

Füessl, H.S., Yiangou, Y., Ghatei, M.A., Goebel, F.D. and Bloom, S.R. (1988) *Diabetologia*, **31**, 492A

Furness, J.B. and Costa, M. (1987) *The enteric nervous system*, New York, Churchill Livingstone

Gibbs, J., Young, R.C. and Smith, G.P. (1973) *Journal of Comparative Physiology and Psychology*, **84**, 488-495

Hampton, S.M., Kwasowski, P., Tan, K.S., Morgan, L.M. and Marks, V. (1983) *Diabetologia*, **24**, 278-281

Hampton, S.M., Morgan, L.M., Tredger, J.A., Cramb, R. and Marks, V. (1986) *Diabetes*, **35**, 612-616

Hermansen, K. (1984) *Endocrinology*, **114**, 1770-1775

Holst, J.J., Ørskov, C., Nielson, O.V. and Schwartz, T.W. (1987) *FEBS Letters*, **211**, 169-174

Jones, I.R., Owens, D.R., Moody, A.J., Luzio, S.D., Morris, T. and Hayes, T.M. (1987) *Diabetologia*, **30**, 707-712

Keast, J.R., Furness, J.B. and Costa, M. (1985) *Journal of Comparative Neurology*, **236**, 403-422

Kissebah, A., Peiris, A. and Evans, D. (1989) In *Proceedings of the 18th Steenbock symposium on hormones, thermogenesis and obesity*, (eds H. Lardy and F. Stratman), New York, Elsevier Science Publishing Company, pp 77-91

Kreymann, B., Williams, G., Ghatei, M.A. and Bloom, S.R. (1987) *Lancet*, ii, 1300-1304

Kreymann, B., Yiangou, Y., Kanse, S., Williams, G., Ghatei, M.A. and Bloom, S.R. (1988) *FEBS Letters*, **242**, 167-170

Kwasowski, P., Flatt, P.R., Bailey, C.J. and Marks, V. (1985) *Bioscience Reports*, **5**, 701-705

Lithell, H. (1987) In *Recent Advances in Obesity Research: V* (eds E. Berry, S.H Blondheim, H.E. Eliahou and E. Shafrir) London, John Libbey, pp 77-81

Lotter, E.C., Krinsky, R., McKay, J.M., Treneer, C.M., Porter, D. and Woods, S.C. (1981) *Journal of Comparative Physiology and Psychology*, **95**, 278-287

Makhlouf, G.M. and Grider, J.R. (1989) In *Handbook of Physiology*, **2** (6) (eds S.G. Schultz, G.M. Makhlouf and B.B. Rauner), Bethesda, American Physiology Society, pp 123-132

McIntyre, N., Holdsworth, D. and Turner, D.S. (1964) *Lancet*, ii, 20-21

Mojsov, S., Heinrich, G., Wilson, I.B., Ravazzola, M., Orci, L. and Habener, J.F. (1986) *Journal of Biological Chemistry*, **261**, 11880-11889

Morgan, L.M. (1979) *Annals of Clinical Biochemistry*, **16**, 6-14

Morgan, L.M., Flatt, P.R. and Marks, V. (1988) *Nutrient Research Reviews*, **1**, 79-97

Morgan, L.M., Hampton, S.M., Tredger, J.A., Cramb, R. and Marks, V. (1988) *British Journal of Nutrition*, **59**, 373-380

Morley, J.E. (1980) *Life Sciences*, **27**, 355-368

Morley, J.E. (1987) *Endocrine Reviews*, **8**, 256-287

Morley, J.E. and Flood, J.F. (1987) *Life Sciences*, **41**, 2157-2165

Morley, J.E., Levine, A.S., Bartness, T.J., Nizielski, S.E., Shaw, M.J. and Hughes, J.J. (1985) *Annals of the New York Academy of Science*, **488**, 413-416

Oben, J., Morgan, L., Fletcher, J. and Marks, V. (1989) *Journal of Endocrinology*, **123**, 165

Ørskov, C., Holst, J.J., Knuhtsen, S., Baldiserra, F.G.A., Poulsen, S.S. and Vagn Nielsen, O. (1986) *Endocrinology*, **119**, 1467-1475

Ørskov, C., Holst, J.J., Poulsen, S.S. and Kirkegaard, P. (1987) *Diabetologia*, **30**, 874-881

Pearse, A.G.E. (1984) In *Interdisciplinary Neuroendocrinology*, (eds M. Ratzenhofer, H. Höfler and G.F. Walter), Basel, Karger, **12**, p 1-7

Peiris, A., Sothmann, M., Hoffmann, R., Hennes, M., Wilson, C.R., Gustafson, A. and Kissebah, A. (1989) *Annals of Internal Medicine*, **110**, 867-872

Penman, E., Wass, J.A.H., Medback, S., Morgan, L.M., Lewis, J., Besser, G.M. and Rees, L.H. (1981) *Gastroenterology*, **81**, 692-699

Ponter, A.A., Salter, D.N., Morgan, L.M. and Flatt, P.R. (1990) *Animal Production*, **50**, 571

Rebbuffe-Scrive, M. (1987) In *Recent Advances in Obesity Research V*, (eds E. Berry, S.H. Blondheim, H.E. Eliahou and E. Shafrir), London, John Libbey, pp 82-91

Richter, W.O. and Schwandt, P. (1989) *Hormone and Metabolic Research*, **21**, 216-217

Romsos, D.R. and Leveille, G.A. (1975) In *Modification of lipid metabolism*, (eds E.G. Perkins and L.A. Witting), New York, *Academic Press*, pp 127-141

Salminen, S., Salminen, E. and Marks, V. (1982) *Diabetologia*, **22**, 480-482

Samols, E., Tyler, J., Marri, G. and Marks, V. (1965) *Lancet* i, 415-416

Shaw, M.J., Hughes, J.J., Morley, J.E., Levine, A.S., Silvis, S.E. and Shafer, R.B. (1985) *Annals of the New York Academy of Sciences*, **448**, 640-641

Shima, K., Hirota, M. and Ohboshi, C. (1988) *Regulatory Peptides*, **22**, 245-252

Smith, G.P., Greenberg, D., Falusco, J.D., Gibbs, J., Liddie, R.A. and Williams, J.A. (1985) *Proceedings of the Society of Neuroscience*, **11**, 55

Smith, G.P., Jerome, C., Cushin, B.J., Eterno, R. and Simansky, K.J. (1981) *Science*, **213**, 1036-1077

Smith, U. (1985) In *Recent Advances in Obesity Research IV*, (eds J. Hirsch and T.B. van Itallie), London, John Libbey, pp 33-36

Sykes, S., Morgan, L.M., English, J. and Marks, V. (1980) *Journal of Endocrinology*, **85**, 201-207

Takahashi, H., Manak, H., Katsuyuki, S., Fukase, N., Tominaga, M., Sasaki, H., Kawai, K. and Ohashi, S. (1990) *Biomedical Research*, **11**, 99-108

Thim, L. and Moody, A.J. (1981) *Regulatory Peptides*, **2**, 139-150

Unger, R.H. and Eisentraut, A.M. (1969) *Archives of Internal Medicine*, **123**, 261-266

Valet, P., Berlan, M., Beauville, M., Crampes, F., Montastruc, J.L. and Lafontan, M. (1990) *Journal of Clinical Investigation*, **85**, 291-295

Wasada, T., McCorkle, K., Harris, V., Kawai, K., Howard, B. and Unger, R.H. (1981) *Journal of Clinical Investigation*, **68**, 1106-1107

West, D.B., Fey, D. and Woods, S.C. (1984) *American Journal of Physiology*, **246**, R776-787

Wood, A.J. and Groves, T.D.D. (1965) *Canadian Journal of Animal Science*, **45**, 8-13

Zarbin, M.A., Wamsley, J.K., Innis, R.B. and Kuhar, M.J. (1981) *Life Sciences*, **29**, 697-705

10
CONTROL OF FAT AND LEAN DEPOSITION IN FORAGE FED CATTLE

D.E. BEEVER
AFRC Institute of Grassland and Environmental Research, Hurley

J.M. DAWSON and P.J. BUTTERY
University of Nottingham School of Agriculture, Sutton Bonington, Loughborough, LE12 5RD

Introduction

Both grazed and conserved forage make a significant contribution to the nutrition of beef cattle within the United Kingdom, although it is recognised that such systems are inherently disadvantaged compared with those based on high cereal usage in terms of feed intake, the overall efficiency of nutrient utilisation and the composition of the final carcass with respect to the relative yields of fat and protein. Such issues are of considerable importance in a community which demonstrates increasing awareness towards environmental pollution as well as quality of the human diet in relation to both quantity and type of fat consumed. Equally there is a growing demand to develop more publicly acceptable methods of meat production.

Consequently, the United Kingdom beef industry, like the dairy industry is facing great pressure to provide sustainable and publicly acceptable technologies which will both improve the efficiency of beef production and beneficially manipulate the composition of the resultant carcass. In order to achieve these targets, it will be necessary to examine the nutritional and endocrinological control of muscle and fat deposition in forage fed cattle and, through an improved recognition of the importance of nutritional/endocrinological interactions, suggest alternative feeding and management strategies which the industry may wish to adopt. Perturbations to the nutritional status of the animal are relatively straightforward to effect and usually provoke little or no adverse public opinion. In contrast, perturbation of the endocrinological status of the animal, generally by exogenous means is a more sensitive issue and must be achieved using technologies which do not attract public criticism. It is against this background that current possibilities, including those evolving from the molecular sciences, will be discussed in this paper.

Growth of muscle and fat

In this paper, growth is considered as the net accretion of protein and fat in respective tissues, being the summation of the animals genetic capacity to grow, the nutritional status of the animal and aspects of its environment. The animal's endocrine system is considered to be the overall controller of growth and in this respect the importance of growth hormone (GH) is recognised. However, major interactions between the endocrine system and the nutritional status of the animal are accepted, and this led Breier and Gluckman (1991) to identify two distinct patterns of mammalian growth viz (1) substrate limited, GH-independent growth where nutrient supply is the dominating regulator of growth, with GH considered to play a less important role although the importance of insulin may be increased, and (2) GH-dependent growth which is largely regulated through the GH axis, with the contribution of other hormones being permissive rather than regulatory. Between two such extremes, many intermediate forms of growth can be identified, and it is within the context of nutritional/endocrinological interactions that this paper will seek to establish ways in which the performance of beef cattle fed predominantly forage diets can be beneficially manipulated.

Nutritional control of protein and fat accretion

In ruminant animals, amino acids absorbed from the small intestine, following the enzymatic hydrolysis of ruminally derived microbial protein and ruminally undegraded dietary protein are the main substrates for tissue protein synthesis (Armstrong and Hutton, 1975) with limited evidence suggesting some contribution from low molecular weight peptides, also absorbed from the small intestine (Parker, MacGregor, Finlayson, Stockall and Balias, 1984). The quantity of protein presented to the small intestine and its subsequent availability within that organ is influenced by both the quantity and quality of the diet consumed by the animal, and there is considerable evidence of impaired protein supply on grazed (Beever and Siddons, 1986) and particularly ensiled (Thomas, 1982) forages. Both types of diets are characterised by relatively high quantities of soluble nitrogen as a proportion of total dietary N (Mangen, 1982; Thomas, 1982) and once such feeds enter the rumen, digestion of the nitrogen fractions usually occurs rapidly, at a rate which is likely to exceed energy availability within the rumen. Thus, on many fresh forages, the supply of ruminally degraded N exceeds the capacity for microbial N synthesis due to restricted energy availability, even when dietary soluble carbohydrate concentrations are significant, whilst on fermented (ensiled) forages where most of the soluble carbohydrate has been degraded prior to feeding, the efficiency of N utilisation within the rumen deteriorates further and lower efficiencies of microbial N synthesis are often recorded (McAllan, Siddons and Beever, 1987). The full extent of this was demonstrated by Siddons, Nolan, Beever and MacRae (1985) when rumen ammonia production on a grass silage diet was shown to represent 75% of total dietary N intake, with only 30% of ruminally-produced ammonia being converted into microbial protein, and almost 60% of ingested N being absorbed from the rumen as ammonia, thus making no significant contribution to the animal's amino acid supply.

To consider this phenomenon further, Rooke, Lee and Armstrong (1987) examined the consequence of intraruminal infusions of energy and nitrogen on microbial N synthesis in the rumen of silage fed cattle. The results are summarised in Table 1 where, as expected, an intraruminal infusion of urea had no effect on microbial protein synthesis due presumably to the excess of ammonia already present in the rumen. An isonitrogenous infusion of casein caused marginal responses in the extent of microbial N synthesis, findings which support the earlier suggestion of Maeng and Baldwin (1975) of an obligatory requirement for preformed amino acids for microbial protein synthesis. The response to an isoenergetic infusion of glucose however was much greater, confirming the earlier suggestion of an imbalance of readily fermentable energy and N with silage diets and finally, when glucose and casein were co-infused into the rumen, further increases in the extent and efficiency of microbial synthesis were detected, comparable to those expected with fresh forage (Beever and Siddons, 1986). Thus from this one study, the pronounced negative effect of ensiling forage on protein supply was clearly demonstrated and the potential to stimulate microbial metabolism and the need to pay greater attention to the preservation of forage protein were all established.

Table 1 EFFECT OF INTRARUMINAL INFUSIONS OF ENERGY AND/OR NITROGEN ON MICROBIAL NITROGEN SYNTHESIS IN THE RUMEN OF SILAGE FED CATTLE

	Control	Urea	Casein	Glucose	Glucose + Casein
Intake:					
Total Nitrogen (g/d)	94	119	116	86	103
Soluble carbohydrate (kg/d)	0.32	0.28	0.28	1.00	0.93
Microbial N flow (g/d)	63	68	75	81	109
Microbial N synthesis (g/kg OMADR)	22	25	25	27	38

OMADR = organic matter apparently digested in the rumen
From Rooke *et al*, 1987

In this regard, considerable success has been achieved with the use of aldehyde-containing silage additives. Through their capacity to 'protect' forage protein during the ensiling process and to a lesser, or at least more variable, extent during rumen fermentation, substantial improvements in duodenal protein supply have been reported. When Siddons, Arricastres, Gale and Beever (1984) compared both formaldehyde and glutaraldehyde, duodenal non-ammonia nitrogen supply (g/gN intake) increased from 0.61 to 0.86 and 0.94 respectively, with increases in both microbial protein synthesis and undegraded dietary protein flow contributing to these effects. Data in Table 2 relate to similar experimentation undertaken by Thomson, Beever, Lonsdale, Haines, Cammell and Austin (1981) with growing cattle where the effect of formaldehyde on duodenal protein supply was substantial and in a companion growth study, was shown to enhance net protein deposition per unit ME intake by almost 50%, with a concomitant reduction in net fat deposition such that overall energy retention remained unchanged.

Table 2 EFFECT OF FORMALDEHYDE TREATMENT OF GRASS PRIOR TO ENSILING ON DUODENAL AMINO ACID SUPPLY AND CARCASS PROTEIN AND FAT RETENTION IN GROWING CATTLE

	Control Silage	Formaldehyde Silage
Total amino acid flow to small intestine g/kg DMI	123	164
Absorbed protein# (g/MJ ME intake)	7.5	10.5
Body retention (g/MJ):		
Protein	1.9	2.8
Fat	3.3	2.8

# Estimated assuming 68% availability of amino acids in small intestine, and metabolisable energy (ME) content derived from digestible organic matter content.
DMI = dry matter intake.
From Thomson *et al*, 1981

As an alternative to manipulation of ruminal events in order to enhance protein supply from forage diets, the use of protein supplements particularly those of relatively low susceptibility to rumen degradation has been examined by many research workers. Data in Table 3 demonstrate the effect of increasing levels of fishmeal as a supplement to grass silage in young cattle on growth and carcass composition. Despite a relatively high rate of liveweight gain on the control diet (0.77 kg/d), substantial improvements in liveweight were recorded up to the second level of fishmeal inclusion. Only those animals receiving either the control diet or the highest level of supplement (*ie* 150 g/kg silage dry matter) were slaughtered but from that data it is clear that almost all of the increase in liveweight and carcass weight recorded between the two extreme treatments could be accounted for by increased protein accretion (Gill, Beever, Buttery, England, Gibb and Baker, 1987). Complementary to this study, the net disappearance of

Table 3 EFFECT OF FISHMEAL SUPPLEMENTATION ON LIVEWEIGHT GAIN AND CARCASS COMPOSITION OF GROWING CATTLE OFFERED GRASS SILAGE

	Control	Fishmeal Supplement#		
		50	100	150
Liveweight gain (g/d)	0.77	0.94	1.02	1.01
Protein gain (g/d)	94.6	ND	ND	145.2
Fat gain (g/d)	92.1	ND	ND	97.5

g fishmeal/kg silage dry matter intake
ND, not determined
From Gill *et al*, 1987

amino acids from the small intestine was examined in similar animals fed the same diets at similar levels of intake (Beever, Gill, Dawson and Buttery, 1990). As expected, both duodenal supply and small intestinal disappearance of amino acids were elevated by fishmeal inclusion in the diet but irrespective of the method of calculation used (see Table 4), it was found that the efficiency of utilisation of absorbed amino acids was considerably lower (52%) than expectations (ARC, 1980), and appeared to be unaffected by either the level of amino acid supply or possible effects of the fishmeal supplement on the composition of absorbed amino acids.

Table 4 EFFECT OF FISHMEAL INCLUSION IN A GRASS SILAGE DIET OFFERED TO GROWING CATTLE, ON THE SMALL INTESTINAL ABSORPTION OF AMINO ACIDS AND THE EFFICIENCY OF UTILISATION OF ABSORBED AMINO NITROGEN, CALCULATED BY TWO METHODS (a, b)

	Control	+ Fishmeal[#]	Incremental Response
(a)			
Amino acid N absorbed (g/d)	37.8	53.0	15.2
N retention (g/d)	15.1	23.2	8.1
Partial efficiency (g retained/g supplied)			0.53
(b)			
Amino acid N available for production (g/d)	29.8	45.0	
Efficiency of utilisation (g retained/g supplied)	0.51	0.52	

[#] 150 g/kg total diet dry matter
For details see Beever *et al*, 1990

If such findings of low efficiency of amino acid utilisation are confirmed in other studies, and the suggestion that the efficiency may be higher on cereal diets than those based on high forage use (Beever, Cammell, Thomas, Spooner, Haines and Gale, 1988; Thomas, Gibbs, Beever and Thurnham, 1988) is proved, then reasons for this phenomenon need to be elucidated. The importance of gut:tissue interactions and metabolism in the splanchnic tissues *per se* needs to be established as there are already strong indications that both gut and liver metabolism can have a major effect on the ultimate supply of metabolites to peripheral tissues. Initial studies (Wilton, Gill and Lomax, 1988) which considered the consequence of elevated ammonia fluxes to the liver on hepatic metabolism, demonstrated that whilst hepatic removal of ammonia was complete, the incremental production of urea by the liver generally exceeded stoichiometric predictions. Subsequently it was confirmed by Fitch, Gill, Lomax and Beever (1989) and Maltby, Lomax, Beever and Pippard (1991) that hepatic amino acid utilisation was increased in response to an increased load of ammonia, with a concomitant reduction in hepatic output of free amino acids to peripheral tissues. As yet there are insufficient data on the hepatic output of peptides and proteins for a satisfactory balance of nitrogen-containing moieties across the liver to be established. Nonetheless, the enhanced synthesis of urea, which may be diet related as its occurrence has not been established in all dietary situations, is of sufficient concern to merit further research.

The principal energy-yielding (*ie* non-protein) components of fresh forages are water-soluble carbohydrates and fibre, whilst ensiled forages generally have little or no soluble carbohydrate but have high levels of lactic acid and other fermentation acids. Both types of diets give relatively high yields of rumen acetate and butyrate, whilst yields of ruminally-derived propionate and small intestinally-absorbed glucose and lipid are generally low. Consequently, most forage diets tend to promote the synthesis of body fat with a large proportion being derived via *de novo* synthesis from acetate and β-hydroxy butyrate. Attempts to reduce the ruminal yields of acetate and butyrate and enhance propionate through the use of feed additives such as monensin have been remarkably successful (Chalupa, 1980), and associated reductions in methane production through a significant repartition of ruminally-derived hydrogen have been observed. Use of such products in practice have invariably enhanced the efficiency of feed utilisation, although in some instances reductions in feed intake have been observed. Evidence for a substantial shift in the ratio of carcass protein:fat has, however, not been established.

Endocrinological manipulation of protein and fat accretion

ANABOLIC STEROIDS

In the study on the effect of dietary protein supplementation on the performance of grass silage-fed cattle referred to earlier (Gill *et al*, 1987), additional treatments also examined the consequence of ear implantation of oestradiol-17β in the presence or absence of different levels of fishmeal. From the data presented in Table 5 it can be seen that in animals fed the highest fishmeal-supplemented diet, oestradiol caused further enhancement of liveweight and carcass weight gain, which were wholly attributable to increased rates of protein gain. In contrast however, on the unsupplemented diet, no effects of oestradiol implantation on liveweight, carcass weight or protein and fat gain were discernable, indicating an important nutritional/endocrinological interaction. This lack of response on the unsupplemented diet confirms observations which have been made in practice, and indicates that the nutritional demands of endocrinologically-manipulated animals need to be recognised and established if the potential improvements in animal performance are to be achieved.

Table 5 THE EFFECT OF FISHMEAL SUPPLEMENTATION AND OESTRADIOL IMPLANTATION ON THE PERFORMANCE OF GROWING CATTLE OFFERED GRASS SILAGE *AD LIBITUM*

	Control Silage		Fishmeal Supplemented Silage#	
	No Implant	+ Oestradiol	No Implant	+ Oestradiol
Liveweight gain (g/d)	0.77	0.81	1.01	1.16
Protein gain (g/d)	94.6	93.7	145.2	167.8
Fat gain (g/d)	92.1	101.6	97.5	100.4

150 g fishmeal/kg silage dry matter intake
From Gill *et al*, 1987

There are several reports (Davis, Ohlson, Klind and Anfinson, 1977; Gopinath and Kitts, 1984) which indicate that oestrogens enhance plasma growth hormone (GH) concentrations in ruminants, suggesting that this mechanism could be involved in their mode of action. Hepatic GH receptor numbers and their binding affinities have been shown to be influenced by both plane of nutrition and oestradiol treatment (Breier, Gluckman and Bass, 1988a). The data summarised in Table 6 show that cattle fed on a high plane of nutrition have increased receptor affinity (Kd) compared with those fed at maintenance and they also exhibit a second, high-affinity binding site with 10-fold greater affinity for GH which was not detectable in hepatic membranes from animals fed on the low plane of nutrition. In contrast, oestradiol had no effect on Kd but increased receptor number with a relatively greater increase in the number of high-affinity sites, detectable only in the well-fed animals. This differential response of oestradiol with respect to plane of nutrition may explain the lack of response to oestradiol observed in cattle fed unsupplemented grass silage (Gill *et al*, 1987) compared with the increase in protein accretion observed when the implanted animals received protein supplementation. The capacity of the high-affinity GH binding site has been shown to correlate with plasma IGF-1 concentrations and body weight gain (Breier *et al*, 1988a,b). Exogenous administration of bovine GH

Table 6 THE EFFECT OF PLANE OF NUTRITION AND OESTRADIOL IMPLANTATION ON THE NUMBER AND BINDING CHARACTERISTICS OF HEPATIC GH RECEPTORS OF GROWING STEERS

	Control		Oestradiol	
Plane of Nutrition	Low	High	Low	High
Receptor numbers (pmol/100mg liver)				
High affinity	ND	1.87	ND	6.56
Low affinity	22.6	20.1	32.8	30.1
Dissociation constants (Kd - pmol/l)				
High affinity	ND	11.6	ND	10.8
Low affinity	197	106	193	111

ND = not detectable From Breier *et al*, 1988a

was found to stimulate plasma IGF-1 concentrations only in well-fed animals, with the response being greater when the animals had been previously implanted with oestradiol (Breier *et al*, 1988b), presumably reflecting the increased GH binding to hepatic high-affinity sites, resulting in enhanced IGF-1 levels and consequently improved growth rate.

EXOGENOUS GROWTH HORMONE ADMINISTRATION

The importance of GH in modifying nutrient utilisation in both growing and lactating animals has been known for many years and its ubiquitous action on glucose, fatty acid and amino acid metabolism as well as cell proliferation is widely recognised (Pell and Bates, 1990). With respect to the growing animal, most interest has centred on its potential to increase nitrogen retention and, with increased availability of GH through recombinant DNA technology, considerable experimentation has been undertaken to examine its efficacy in various species. Responses with porcine GH administration to growing pigs have been most dramatic, with major shifts in the protein:fat ratio of the resultant carcass (see for example, Campbell, Steele, Caperna, McMurty, Soloman and Mitchell, 1989). Use of bovine GH in growing ruminants however has given variable results and in some studies (*eg* Johnson, Hathorn, Wilde, Treacher and Butler-Hogg, 1987) no anabolic effects of GH were observed on either liveweight gain or muscle growth in young lambs. In contrast, Pell (1989) using twin lambs in order to reduce animal variability with one serving as the control while the other was treated with recombinant GH from weeks 9 to 20 of age, demonstrated a 36% improvement in liveweight gain (C, 225; GH 307 g/d). Tissue weights at slaughter, expressed as a fraction of final liveweight in order to allow comparison between the treated and control lambs, indicated that both carcass and visceral fat were significantly reduced in the GH-treated lambs, whilst significant increases in the fractional weights of intestines, liver and heart were also observed. Overall increases in the protein content of the carcass were small, but specific muscle weights were significantly increased in the GH-treated lambs when expressed as a fraction of cold carcass weight. Sustained responses in plasma IGF-1 concentrations were demonstrated in the GH-treated lambs which increased more than three-fold during the experiment, whilst IGF-1 concentrations in the control lambs declined gradually with time. As yet the relationships between GH, IGF-1 and other anabolic hormones are not fully understood, but there is evidence to suggest that GH may act both directly on some tissues such as liver and adipose and indirectly through IGF-1 on other tissues such as muscle and bone (Vernon and Flint, 1989).

Clearly, the response to exogenous GH obtained by Pell (1989) does not reconcile with those studies where only minimal effects of GH have been reported; reasons for which are not immediately apparent. The specificity of recombinant GH has been questioned but equally there is some suggestion that the responsiveness of the animal to exogenous GH may be related to dose, sex and age as well as the dietary intake of energy and protein. In consideration of this latter respect, Pell, Gill, Beever, Jones and Cammell (1989) examined the growth of 72 lambs allocated to one of three dietary protein levels (120, 160 and 200 g/kg DM) which were fed at two energy levels (30 g/kg LW or *ad libitum*, approximately 53 g/kg LW) and within each dietary treatment lambs received either buffered saline (control) or bovine GH injections daily from 9 weeks of age until slaughter at week 20. On the restricted energy intake, GH induced a significant increase in liveweight gain irrespective of dietary protein content and these changes were accompanied by significantly increased muscle weights but no consistent changes in visceral fat weights were detected. In contrast, at *ad libitum* feeding GH treatment had no effect on liveweight gain, but a major increase in muscle weight and a decrease in visceral fat weight was detected. Thus it appears from this study that the anabolic effect of GH on protein accretion was independent of total energy intake and crude protein concentration of the diet, whilst the catabolic effect of GH on visceral fat was related to energy intake. Plasma IGF-1 concentrations were measured during the last week of this study and significant positive effects of GH treatment, dietary energy intake and dietary protein level were established.

B-AGONISTS

β-adrenergic agonists comprise a group of related compounds which are structural analogues of the naturally occurring catecholamines, adrenaline and noradrenaline. Studies with ruminants receiving β-agonists have consistently shown reductions in carcass fat contents, generally associated with significant increases in net protein accretion.

The principal mode of action of β-agonists appears to be on adipose tissue via the β-receptors and, through a cascade of events occurring at the enzymic level involving the synthesis and subsequent release of cyclic AMP into the cytosol, hormone-sensitive lipase is activated and lipolytic rate increases. *In vitro* incubations of adipose tissue from β-agonist-treated animals have failed to establish the relative importances of increased lipolysis or reduced lipogenesis in the overall reduction of carcass fat but elevated concentrations of non-esterified fatty acids (NEFA) in plasma have been reported in most studies, which is consistent with an increase in lipolysis (see Buttery, Dawson, Beever and Bardsley, 1991).

At the same time, β-agonists increase net muscle accretion, which appears to be mediated through increased fibre diameter rather than increased cell number. There is some suggestion that type II fibres respond more consistently than type I fibres, although uncertainty remains as to whether the effect on net protein accretion is due largely to increased synthetic or reduced degradation rates. Studies by Emery, Rothwell, Stock and Winter (1984) with rats and Dawson, Buttery, Lammiman, Soar, Essex, Gill and Beever (1991) with cattle have measured increases in *in vivo* rates of protein synthesis in ß-agonist-treated animals. The concentration of mRNA for α-actin in skeletal muscle has been shown to increase two-fold in pigs fed ractopamine (Helferich, Jump, Anderson,

Skjaerlund, Merkel and Bergen, 1990) and this is consistent with the increase in muscle protein synthesis seen in pigs similarly treated (Bergen, Johnson, Skjaerlund, Babiker, Ames, Merkel and Anderson, 1989). In contrast however, Reeds, Hay, Dorwood and Palmer (1986) with rats and Bohorov, Buttery, Correia and Soar (1987) with sheep showed no increase in muscle protein synthetic rate. With young rats receiving an oral dose of clenbuterol, Reeds *et al* (1986) reported significant reductions in protein degradation rates of *gastrocnemius* and *soleus* muscles after 4, 11, 21 and 25 days of treatment, and concluded that this was the predominant mechanism of action of β-agonist on net protein accretion.

In support of this suggestion, Higgins, Lasslett, Bardsley and Buttery (1988) investigated the activities of the calpain system in control and β-agonist-treated sheep. This system in mammalian species comprises two calcium-activated neutral protease enzymes with differing calcium requirements, together with a naturally occurring inhibitor, calpastatin. Despite no change in liveweight gain, weight of the target muscle, *longissimus dorsi*, was increased by almost 25% by clenbuterol. Accompanying such changes, an increase in calpastatin activity was observed together with a small increase in calpain II activity and a reduction in calpain I activity. Similar observations were made by Kretchmar, Hathaway, Epley and Dayton (1989) and Kretchmar, Hathaway, Epley and Dayton (1990) using the beta-agonist L-644,969. The major species of calpastatin mRNA was increased by 16% and the calpain II large subunit mRNA was increased by 30% in the *L.dorsi* of cimaterol-treated cattle (Parr, Bardsley, Gilmour and Buttery, 1991). Calpains have been implicated in the catabolism of the Z-disk, an initial step in the catabolism of the myofibril (Goll, Kleese and Szpacenko, 1989). When all the data available in the literature and from unpublished experiments conducted at the University of Nottingham were assembled, there was evidence for some interesting correlations between the effect of ß-agonists on muscle growth and the activity of the calpain systems. Despite the data coming from a wide range of experiments conducted under different conditions in different species and using a variety of agonists, calpain I activity was negatively correlated with the extent of ß-agonist-induced muscle growth while both calpain II and calpastatin were positively correlated (Bardsley, Allcock, Dawson, Dumelow, Higgins and Buttery, 1992). These changes in the activity of the calpain system are however not seen when muscle growth is stimulated by diet, at least in chickens and sheep (Ballard, Bardsley and Buttery, 1988; Higgins *et al*, 1988). Clearly such responses need to be interpreted with caution, especially as the calpain system may be more specifically associated with the degradation of hormone receptors rather than the structural proteins.

Additionally there are suggestions that β-agonists may be acting at least in part by a redirection of blood flow towards muscle at the expense of adipose blood supply (Rothwell, Stock and Sudera, 1987), but confirmatory evidence of this is not available. In one experiment reported by Dawson, Buttery, Gill and Beever (1989), cimaterol reduced liveweight gain in silage-fed steers. On the control silage, which was of modest quality, gain averaged 0.21 kg/d, but this fell to 0.03 kg/d in animals treated with the β-agonist, whilst fishmeal supplementation alone increased gain to 0.61 kg/d. Accompanying the reduction due to β-agonist were small increases in plasma GH and NEFA concentrations, and a significant reduction in plasma insulin concentration. It is interesting to note however that this experiment was undertaken at a time of very cold weather and it was concluded that possible vasodilatory effects of the β-agonist together with the fat mobilisation exacerbated the environmental conditions. This strongly emphasises the need for considerable caution in predicting the response of animals to metabolic stimulants such as the β-agonists and the need to recognise and reconcile the importance of the animals nutritional and endocrinological environment.

In another experiment, using silage-fed beef cattle Dawson, Buttery, Lammiman, Soar, Essex, Gill and Beever (1991) found no change in liveweight gain in response to β-agonist (cimaterol) treatment but reported significant increases in the weight and protein content of the *vastus lateralis* and *longissimus dorsi* muscles. In contrast, the *semitendinosus* muscle showed no myogenic response to cimaterol, indicating that not all muscles respond identically to carcass manipulating agents. Small increases in protein synthetic rates were observed in the two responding muscles but the ratio of urinary N^r-methylhistidine : creatinine excretion was shown to be significantly reduced in the β-agonist-treated animals, suggesting that changes in both protein synthesis and degradation rates were contributing to the anabolic effect of the β-agonist (Dawson, Buttery, Lammiman, Soar, Essex, Gill and Beever, 1991).

In a further experiment, with growing cattle fed a dried grass:barley diet, Dawson, Beever, Gill and Buttery (1991) reported a substantial increase in liveweight gain when cimaterol was included in the diet, compared with earlier experimentation cited above when cimaterol had been delivered by subcutaneously implanted osmotic minipumps and the animals were fed grass silage. To what extent this variation in response was associated with the different route of administration rather than the diet is not possible to ascertain.

IMMUNISATION AGAINST SOMATOSTATIN

Alternative to the use of exogenous GH, other studies have attempted to increase endogenous GH secretion by either administration of growth hormone releasing factor (GRF) or neutralisation of the inhibitory effect of somatostatin on GH release. Initial studies by Spencer, Garessen and Hart (1983) with Dutch moor sheep showed substantial increases in liveweight gain in response to somatostatin immunisation without any significant changes in GH concentrations, but subsequent studies have reported more modest changes in liveweight gain and the study of Varner, Davis and Reeves (1980) failed to demonstrate any increase in growth rate in immunised sheep despite the occurrence of significantly elevated plasma GH concentrations.

In a recent experiment conducted by Dawson, Beever, Gill and Buttery (1991) the effect of immunisation against somatostatin was examined in growing cattle fed on a forage:concentrate diet, with or without the simultaneous dietary administration of the β-agonist, cimaterol. Two of the objectives of the experiment were to consider whether immunisation could enhance β-adrenergic stimulation of GH release, possibly overcoming some earlier suggestions that β-agonist stimulation of GH release may only be transitory due to somatostatin secretion (Krieg, Perkins, Johnson, Rogers, Arimura and Cronin, 1988) and to examine whether the repartitioning effect could be further enhanced by the combined treatment of immunisation plus cimaterol administration.

The main results of this study for 24 Friesian steers which started the experiment at 12 weeks of age and remained on treatment until week 28 are

Table 7 EFFECT OF IMMUNISATION AGAINST SOMATOSTATIN AND DIETARY INCLUSION OF CIMATEROL ON THE PERFORMANCE OF GROWING CATTLE

	Control	Immun.	Cimaterol	Cim. + Imm.
Liveweight gain (kg/d)	0.86	0.96	1.13	1.03
Empty body weight (kg)	153.7	168.5	179.8	167.8
Muscle weights (kg):				
Longissimus dorsi	2.04	2.22	2.80	2.70
Semitendinosus	0.75	0.83	0.98	0.98
Vastus lateralis	0.82	0.87	1.08	1.10
Fat weights (kg):				
KKCF	0.85	1.00	0.56	0.48
Mesenteric	2.11	2.63	2.23	1.65
Omental	1.27	1.43	1.14	0.90

KKCF = Kidney knob and channel fat
From Dawson, Beever, Gill and Buttery (1991)

presented in Table 7. Daily liveweight gain averaged 0.86 kg/d on the control diet and the response to immunisation was a non-significant increase of 12% - in line with many other studies. In contrast, dietary inclusion of cimaterol caused a significant improvement of 31% in liveweight gain (+0.27 kg/d) whilst the combined treatment gave an intermediate response (+0.17 kg/d). These effects were reflected in increased empty body and carcass weights and from examination of target muscle and fat weights, substantial changes due to the treatments could be observed. Immunisation (I) alone increased the weight of all three target muscles by approximately 10% but this was accompanied by an increase in net fat deposition when either total weight of three target fat depots or their relative contributions to total empty body weight (control 27.5, I 30.0 g/kg EBW) were considered. In contrast, cimaterol (CIM) had a pronounced effect on both total (+35%) and relative (control 23.5, CIM 27.1 g/kg EBW) weights of the target muscles while total and relative (control 27.5, CIM 21.9 g/kg) weights of fat were substantially reduced. The combined treatment (CIM/I) gave virtually identical muscle weights to those obtained with cimaterol alone (CIM 4.86, CIM/I 4.78 kg) but total (CIM 3.93, CIM/I 3.03 kg) and relative (CIM 21.9, CIM/I 18.1 g/kg EBW) fat weights were reduced, and represented a 40% reduction in net fat deposition compared with the immunised treatment which sustained the highest level of fat accretion.

Whilst the dietary inclusion of β-agonist in both immunised and non-immunised animals caused substantial reductions in fat accretion and accompanying increases in muscle growth, evidence from *in vitro* studies on the relative rates of lipogenesis and lipolysis were inconclusive. Both treatments gave elevated serum NEFA concentrations, indicative of a stimulation of lipolysis but a similar response was also observed on the immunisation-only treatment, where actual levels of fat accretion increased. Consequently, elucidation of the mechanisms involved in the responses obtained is difficult, particularly the dramatic differences observed in fat accretion in immunised animals which did or did not receive β-agonist in the diet.

Effect on carcass quality

From the earlier sections of this paper, it is evident that through the application of nutritional and/or endocrinological technology, dramatic improvements in the level and composition of carcass gain and in the efficiency of nutrient utilisation in beef cattle can be achieved. There is concern however that such changes may be associated with deleterious effects on the quality of the product as perceived by the meat trade and the final consumer. There is sufficient evidence to indicate that specific muscles and adipose tissue depots will react differentially to the applied stimuli and if such trends are confirmed, then the consequences of such on the subsequent yield of high quality meat cuts need to be assessed. Furthermore, if the overriding desire to reduce carcass fat content continues, the consequences to the beef industry may be very serious if discernible deterioration in meat eating quality is established.

There is therefore an urgent need for research workers of all disciplines associated with the production and marketing of beef to collaborate if sustained progress is to be made and all possible adverse effects avoided. To date however, such collaboration has been minimal. In a recent experiment Dawson, Buttery, Gill and Beever, 1990 examined the effect of ß-agonist treatment on the type and quantity of collagen in muscle, both of which can affect meat tenderness. Cimaterol was found to reduce the total amount of intramuscular collagen, but also to significantly increase the proportion which was heat insoluble. Clearly these two findings are conflicting in terms of their effects on meat tenderness, but indicate important changes in muscle structure which may affect meat quality. Furthermore, cimaterol significantly reduced the content of intramuscular fat in all target muscles, whilst fishmeal with or without oestradiol had no measurable effect (Dawson, Buttery, Lammiman, Soar, Essex, Gill and Beever, 1991).

All of these aspects are discussed more fully in the subsequent paper by Wood and Warris (1992).

Conclusions

It is evident from this paper, that considerable opportunities do exist to manipulate the performance of forage-fed beef cattle. Through appropriate use of nutritional and metabolic manipulations, and the recognition of the important nutritional/endocrinological interactions, major changes in the rate and composition of carcass gain can be achieved. In this respect both exogenous administration of growth hormone, and the dietary inclusion of β-agonists have given very encouraging results, but the different modes of action of such agents needs to be recognised. As illustrated by MacRae and Lobley (1991), it appears that the effects of β-agonists on protein metabolism are largely restricted to skeletal muscles, whilst GH, in both bovine and porcine livestock has been shown to increase protein accretion in both skeletal and visceral tissues. At the same time it is important to recognise the need to avoid any impairment of eating quality, whilst the differential response of different tissue types is open for further exploitation. Currently, however many of the metabolic agents which are used experimentally do not have the necessary approval for use within the Industry, and all those associated with the production of meat from all classes of farm livestock must be aware of the growing public concern over acceptable farming practices. In this respect, hormone administration is unlikely to regain approval within the EEC, but alternative approaches which rely on immunological technology may be more successful. At the same time, well-defined animal breeding programmes designed to provide progeny which are more suited to current markets need to be sustained, and above all else, there will never be a substitute for good animal husbandry, and in particular the provision of a nutritionally-balanced and adequate diet.

References

Agricultural Research Council (1980) *Farnham Royal: Commonwealth Agricultural Bureaux.*

Armstrong, D.G. and Hutton, K. (1975) In '*Digestion and Metabolism in the Ruminant*' (eds I.W. McDonald and A.C.I. Warner), Armidale, Australia, University of New England Publishing Unit, pp 432-447

Ballard, R., Bardsley, R.G. and Buttery, P.J. (1988) *British Journal of Nutrition* **59**, 141-147

Bardsley, R.G., Allcock, S.M.J., Dawson, J.M., Dumelow, N.W., Higgins, J.A., Lasslett, Y.V., Lockley, A.K., Parr, T. and Buttery, P.J. (1992) *Biochimie* **74** (In press)

Beever, D.E. and Siddons, R.C. (1986) In '*Control of Digestion and Metabolism in Ruminants*' (eds L.P. Milligan, W.L. Grovum and A. Dobson), Englefield Cliffs, N.J., Prentice Hall, pp 479-97

Beever, D.E., Cammell, S.B., Thomas, C., Spooner, M.C., Haines, M.J. and Gale, D.L. (1988) *British Journal of Nutrition*, **60**, 307-319

Beever, D.E., Gill, M., Dawson, J.M. and Buttery, P.J. (1990) *British Journal of Nutrition*, **63**, 489-502

Bergen W.G., Johnson, S.E., Skjaerlund, D.M., Babiker, A.S., Ames, S.K., Merkel, R.A. and Anderson, D.B. (1989) *Journal of Animal Science*, **67**, 2255-2262

Bohorov, O., Buttery, P.J., Correia, T.H.R.D. and Soar, J.B. (1984) *British Journal of Nutrition*, **57**, 99-107

Breier, B.H., Gluckman, P.D. and Bass, J.J. (1988a) *Journal of Endocrinology*, **116**, 169-177

Breier, B.H., Gluckman, P.D. and Bass, J.J. (1988b) *Journal of Endocrinology*, **118**, 243-250

Breier, B.H. and Gluckman P.D. (1991) *Livestock Production Science*, **27**, 77-94

Buttery, P.J., Dawson, J.M., Beever, D.E. and Bardsley, R.G. (1991) *Proceedings of the 6th European Association for Animal Production Meeting on Protein Metabolism and Nutrition*, pp 88 102

Campbell, R.G., Steele, N.C. Caperna, T.J. McMurty, J.P., Soloman, M.B. and Mitchell, A.D. (1989) *Journal of Animal Science*, **64**, 1265-1271

Chalupa, W. (1980) In *Digestive Physiology and Metabolism in Ruminants*, (eds Y. Ruckebush and P. Thivend), Lancaster, M.T.P. Press, pp 325-347

Davis, S.L., Ohlson, D.L., Klind J. and Anfinson, N.S. (1977) *American Journal of Physiology*, **233**, E519-E523

Dawson, J.M., Buttery, P.J., Gill, M. and Beever, D.E. (1989) *Asian-Australasian Journal of Animal Science*, **2**, 243-245

Dawson, J.M., Buttery, P.J., Gill, M. and Beever, D.E. (1990) *Meat Science*, **28**, 289-297

Dawson, J.M., Buttery, P.J., Lammiman, M.J., Soar, J.B., Essex, C.P., Gill, M. and Beever D.E. (1991) *British Journal of Nutrition*, **66**, 171-185

Dawson, J.M., Beever,D.E., Gill, M. and Buttery, P.J. (1991) *Animal Production*, **52**, 564

Emery, P.W., Rothwell, N.J. Stock, M.J. and Winter, P.D. (1984) *Bioscience Reports*, **4**, 83-91

Fitch, N.A., Gill, M., Lomax, M.A. and Beever, D.E. (1989) *Proceedings of Nutrition Society*, **48**, 76A

Gill, E.M., Beever, D.E., Buttery, P.J., England, P., Gibb, M.J. and Baker, R.D.(1987) *Journal of Agricultural Science, Cambridge*, **108**, 9-16

Goll, D.E., Kleese W.C. and Szpacenko, A. (1989) In *Animal Growth Regulation*, (eds D.R. Campion, G.J. Hausman and R.J. Martin), Plenum, New York, p 141-182

Gopinath, R. and Kitts, W.D. (1984) *Growth*, **48**, 499-514

Helferich, W.G., Jump, D.B., Anderson, D.B., Skjaerlund, D.M., Merkel, R.A. and Bergen, W.G. (1990) *Endocrinology*, **126**, 3096-3100

Higgins, J.A., Lasslett, Y.V., Bardsley, R.G. and Buttery, P.J. (1988) *British Journal of Nutrition*, **60**, 645-652

Johnson, J.D., Hathorn, D.J., Wilde, R.M., Treacher, T.T. and Butler-Hogg, B.W. (1987) *Animal Production*, **44**, 405-414

Kretchmar, D.H., Hathaway, M.R., Epley, R.J. and Dayton, W.R. (1989) *Archives of Biochemistry and Biophysics*, **275**, 228-235

Kretchmar, D.H., Hathaway, M.R., Epley, R.J. and Dayton, W.R. (1990) *Journal of Animal Science*, **68**, 1760-1772

Krieg, R.J., Perkins, S.N., Johnson, J.H., Rogers, J.P., Arimura, A. and Cronin, M.J. (1988) *Endocrinology*, **122**, 531-537

MacRae, J.C. and Lobley, G.E. (1991) *Livestock Production Science*, **27**, 43-59

Maeng W.J. and Baldwin, R.L. (1975) *Journal of Dairy Science*, **59**, 636-642

Maltby, S.A., Lomax, M.A., Beever, D.E. and Pippard, C.J. (1991) In *Energy Metabolism of Farm Animals*, No. 58, (eds C. Wenke and M. Boessinger, eds), *Proceedings of the 12th Symposium, Zurich, Switzerland*, EAAP Publication, pp 20-23

Mangen, J.L. (1982) In *Forage protein in Ruminant Animal Production*, BSAP Occasional Publication No. 6. (eds D.J. Thomson, D.E. Beever and R.G. Gunn), Thames Ditton, BSAP, pp 25-40

McAllan, A.B., Siddons, R.C. and Beever, D.E. (1987) In *Feed Evaluation and Protein Requirement Systems for Ruminants*, (eds R. Jarrige and G. Alderman), Luxembourg, CEC, pp 111-128

Parker, D.S., MacGregor, R.C., Finlayson, H.J., Stockall, P. and Balias, J. (1984) *Canadian Journal of Animal Science*, **64**, (Suppl) 136-137

Parr, T., Bardsley, R.G., Gilmour, R.S. and Buttery, P.J. (1991) *Journal of Muscle Research and Cell Motility*, **12**, 82

Pell, J.M. (1989) In *Biotechnology in Growth Regulation* (eds R.B. Heap, C.G. Prosser and G.E. Lamming), London, Butterworths, pp 85-96

Pell, J.M., Gill, M., Beever, D.E., Jones, A.R. and Cammell, S.B. (1989) *Proceedings of the Nutrition Society*, **48**, 83A

Pell, J.M. and Bates, P.C. (1990) *Nutrition Research Reviews*, **3**, 169-192

Reeds, P.J., Hay, S.M., Dorwood, P.M. and Palmer, R.M. (1986) *British Journal of Nutrition*, **56**, 249-258

Rooke, J.A., Lee, N.H. and Armstrong, D.G. (1987) *British Journal of Nutrition*, **57**, 89-98

Rothwell, N.J., Stock, M.J., and Sudera, D.K. (1987) *British Journal of Phamacology*, **90**, 601-607

Siddons, R.C., Arricastres, C. Gale, D.L. and Beever, D.E. (1984) *British Journal of Nutrition*, **52**, 391-401

Siddons, R.C., Nolan, J.V., Beever, D.E. and MacRae, J.C. (1985) *British Journal of Nutrition*, **54**, 175-187

Spencer, G.S.G., Garessen, G.J. and Hart, I.C. (1983) *Livestock Production Science*, **10**, 25-37

Thomas, C., Gibbs, B.G., Beever, D.E. and Thurnham, B.R. (1988) *British Journal of Nutrition*, **60**, 297-353

Thomas, P.C. (1982) In *Forage in Ruminant Animal Production BSAP Occasional Symposium No. 6.* (eds D.J. Thomson, D.E. Beever and R.G. Gunn), Thames Ditton-BSAP, pp 67-76

Thomson, D.J., Beever, D.E., Lonsdale, C.R., Haines, M.J., Cammell, S.B. and Austin, A.R. (1981) *British Journal of Nutrition*, **46**, 193-207

Varner, M.A., Davis, S.L. and Reeves, J.J. (1980) *Endocrinology*, **106**, 1027-1032

Vernon, R.G. and Flint, D.J. (1989) In *Biotechnology in Growth Regulation* (eds R.B. Heap, C.G.Prosser and G.E. Lamming), London, Butterworths, pp 57-71

Wilton, J.C., Gill, M. and Lomax, M.A. (1988) *Proceedings of Nutrition Society*, **47**, 153A

Wood, J.D. and Warriss, P.D. (1992) In *The Control of Fat and Lean Deposition*, (eds P.J. Buttery, K.N. Boorman, D. K. Lindsay), London, Butterworths

11

CONTROL OF LEAN AND FAT DEPOSITION IN BIRDS

H.D. GRIFFIN, C. GODDARD and S.C. BUTTERWITH
Department of Cellular and Molecular Biology
AFRC Institute of Animal Physiology and Genetics Research
Edinburgh Research Station, Roslin, Midlothian EH25 9PS, UK

Introduction

The economic importance of poultry and their ease of study has meant that much more is known about the control of fat and lean deposition in birds than in mammalian species of livestock. This paper will review some of the more recent findings, with the aim of highlighting the areas where results of studies on birds are relevant to other species and the areas where differences between birds and mammals are particularly important.

The literature cited is meant to be illustrative rather than comprehensive and readers are referred to the following reviews for further details of the endocrinological control of growth in birds: on the role of growth hormone and other growth factors (Scanes, 1987; Decuypere and Boyse, 1988; Scanes, Aramburo and Campbell, 1990); on hormonal control of embryonic development (Scanes, Hart, Decuypere and Kuhn, 1987; De Pablo, 1991); on molecular biology of avian growth factors (Goddard and Bulfield, 1990; Goddard and Boswell, 1991) and on endocrine-nutrition interactions (Scanes and Griminger, 1990).

Much of the present knowledge on the endocrine control of growth in birds has been gained from studies on the *in vivo* effects of ablation (eg hypophysectomy), the administration of exogenous hormones and from studies *in vitro*. These approaches have clearly implicated a large number of hormones in the control of growth in birds, including growth hormone, insulin, the insulin-like growth factors, corticosterone and the thyroid hormones. The extent to which each is simply required for growth or is actively involved in its control is uncertain.

Discussion in this paper will concentrate on growth in meat-type poultry and on the developmental and genetic differences in the regulation of lean and adipose tissue growth. The control of fattening in laying hens and in migratory birds is of considerable interest but will not be reviewed here. The paper concludes with a brief discussion of approaches to the manipulation of growth in poultry.

Muscle cell proliferation and protein accretion

The structure and origins of skeletal muscle are very similar in all vertebrates (Swatland, 1984). Muscle tissue arises during embryonic development from myoblast cells that proliferate, differentiate and fuse to form multinucleate myotubes that are then assembled into myofibrils. The number of muscle fibres is fixed at birth or hatch and the nuclei that have been incorporated into a myotube are no longer capable of cell division. However the DNA content of muscle fibres increases during post-natal growth in all species. This occurs by proliferation of myogenic satellite cells found between the plasmalemma and basal lamina of each muscle fibre and their fusion with existing myotubes.

Commercial selection of broiler chickens by the method of 'weight for age' for more than 40 generations has produced strains that reach marketable weight (in the UK) of 2 to 2.5 kg within 45 d. At this age, broiler chickens are about 4-fold heavier than chicks of layer strains selected for egg size and persistency of lay and they can contain up to 8 times as much breast muscle (Bulfield, Isaacson and Middleton, 1988).

There is no difference in the weight of broiler and layer chicks at 1 day of age (Goddard, Wilkie and Dunn, 1988). This is perhaps not surprising considering the similarity in weight of the eggs laid by the two strains and the physical constraints of egg size on avian embryonic development. This includes restrictions on CO_2 and 0_2 exchange as well as limitations on the space available within the shell. Day-old broiler chicks have a greater number of muscle fibres than layer chicks, but the fibres are of significantly smaller diameter (Tinch, 1990).

The growth of the newly-hatched broiler chick is very much faster than that of the layer chick. The fractional rate of growth (% increase/d) in the first week after hatch is about 1.7 times greater in broilers than in layers but this difference declines slowly thereafter so that at 6 weeks of age there is no significant difference between strains (Goddard *et al*, 1988). As in other species (Swatland, 1984), growth is accompanied by dramatic increase in muscle fibre diameter but no change in fibre number. The increase in muscle fibre diameter and in DNA content of muscle in broilers is much faster than in layers (Tinch, 1990), but there are only small differences in muscle fibre type between the two strains (Aberle, Addis and Shoffner, 1979).

Studies on cell proliferation *in vivo* using bromodeoxyuridine indicate that the proliferation of cells closely associated with muscle fibres (presumably satellite cells) at 5 d of age is 3 to 4-fold greater in broiler chicks than in those of a layer strain (Goddard, Loveridge and Farquaharson, unpublished data). Satellite cells isolated from 5-d-old broiler chicks appear to proliferate faster *in vitro* than do satellite cells from layer chicks (Goddard, unpublished data) and this suggests that differences in rate of proliferation *in vivo* are in part a consequence of intrinsic differences in the satellite cells themselves.

The relative importance of protein synthesis and degradation to muscle protein accretion during growth has been the subject of several studies in birds. As with other species, rates of both synthesis and degradation decline rapidly with age (see

Scanes, 1987). Differences in protein accretion between broiler and layer strains appear to be largely due to the slower rate of protein dradation in the more rapidly growing strain (Kang, Sunde and Swick, 1985). The mechanisms controlling protein degradation are not clear in any species. Lysosomal proteases may be involved, but Saunderson and Leslie (1987) found no significant differences in muscle cathepsin D activity between fast and slow growing strains. Calpain I and II and calpastatin have been identified in birds (*eg* Murachi, Murakami, Udea, Fukui, Hamakubo, Adachi and Hatanaka, 1989) but no comparisons of their activities with age or between strains have been reported.

Studies on the endocrine control of lean tissue growth in birds have concentrated on the role of growth-hormone (GH), insulin-like growth factor-I (IGF-I) and the thyroid hormones (T3 and T4). Growth hormone is detectable in the chick embryo at 12 d of incubation (Kikuchi, Buonomo, Kajimoto and Rotwein, 1991). Its concentration in the plasma increases rapidly after hatch, remains high for 2-3 weeks and declines thereafter (see Scanes, 1987; Goddard *et al*, 1988). GH is secreted in a pulsatile fashion in young male birds with peaks every 1 to 1.5 h, but this pattern is much less evident as plasma GH levels decline with age and in female chicks. It is tempting to associate the rapid early growth of birds with the high circulating levels of growth hormone. However, strains of chicken with low growth rate have higher plasma concentrations of GH than more rapidly growing strains (see Scanes, 1987; Goddard *et al*, 1988). Differences in plasma GH are therefore not the prime cause of genetic differences in growth rate produced by selection.

The effect of exogenous GH on chicken growth appears to be very dependent on the method of administration. Daily subcutaneous injections have little or no effect on growth but pulsatile infusion, at least into 8-week-old birds, produced marked improvement in growth body composition and feed efficiency (Vasilatos-Younken, Cravener, Cogburn, Mast and Wellenreiter, 1988). The effects of GH on muscle are believed to be mediated principally via IGF-I, although the relative importance of an endocrine role of IGF-I synthesised in the liver and the autocrine/paracrine role of IGF-I synthesised locally is uncertain. Neither the avian nor amphibian mannose-6-phosphate receptor binds IGF-II and this allowed Duclos, Wilkie and Goddard (1991) to show that IGF-I and -II stimulate the proliferation of chicken satellite cells solely via the IGF-I receptor.

Plasma IGF-I concentrations rise steadily over the period of growth in chickens, but differences between fast and slow-growing strains are small and inconsistent in spite of marked differences in plasma GH concentrations (Scanes, Dunnington, Buonomo, Donoghue and Siel, 1989; Goddard *et al*, 1988). However, the significance of measurements of circulating IGF-I is uncertain. Specific IGF binding proteins have been identified in chicken serum similar to those in mammals (Armstrong, McKay, Morrell and Goddard, 1989) and changes in their concentration presumably can have a major effect on plasma IGF-I concentrations and/or activity. Much of the circulating IGF-I is probably of hepatic origin and its importance in influencing growth in extra-hepatic tissues is not known.

The effects of GH on local IGF-I production and growth is clearly likely to be dependent on tissue levels of GH receptors, on IGF-I and IGF-I binding protein synthesis and IGF-I receptor gene expression and function. A systematic examination of all these factors is needed to established the role of the GH/IGF-I axis in growth, but to date attention in chickens has been directed primarily at the liver. Developmental studies on young birds have shown major changes in IGF-I binding to chick liver during early growth (Duclos and Goddard, 1990) and a gradual increase in GH binding to turkey liver with age (Vasilatos-Younken, Gray, Bacon, Nestor, Long and Rosenburger, 1990).

The sex-linked mutant (*dw*) gene used commercially in maternal lines of broilers represents an alternative model to study growth in the chicken (Tixier-Boichard, Huybrechts, Kuhn, Decuypere, Charrier and Mongin, 1989). Dwarf (*dw*) birds are characterised by low circulating levels of IGF-I and T3 in spite of normal levels of GH and T4. The low plasma T3 concentration in dwarf (*dw*) birds is explained by a low peripheral mono-deiodination. Both this and low IGF-I production may be due to reduced GH receptor levels in the liver of dwarf birds, since binding of GH is low in young dwarf birds and does not increase with age as in normal broilers. The effect of the *dw* gene on GH binding and IGF-I production by muscle and other tissues is as yet unknown, although information on this would be very useful in determining the relative importance of hepatic or local production of IGF-I in control of muscle growth. The thyroid hormones are clearly important in avian growth, but no consistent differences in circulating T3 and T4 concentrations have been reported in selected lines (Lilburn, Leung, Ngiam-Rilling and Smith, 1986; Lauterio, Decuypere and Scanes, 1986; Goddard *et al*, 1988).

Many other factors have been implicated in control of muscle growth and development, including fibroblast growth factor and the transforming growth factor-β family, but little is known about their developmental expression in any species. In general growth factors appear to be highly conserved across species and the local paracrine and autocrine mechanisms controlling growth are likely to be basically similar in all vertebrates.

Fat deposition in birds

ADIPOCYTE HYPERPLASIA

Hood (1984) found that much of the growth of the abdominal fat pad in broiler chickens up to about 14 weeks of age involved an increase in cell number, whereas almost all the growth of this depot after this age was due to an increase in adipocyte size. Different depots in the chicken increase at different rates (Butterwith, 1989) and studies using [^{3}H]-thymidine suggest that this is in part due to differences in the rate of adipocyte hyperplasia (March and Hansen, 1977). Differences in adipocyte cell number have been reported between broiler and layer chicks (March and Hansen, 1977; Griffin, Butterwith and Goddard, 1987) and between lean and fat lines produced by divergent selection for abdominal fat pad weight (Hermier, Quinard-

Boulange, Dugail, Guy, Salichon, Brigant, Ardouin and Leclercq, 1989). Comparisons of adipocyte cell number have to be treated with caution, since none of the methods for measuring adipocyte cell number is without criticism (see Gurr and Kirtland, 1978).

The mature adipocyte has no capacity for cell proliferation and adipocyte hyperplasia results from the proliferation of fibroblast-like adipocyte precursor cells and their differentiation into mature adipocytes. Adipocyte precursor proliferation and differentiation have been extensively studied in cell lines (*eg* 3T3-LI and ob 1771) but these may not be typical of precursors *in vivo* (see Butterwith, 1988). Primary cultures of adipocyte precursors isolated from adipose tissue are likely to be more similar to the native precursor and a distinct advantage of using precursors derived from chicken adipose tissue is the ability to produce essentially homogenous cultures using either medium containing foetal calf serum (Cryer, Woodhead and Cryer, 1987) or a serum substitute (Butterwith and Griffin, 1989). This has not yet proved possible in some other livestock species. In the pig, for example, precursor cultures appear to contain only a few cells that differentiate into adipocytes (Hausman, Novakofski, Martin and Thomas, 1984; Hausman, 1989). It is not clear whether this is due to the presence of contaminating cells other than adipocyte precursors or to the culture conditions not being appropriate to induce differentiation in all the cells present.

Precursors isolated from the adipose tissue of broiler chickens proliferate at a much faster rate *in vitro* than do those from layer-strain chicks (Donelly, Cryer and Butterwith, unpublished data) and this indicates that at least part of the genetic difference between the two strains is due to an intrinsic difference at the cellular level.

The role of circulating hormones in stimulating adipocyte hyperplasia is uncertain. Insulin stimulates the growth of some precursor cell lines, but may be acting through the IGF-I receptor (Grimaldi, Djian, Forest, Poli, Nrel and Ailhaud, 1983) and corticosteroids and growth hormone have been shown to stimulate the differentiation of precursors into mature fat cells (Morikawa, Nixon and Green, 1982; Ringold, Chapman and Knight, 1986). However, whereas some preadipocyte cell lines require growth hormone for differentiation, this is not the case for primary cultures of pig preadipocytes where growth hormone inhibits differentiation (Hausman and Martin, 1989). No information is available on the effects of these hormones on proliferation of avian precursors.

One area which has received relatively little study is the role of autocrine/paracrine acting factors in the regulation of adipocyte hyperplasia. For example IGF-I is expressed in preadipocytes during differentiation (Doglio, Dani, Fredrikson, Grimaldi and Ailhaud, 1987; Gaskins, Kim, Wright, Rund and Hausman, 1990) and TGF-β is expressed during both proliferation and differentiation (Weiner, Shah, Smith, Rubin and Zern, 1989). TGF-β stimulates the proliferation of chicken adipocyte precursors (Butterwith and Goddard, unpublished) and inhibits their differentiation (Butterwith and Gilroy, 1991; Torti, Torti, Larrick and Ringold, 1989; Ignotz and Massague, 1985). Both IGF-I and TGF-β might therefore be important autocrine/paracrine regulators of adipocyte hyperplasia. This may also be the case for other growth factors. Recent studies (Butterwith, unpublished data) have shown that the inclusion of very low density lipoproteins in incubation media can considerably increase the effectiveness of PDGF, TGF-α, IGF-I and, to a limited extent, TGF-β as mitogens for chicken precursors. This observation raises the intriguing possibility of a link between adipocyte hyperplasia and the mechanisms supplying fatty acids to adipose tissue that would be clearly relevant to mammalian species.

METABOLIC CONTROL OF FAT DEPOSITION

Early studies in chickens identified the liver as the major site of lipogenesis in birds. O'Hea and Leveille (1969), for example, calculated that the liver accounted for 95% of *de novo* fatty acid synthesis in young chicks. Subsequent studies by a number of different groups confirmed that the rate of hepatic lipogenesis in birds was very much greater than that in adipose tissue and this has led to the general assumption that almost all the fat that accumulates in avian adipose tissue is synthesised in the liver or derived from the diet (see Griffin and Hermier, 1988). More recent evidence indicates extra-hepatic lipogenesis becomes increasingly important as birds grow older (Saadoun and Leclercq, 1986) although the tissues responsible have not been identified.

Transport of lipid from the liver to the extrahepatic tissues in immature females and male birds is very similar to that in rat and man (see Griffin and Hermier, 1988). The plasma of these birds contains very low density lipoproteins, low density lipoproteins and high density lipoproteins with similar sizes, lipid composition and functions to their counterparts in mammals. Avian apolipoproteins are not identical to those in mammals: for example, birds do not have the direct equivalent of mammalian apo-E or apo-A-2 and synthesise only the apo-100 form of apo-B. In laying hens, the liver is responsible for synthesis of the major yolk precursors and plasma lipoprotein metabolism and apolipoprotein content are very different from that in immature hens.

The route of absorption of dietary lipid in birds is very different from that in mammals. Dietary fat is secreted by the avian intestine directly into the portal vein as large triglyceride-rich lipoproteins: these were originally called portomicrons by Bensadoun and Rothfield (1972) by analogy with mammalian chylomicrons. Re-esterification of absorbed fatty acids in the intestinal mucosa is not complete in birds and up to 50% of absorbed fat may be released into the portal vein as non-esterified fatty acids (Sklan, Geva, Budowsky and Hurwitz, 1984). This is a much higher proportion than in the rat, where at least 90% of dietary fat is incorporated into chylomicrons.

The absorption of fat via the portal vein does not mean that dietary fatty acids are all taken up directly by the liver. The endothelial cells forming the capillary bed in avian liver are much more closely aligned than those in mammalian liver (Fraser, Heslop, Murray and Day, 1986) and portomicrons are thought to pass directly through the liver into the general circulation. However, most of the non-esterified fatty acids passing into the portal blood are probably removed from the circulation during their initial passage through the liver. In addition, a proportion of portomicron triglyceride will be taken up by the liver after hydrolysis releasing portomicron remnants into the peripheral circulation. Hepatic synthesis of fatty acids is strongly inhibited by dietary fat in birds (Saadoun and Leclercq, 1987) and this probably accounts for the lack of effect of dietary fat content *per se* on body composition. Body fat content of chickens is influenced by the energy content of the diet.

Studies in mammals have shown that a substantial proportion of lipoprotein triglyceride is taken up and oxidised by muscle. If this also occurred in birds, it could have a significant effect on the relative importance of liver as a source of fatty acids for adipose tissue growth. However, *in vivo* measurement of the rate of lipoprotein triglyceride secretion and of the fate of biologically labelled [^{14}C]-VLDL (Griffin, unpublished) indicated that lipoprotein metabolism accounted for about 85% of the fatty acids deposited in the abdominal fat pad of 6-week-old broilers. Adipose tissue lipogenesis was sufficient to account for the remaining 15%. Only 25% of intravenously injected ^{14}C-VLDL was oxidised within 8 h, indicating that only a relatively small proportion was taken up by muscle.

Lipogenic enzyme activity is measurable in avian adipose tissue and the low rate of adipose tissue lipogenesis in birds could be due to its suppression by exogenous fatty acids from the liver or intestine. However low rates of secretion of VLDL into the circulation produced by genetic selection or by feeding diets with high protein/energy ratio were not found to stimulate a compensatory increase in incorporation of [^{3}H]-water into adipose tissue fatty acids (Griffin, unpublished). The importance of adipose tissue lipogenesis is not, therefore, greater in leaner birds.

Plasma VLDL concentrations in fully-fed broiler chickens and turkeys are sufficiently well correlated with body fat content to act as an indirect measure of fatness (Griffin and Whitehead, 1982). The correlation coefficients found (r=0.6-0.7) indicate that approximately 50% of the genetic variation in fatness in normal commercial populations is attributable to differences in plasma VLDL concentration. Divergent selection for high and low plasma lipoprotein concentration has produced lean and fat lines with marked differences in feed and protein conversion efficiencies (see Whitehead,1988). After 10 generations of selection, the difference in VLDL concentrations between lines was about 20-fold and this was largely due to a 4 to 5-fold difference in the rate of VLDL secretion from the liver (Griffin, Acamovic, Guo and Peddie, 1989; Griffin, Windsor and Whitehead, 1991). However, the rates of hepatic lipogenesis between strains differ by only 2-fold (Griffin, Windsor and Zammit, 1990). β-hydroxybutyrate concentrations in the plasma of *ad libitum*-fed chickens are much higher than in fed rats (0.5 vs 0.1 μmole/ml) and this implies that β-oxidation is a significant fate for fatty acids in chicken liver, even in the fully-fed state. Malonyl CoA is believed to be the major regulator of β-oxidation in mammals through its inhibition of mitochondrial palmitoyltransferase I, but malonyl CoA concentrations are low in chicken liver despite a high rate of lipogenesis (Griffin *et al*, 1990). The combination of a relatively high rate of β-oxidation in chicken liver with a high rate of fatty acid synthesis would explain how relatively small changes in hepatic lipogenesis can have a large effect on rate of VLDL secretion and provides a particularly sensitive method for controlling rate of lipoprotein secretion.

Lipoprotein lipase (LPL) is believed to have a key role in controlling the partitioning of plasma triglyceride between the mammary gland and adipose tissue during lactation (Cryer, 1981), but the evidence for a major role for lipoprotein lipase in partitioning lipid between skeletal muscle and adipose tissue during growth is much less certain. Divergent selection of broiler chickens for plasma VLDL concentration has created significant differences in plasma post-heparin, adipose tissue and muscle LPL activity, all of which are greater in the leaner, low-VLDL line (Griffin *et al*, 1989). The higher activity in muscle in the lean line is not accompanied by any significant difference in the proportion of VLDL-TG that is oxidised to CO_2 (Griffin *et al*, 1991).

The difference in the relative proportion of whole body LPL activity in muscle and adipose tissue is much greater between broiler and layer strain chicks, but this is accompanied by only a small difference in the proportion of [^{14}C]-VLDL-TG oxidised to CO_2 in the two strains (Griffin *et al*, 1991). Differences in the proportion of plasma VLDL-TG taken up by the abdominal fat appear to be driven by differences in VLDL secretion rate independently of depot LPL activity (Griffin *et al*, 1989) and, perhaps as expected, by differences in depot LPL activity (Leclercq, Hermier and Guy, 1990; Hermier *et al*, 1989). This limited evidence appears to indicate that avian adipose tissue is particularly effective at sequestering circulating triglyceride and that the partitioning of the latter between skeletal muscle and adipose tissue is not a simple function of the relative LPL activities of the two tissues. Clarification of the role of LPL during growth will require the development of methods for measuring functional LPL activity in tissues (*ie* the LPL activity at the capillary bed) but the broiler chicken does provide a very appropriate model for the *in vivo* study of lipoprotein metabolism.

Hormonal control of fat deposition in birds

LIPOGENESIS

The relative importance of insulin and glucagon in the control of metabolism is very different between birds and mammals, with short-term control of hepatic metabolism in birds being primarily under the control of glucagon (Cramb and Langslow, 1984). Plasma insulin concentrations in birds respond to long-term fasting and refeeding in much the same way as mammals, but avian pancreatic β-cells are unresponsive to changes in plasma glucose concentration and the mechanisms responsible for controlling release of insulin into the portal vein are not known (see Rideau, 1988).

Unlike the situation in the rat, insulin has no immediate effects on carbohydrate and lipid metabolism in the isolated avian heptocyte (Cramb and Langslow, 1984).

Long-term induction of lipogenic enzyme synthesis by insulin has been demonstrated *in vitro* (eg Tarlow, Watkins, Reed, Miller, Zwergel and Lane, 1977) and this is important in regulating lipogenesis *in vivo* (see Goodridge, 1987). In contrast, glucagon has an immediate inhibitory effect on both lipogenesis and lipoprotein secretion from the chicken hepatocyte (Tarlow *et al*, 1977). Differences in responsiveness to insulin and glucagon are reinforced by comparisons of receptor occupancy. Cramb and Langslow (1984) estimated that the most of the insulin receptors on chicken hepatocytes would be occupied at the likely concentrations of insulin in the portal blood in fed birds. However, only 10% of glucagon receptors would be occupied at the likely concentrations of glucagon.

Cramb and Langslow (1984) have suggested that these species differences in hormonal control of metabolism are an evolutionary consequence of differences in behaviour and digestive physiology. Granivorous birds tend to eat every 15 -20 min throughout the day and retain much of the food for long periods within the crop. This provides a steady stream of nutrients to the liver and removes the need for regular switching between anabolic and catabolic states characteristic of species eating at less frequent intervals.

Other hormones that have been shown to influence lipogenesis *in vitro* include GH, IGF-I, T3 and corticosterone, prolactin and oestrogen (Leclercq, 1984; Decuypere and Boyse, 1988; Cupo and Cartwright, 1989). However, comparison of lean and fat lines of chicken have revealed no consistent differences in circulating concentrations of glucagon, corticosterone or oestrogen, although T_3 concentrations tend to be higher in the plasma of leaner birds and insulin concentrations tend to be higher in fatter birds (particularly after fasting and refeeding (see Simon, 1988; Leclercq, 1988). Recent studies on the lean and fat lines of broiler chicken produced by divergent selection for plasma VLDL concentration have shown that the low VLDL line has significantly higher GH levels than the fatter, high VLDL line (Griffin and Goddard, unpublished). In contrast, plasma GH concentrations in lean and fat lines produced by divergent selection for abdominal fat were similar or slightly higher in the fat line (Leclercq, 1988).

Daily injection of GH into broiler chickens has little or no effect on growth or body composition. However pulsatile, but not continuous, infusion of GH into 8-week-old birds stimulated growth, reduced body fat content and markedly improved feed conversion efficiency (Vasilatos-Younken *et al*, 1988). This dependence of the response to GH on the method of administration is not peculiar to birds since it is also observed in the rat, but single daily injection of GH are effective in stimulating growth in pigs. These differences in response may be due to differences in control of hepatic GH receptor levels. GH binding to chicken hepatocyte membranes is decreased by daily injection of GH (Leung, Taylor, Wein and Van Iderstine, 1986), whereas that in the rat has been reported to either increase or decrease. In contrast, GH binding to pig liver membranes is increased by daily injection of GH in a dose-dependent manner (Chung and Etherton, 1986).

Pulsatile but not continuous administration of GH to chickens significantly reduces hepatic lipogenic enzyme activity (Roseborough, McMurtry and Vasilatos-Younken,

1990). This response may be due to a direct effect of GH on the liver, since GH antagonises the long-term stimulation of lipogenesis by insulin in avian hepatocytes (Harvey, Scanes and Howe, 1977; Tarlow *et al*, 1977). The mechanisms responsible are not known.

Adipose tissue lipolysis in birds

Adrenalin and nor-adrenalin are the major lipolytic hormones in mammals but the responsiveness of chicken adipose tissue to β-adrenergic stimulation is low and declines after hatch (Langslow and Lewis, 1974). In contrast, the avian adipocyte is exceptionally sensitive to stimulation by glucagon and this appears to be the principal lipolytic hormone in birds (Cramb and Langslow, 1984). GH has both anti-lipolytic and lipolytic activity towards the chicken adipocyte *in vitro* but, unlike the situation in mammals, insulin is not antilipolytic in birds (Scanes, 1987). Somatostatin at normal physiological concentrations inhibits glucagon-stimulated lipolysis in chicken adipocytes and may therefore represent the major anti-lipolytic hormone in birds (Strosser, Di Scala-Guenot, Koch and Miahle, 1983).

Young broilers are much fatter than layer strain chicks at the same age. Measurement of VLDL secretion rate and the proportion of VLDL-TG taken up by the abdominal fat pad indicated that VLDL metabolism accounted for about 65-85% of the overall rate of fat deposition in the depot in broilers. In contrast VLDL metabolism appeared to provide fatty acids to the abdominal fat pad of layers at about 4-fold the net rate of accumulation (Griffin, Windsor and Goddard, 1991). This observation implied that adipose tissue triglyceride was turning over very rapidly in young layers and subsequent studies demonstrated a half-life of 2-3 days. No turnover of adipose tissue triglyceride was detectable in broilers. There was no significant difference in plasma immunoreactive glucagon concentration between strains, but plasma GH concentrations were 8 to 10-fold higher in layers than in broilers (Griffin *et al*, 1991).

This evidence strongly implicates GH in the control of fat deposition in young birds. GH may stimulate lipolysis directly or it may increase the sensitivity of adipocytes to lipolytic hormones, as described in sheep adipose tissue (Watt, Finley, Cork, Clegg and Vernon, 1991). As discussed earlier, GH appears also to reduce hepatic lipogenesis *in vivo*. The relative fatness of broiler chickens may well therefore be a direct consequence of the low plasma GH concentrations that accompany selection for rapid growth.

Manipulation of body composition and/or feed efficiency in poultry

The prospects of manipulating growth in poultry are very much constrained by the relatively low value of individual birds, which tends to favour nutritional or genetic approaches. However, other methods are being investigated, as discussed briefly below.

NUTRITIONAL

Broiler chickens tend to eat to satisfy their protein requirement and as a consequence their energy intake and body fat content are dependent on the energy/protein ratio in the diet. Leaner carcasses can be produced by increasing the protein/energy ratio of commercial diets but this would be expensive (see Jackson, Summers and Leeson, 1982). In contrast, the body composition of turkeys is not very responsive to dietary manipulation (Sell, Hasiak and Owings, 1985).

ß-AGONISTS

The initial studies on the effect of β-agonists on growth in broiler chickens showed smaller effects on growth, feed efficiency and body fat content than in mammalian species (Dalrymple, Baker, Gingher, Ingle, Pensack and Ricks, 1984). Nevertheless, the 5% improvement in feed efficiency reported could be commercially significant because of the intensity of competition within the poultry industry. However, most of the subsequent studies in broiler chickens have shown smaller responses to β-agonists.

The mechanisms involved in the growth promoting activity of β-agonists are not clear in any species (Lafontan, Berlan and Prud'Hon, 1988), but one obvious explanation of the limited effect of β-agonists on body fat in birds is that adrenalin is only weakly lipolytic in avian species (Langslow and Lewis, 1974). Differences in the distribution and/or nature of the β-2 receptors between mammals and birds may also be important. Mammalian erythrocytes contain both β-1 and β-2 receptors, whereas avian erythrocytes contain only β-1 receptors and amphibian erythrocytes only β-2 receptors (Lefkowitz, Stadel and Caron, 1983). Similar variation in receptor distribution in target tissues (*eg* muscle, pancreas or pituitary) between species is likely to influence the overall response to β-agonists. The avian β-1 adrenoreceptor gene has been cloned and its nucleotide sequence shows 52% homology to its mammalian counterpart (Chung, Lentes, Gocayne, Fitzgerald, Robinson, Kerlavage, Fraser and Venter, 1987). No information has yet been published about the structure of an avian β-2 receptor, but differences between avian and mammalian adrenoreceptors may well be sufficient to influence their response to agonists/antagonists. Nevertheless, cimaterol has been shown to increase the synthesis and decrease the dradation of myofibrillar proteins in chick embryo muscle cells in culture (Young, Moriarity, McGee, Farrar and Richter, 1990). Cimaterol reduces fatness in ducks (*eg* Dean and Dalrymple, 1988) and ractopamine stimulates weight gain in turkeys (Wellenreiter and Tonkinson, 1990) and this evidence suggests the responses may vary between poultry species and/or between individual β-agonists, as shown in pigs (Peterla and Scanes, 1990).

IN OVO MANIPULATION

The recent development of machines capable of injecting up to 30 000 eggs/h (Thaxton, 1989) has raised the possibility of routine *in ovo* injection. Initially this was seen as an alternative to immunisation of newly-hatched chicks, but several groups have begun to investigate the possibility of endocrinological manipulation by *in ovo* injection. In view of the lack of knowledge of the relationship between the hormonal status of the embryo and growth performance after hatch, these experiments are inevitably empirical. Preliminary reports have described significantly increased growth of broilers after *in ovo* injection of GH (Hargis, Pardue, Lee and Sandel, 1989) and interleukin-2 (Ford, Morgan, Fredrickson, Thaxton, Tyczkowski and Murray, 1990).

IMMUNOLOGICAL MANIPULATION

The role of somatostatin in control of GH secretion is similar in mammals and birds and passive immunisation with anti-somatostatin antibodies produces a substantial increase in circulating GH levels in birds (Spencer, Harvey, Audsley, Hallet and Kestin, 1986; Harvey and Hall, 1987). Active immunisation of broiler chickens has produced significant growth and reductions in fatness (Spencer *et al*, 1986; Buonomo, Sabacky, Della-Fera and Baile, 1987). However responses were not correlated with changes in circulating GH concentrations, and may be operating through alternative mechanisms (eg by removal of the anti-lipolytic effect of somatostatin).

Reports of successful passive immunisation against adipocytes in rats (Flint, Cosgrave, Futter, Gardner and Clark, 1986) has prompted similar studies in the chicken. However the initial reaction to passive immunisation of broiler chickens with sheep anti-chicken fat cell antibodies was for the flesh to turn a dark shade of green and this response appeared to be a consequence of the production of relatively non-specific antisera with high reactivity towards hepatocytes and erythrocytes as well as adipocytes (Butterwith, Kestin, Griffin, and Flint, 1989). No effect on adipose tissue was found and more detailed studies are needed on the specificity of chicken adipocyte antigens.

ANABOLIC STEROIDS

Implantation with oestrogens has been used routinely in the poultry industry in the past to fatten male broiler chickens (see Weppleman, 1984) but this was accompanied by a decrease in feed conversion efficiency. Growth is stimulated by androgens in older birds, but inhibited in younger birds before epiphyseal closure. The effects of trienbolone acetate are particular striking in older turkeys, stimulating growth rate, decreasing fat content and improving feed efficiency (Wise and Ranaweera,1981). The use of anabolic steroids is currently unacceptable, but these observations provide a useful guide to mechanisms controlling growth in older birds. Interestingly, selection for improved feed efficiency in male turkeys appears to exploit variability size and premature sexual development.

GENETIC SELECTION

Most of the progress in poultry meat production to date has been due to genetic selection and this is likely to be the case in the future. Genetic selection in poultry is favoured by a short generation time, a large number of offspring/dam or sire that allows high selection pressure and the centralisation of poultry breeding on perhaps 6 or 7 major companies worldwide. These same characteristics will favour the adoption of transgenic technology, although methods for the routine production of transgenic chickens remain elusive (Crittenden, 1990).

The present increase in growth is expected to continue in the future at a rate of about 40 g/generation or about 2%/year and this will be accompanied by a 1% improvement in feed conversion efficiency (Anon, 1989). The decrease in time taken to reach marketable weight is itself contributing to a changes in selection priorities, since it is increasingly recognised that it is feed efficiency rather than growth rate that is of prime importance. Up to 70% of chickens and turkeys reared in the UK are now used for meat stripping and further processing and this has generated increased pressure to improving meat yield and body composition. This is likely to be achieved in the foreseeable future by direct selection for the parameter of interest (Leenstra, 1986) rather than selection on indirect (*eg* physiological) indices.

Conclusions

Many of the mechanisms responsible for growth are very similar in birds and other vertebrates but there are important differences, both at the metabolic level and in hormonal control of metabolism. In some cases, these differences are no greater than between mammalian species, but in others they are of very practical importance. Commercial or experimental selection has created lines differing widely in growth rate, body composition and feed efficiency and these provide a very valuable set of models for the study of the processes of growth that can be relevant to growth in all species.

Acknowledgements

Research in the authors' own laboratories was supported in part by commissions from the Ministry of Agriculture, Fisheries and Food.

References

Aberle, E.D., Addis, P.B. and Schoffner, R.N. (1979) *Poultry Science*, **58**, 1210-1212

Anon (1989) *Poultry International*, **28**, (11), 48-50

Armstrong, D.G., McKay, C.O., Morell, D.J. and Goddard, C. (1989) *Journal of Endocrinology*, **120**, 373-378

Bensadoun, A. and Rothfield, A. (1972) *Proceedings of the Society for Experimental Biology and Medicine*, **41**, 814-817

Buonomo, F.C., Sabacky, M.J., Della-Fera, M.A. and Baile, C.A. (1987) *Domestic Animal Endocrinology*, **4**, 191-200

Butterwith, S.C. (1988) In *Leanness in Domestic Birds*, (eds B. Leclerq and C. Whitehead), London, Butterworths, pp 203-222

Butterwith, S.C. (1989) *British Poultry Science*, **30**, 927-933

Butterwith, S.C. and Gilroy, M. (1991) *Comparative Biochemistry and Physiology*, (in press)

Butterwith, S.C. and Griffin, H.D. (1989) *Comparative Biochemistry and Physiology*, **94A**, 721-724

Butterwith, S.C., Kestin, S., Griffin, H.D. and Flint, D.J. (1989) *British Poultry Science*, **30**, 371-378

Bulfield, G., Isaacson, J.H. and Middleton, R.J. (1988) *Theoretical and Applied Genetics*, **75**, 432-437

Chung, C.S. and Etherton, T.D. (1986) *Endocrinology*, **119**, 780-786

Chung, F.Z., Lentes, K.U., Gocayne, J., Fitzgerald, M., Robinson, D., Kerlavage, A.R., Fraser, C.M. and Venter, J.C. (1987) FEBS *Letters*, **211**, 200-206

Cramb, G. and Langslow, D.R. (1984) In *Physiology and Biochemistry of the Domestic Fowl*, (ed B.M. Freeman), London, Academic Press, Vol 5, pp 94-125

Crittenden, L.B. (1990) In *Proceedings of the 4th World Congress on Genetics Applied to Livestock Production*, **16**, 93.

Cryer, J., Woodhead, B.G. and Cryer, A. (1987) *Comparative Biochemistry and Physiology*, **86A**, 515-521

Cryer, A. (1981) *International Journal of Biochemistry*, **13**, 525-541

Cupo, M.A. and Cartwright, A.L. (1989) *Comparative Biochemistry and Physiology*, **94B**, 355-360

Dalrymple, R.H., Baker, P.E., Gingher, D.L., Ingle, J., Pensack, M. and Ricks, C.A. (1984) *Poultry Science*, **63**, 2376-2383

De Pablo, F. (1991) *CRC Critical Reviews in Poultry Biology*, (in press)

Decuypere, E. and Boyse, J. (1988) In *Leanness in Domestic Birds*, (eds B. Leclerq and C.C. Whitehead), London, Butterworths, pp 295-312

Dean, W.F. and Dalrymple, R.H. (1988) *Poultry Science*, **67**, Supplement 1,73

Doglio, A., Dani, C., Fredrikson, G., Grimaldi, P. and Ailhaud, G. (1987) *EMBO Journal*, **6**, 4011-4016

Duclos, M.J. and Goddard, C. (1990) *Journal of Endocrinology*, **125**, 199-206

Duclos, M.J., Wilkie, R.S. and Goddard, C. (1991) *Journal of Endocrinology*, **128**, 35-42

Flint, D.J., Coggrave, H., Futter, C.E., Gardner, M.J. and Clark, T.J. (1986) *International Journal of Obesity*, **10**, 69-77

Ford, G.A., Morgan, W., Fredrickson, T., Thaxton, J.P., Tyczkowski, J. and Murray, L. (1990) *Poultry Science*, Supplement 1, 167

Fraser, R., Heslop, V.R., Murray, F.E. and Day, W.A. (1986) *British Journal of Experimental Pathology*, **67**, 783-791

Gaskins, H.R., Kim, J-W., Wright, J.T., Rund, L.A. and Hausman, G.J. (1990) *Endocrinology*, **126**, 622-630

Goddard, C. and Boswell, J. (1991) *CRC Critical Reviews in Poultry Biology*, (in press)

Goddard C. and Bulfield, G. (1990) In *Proceedings of the 4th World Congress on Genetics Applied to Livestock Production*, **16**, 235-245

Goddard, C., Wilkie, R. and Dunn, I.C. (1988) *Domestic Animal Endocrinology*, **5**, 165-176

Goodridge, A.G. (1987) *Annual Review of Nutrition*, **7**, 157-185

Griffin, H.D. and Hermier, D. (1988) In *Leanness in Domestic Birds* (eds B. Leclercq and C. Whitehead), London, Butterworths, pp 175-201

Griffin, H.D. and Whitehead, C.C. (1982) *British Poultry Science*, **23**, 307-313

Griffin, H.D., Butterwith, S.C. and Goddard, C. (1987) *British Poultry Science*, 28, 197-206

Griffin, H.D., Acamovic, F., Guo, K. and Peddie, J. (1989) *Journal of Lipid Research*, **30**, 1243-1250

Griffin, H.D., Windsor, D. and Zammit, V. (1990) *Biochemical Society Transactions*, **18**, 981-982

Griffin, H.D., Windsor, D. and Goddard, C. (1991) *Comparative Biochemistry and Physiology*, (in press)

Griffin, H.D., Windsor, D. and Whitehead, C.C. (1991) *British Poultry Science*, **32**, 195-201

Grimaldi, P., Dijan, P., Forest, C., Poli, P., Negrel, R. and Ailhaud, G. (1983) *Molecular and Cellular Endocrinology*, **29**, 271-285

Gurr, M.I. and Kirtland, J. (1978) *International Journal of Obesity*, **2**, 401-427

Hargis, P.S., Pardue, S.L., Lee, A.M. and Sandel, G.W. (1989) *Growth, Development and Aging*, **53**, 93-99

Harvey, S. and Hall, T. (1987) *General and Comparative Endocrinology*, **65**, 111-116

Harvey, S., Scanes, C.G. and Howe, T. (1977) *General and Comparative Endocrinology*, **33**, 322-328

Hausman, G.J. (1989) *Journal of Animal Science*, **67**, 3136-3143

Hausman, G.J. and Martin, R.J. (1989) *Domestic Animal Endocrinology*, **6**, 331-337

Hausman, G.J., Novakofski, J.E., Martin, R.J. and Thomas, G.B. (1984) *Cell and Tissue Research*, **236**, 459-464

Hermier, D., Quinard-Boulange, A., Dugail, I., Guy, G., Salichon, M.R., Brigant, L., Ardoiun, B. and Leclercq, B. (1989) *Journal of Nutrition*, **119**, 1369-1375
Hood, R.L. (1984) *World's Poultry Science Journal*, **40**, 160-169
Ignotz, R.A. and Massague, J. (1985) *Proceedings of the National Academy of Sciences*, **82**, 8530-8534
Jackson, S., Leeson, S. and Summers. J.D. (1982) *Poultry Science*, **61**, 2232-2240
Kang, C.W., Sunde, M.L. and Swick, R.W. (1985) *Poultry Science*, **64**, 370-379
Kikuchi, K., Buonomo, F., Kajimoto, Y. and Rotwein, P. (1991) *Endocrinology*, **128**, 1323-1328
Lafontan, M., Berlan, M. and Prud'Hon, M. (1988) *Reproduction, Nutrition, Development*, **28**, 61-84
Langslow, D.R. and Lewis, R.J. (1974) *British Poultry Science*, **15**, 267-273
Lauterio, T.J., Decuypere, E. and Scanes, C.G. (1986) *Comparative Biochemistry and Physiology*, **83A**, 627-632
Leclercq, B. (1984) *Poultry Science*, **63**, 2044-2054
Leclercq, B. (1988) In *Leanness in Domestic Birds*, (eds B. Leclercq and C.C. Whitehead, London, Butterworths, pp 25-40
Leclercq, B., Hermier, D., and Guy, G. (1990) *Reproduction, Nutrition and Development*, **30**, 701-715
Lefkowitz, R.J., Stadel, J.M. and Caron, M.G. (1983) *Annual Review of Biochemistry*, **52**, 159-186
Leenstra, F. (1986) *World's Poultry Science Journal*, **42**, 12-52
Leung, F.C., Taylor, J.E., Wien, S. and Van Iderstine, A. (1986) *Endocrinology*, **118**, 1961-1965
Lilburn, M.S., Leung, F.C., Ngiam-Rilling, K. and Smith, J.H. (1986) *Proceedings of the Society for Experimental Biology and Medicine*, **182**, 336-343
March, B.E. and Hansen, G. (1977) *Poultry Science*, **61**, 886-894
Morikawa, M., Nixon, R. and Green, H. (1982) *Cell*, **29**, 783-789
Murachi, T., Murakami, T., Ueda, M., Fukui, I., Hamakubo, T., Adachi, Y. and Hatanaka, M. (1989) *Advances in Experimental Medicine and Biology*, **255**, 445-454
O'Hea, E.K. and Leveille, G.A. (1969), *Comparative Biochemistry and Physiology*, **30**, 149-159
Peterla, T.A. and Scanes, C.G. (1990) *Journal of Animal Science*, **68**, 1024-1029
Rideau, N. (1988) In *Leanness in Domestic Birds* (eds B. Leclercq and C.C. Whitehead, London, Butterworths, pp 269-294
Ringold, G.M., Chapman, A.B. and Knight, D.M. (1986) *Journal of Steroid Biochemistry*, **24**, 69-75
Rosebrough, R.W., McMurtry, J.P. and Vasilatos-Younken, R. (1990) *Poultry Science*, **69**, Supplement 1, 114
Saadoun, A. and Leclercq, B. (1986) *Comparative Biochemistry and Physiology*, **83B**, 607-611
Saadoun, A. and Leclercq, B. (1987) *Journal of Nutrition*, **117**, 428-435

Saunderson, C.L. and Leslie, S. (1987) *Biochemistry Society Transactions*, **15**, 961
Scanes, C.G. (1987) *CRC Critical Reviews in Poultry Biology*, **1**, 51-105
Scanes, C.G. and Griminger, P. (1990) *Journal of Experimental Zoology*, Supplement 4, 98-105
Scanes, C.G., Hart, L.E., Decuypere, E. and Kuhn, E.R. (1987) *Journal of Experimental Zoology*, Supplement, **1**, 253-264
Scanes, C.G., Dunnington, E.A., Buonomo, F.C., Donoghue, D.J. and Siegel, P.B. (1989) *Growth, Development and Aging*, **53**, 151-157
Scanes, C.G., Aramburo, C. and Campbell, R.L. (1990) In *Endocrinology of Birds: Molecular to Behavioural*, (eds M. Wada, S. Ishii and C. Scanes, Japanese Scientific Societies Press, Tokyo/Springer-Verlag, Berlin, pp 92-105
Sell, J.L., Hasiak, R.J. and Owings, W.J. (1985) *Poultry Science*, **64**, 1527-1535
Sklan, D., Geva, A., Budowsky, P. and Hurwitz, S. (1984) *Comparative Biochemistry and Physiology*, **78A**, 507-510
Simon, J. (1988) In *Leanness in Domestic Birds*, (eds B. Leclercq and C. Whitehead, London, Butterworths, pp 253-268
Spencer, G.S.G., Harvey, S., Audsley, A.R.S., Hallett, K.G. and Kestin, S. (1986) *Comparative Biochemistry and Physiology*, **85A**, No 3, pp 553-556
Strosser, M.-T., Di Scala-Guenot, D., Koch, B. and Miahle, P. (1983) *Biochimica Biophysica Acta*, **763**, 191-196
Swatland, H.J. (1984) The Structure and Development of Meat Animals, Prentice Hall
Tarlow, D.M., Watkins, P.A., Reed. R.E., Miller, R.S., Zwergel, E.E. and Lane, D. (1977) *Journal of Cell Biology*, **73**, 332-353
Thaxton, P. (1989) *Poultry International*, **28** (12), 14-15
Tinch, A. (1990) PhD Thesis, University of Edinburgh
Tixier-Boichard, M., Huybrechts, L.M., Kuhn, E., Decuypere, E., Charrier, J. and Mongin, P. (1989) *Genetics, Selection, Evolution*, **21**, 217-234
Torti, F.M., Torti, S.V., Larrick, J.W. and Ringold, G.M. (1989) *Journal of Cell Biology*, **108**, 1105-1113
Vasilatos-Younken, R., Cravener, T.L., Cogburn, L.A., Mast, M.G. and Wellenreiter, R.H. (1988) *General and Comparative Endocrinology*, **71**, 268-283
Vasilatos-Younken, R., Gray, K.S., Bacon, W.L., Nestor, K.E., Long, D.W. and Rosenberger, J.L. (1990) *Journal of Endocrinology*, **126**, 131-139
Watt, P.W., Finley, E., Cork, S., Clegg, R.A. and Vernon, R.G. (1991) *Biochemical Journal*, **273**, 39-42
Weiner, F.R., Shah, A., Smith, P.J., Rubin, C.S. and Zern, M.A. (1989) *Biochemistry*, **28**, 4094-4099
Wellenreiter, R.H. and Tonkinson, L.V. (1990) *Poultry Science*, **69**, Supplement 1, 143
Weppleman, R.M. (1984) *Journal of Experimental Zoology*, **232**, 461-464
Whitehead, C.C. (1988). In *Leanness in Domestic Birds*, (eds B. Leclercq and C.C. Whitehead), London, Butterworths, pp 41-58
Wise, D.R. and Ranaweera, K.N.P. (1981) *British Poultry Science*, **22**, 93-104

Young, R.B., Moriarity, D.M., McGee, C.E., Farrar, W.R. and Richter, H.E. (1990) *Journal of Animal Science*, **68**, 1158-1169

12

RECENT DEVELOPMENTS IN THE USE OF EXOGENOUSLY ADMINISTERED PEPTIDES AND POLYPEPTIDES TO INFLUENCE ANIMAL GROWTH

D.H. BEERMANN and R.D. BOYD
Cornell University, Ithaca, New York, USA

Introduction

Consumer demand for reduced fat content of meat and the economic pressure to increase the efficiency of animal growth provide the incentive for continued efforts to manipulate the allometric patterns of tissue growth. Discovery of the key roles that peptide and polypeptide hormones or growth factors play in the regulation of cellular hyperplasia and (or) metabolism of skeletal muscle, adipose tissue and bone provided the basis for exogenous administration experiments in farm animals. Metabolic hormones such as insulin, thyroid hormones and somatotropin (ST) were first targeted for investigation in normal animals. Early efforts were hindered by two things: 1. a general lack of knowledge concerning the role and mechanisms by which these hormones impact processes such as cell proliferation, growth and metabolism and, 2. limited availability of highly purified hormone preparations. Recent technological advances have made important contributions to the resolution of each, thereby permitting scientists to use new techniques which dramatically alter growth.

Endocrine influences of insulin and the thyroid hormones have been characterized as primarily homeostatic regulators of energy, protein and lipid metabolism. Somatotropin however, exerts its influence in a chronic, coordinated way to regulate metabolism and somatotropic development of major tissues in the body during postnatal growth. This homeorhetic control of metabolism determines how absorbed nutrients are partitioned between protein and lipid deposition (Bauman, Eisemann and Currie, 1982), and therefore, directs to a significant degree the allometric patterns of skeletal muscle and adipose tissue accretion. Investigations conducted in the mid-1980's with pigs (Chung, Etherton and Wiggins, 1985; Boyd, Bauman, Beermann, DeNeergard, Souza and Butler, 1986; Etherton, Wiggins, Chung, Evock, Rebhun and Walton, 1986) and lambs (Johnsson, Hart and Butler-Hogg, 1985) provided evidence of the dramatic effects of exogenous ST administration on growth performance and carcass composition in growing animals. These studies raised important questions regarding dose-response relationships among response variables, the importance of nutritional status in relation to expression of the biological response and responsiveness with

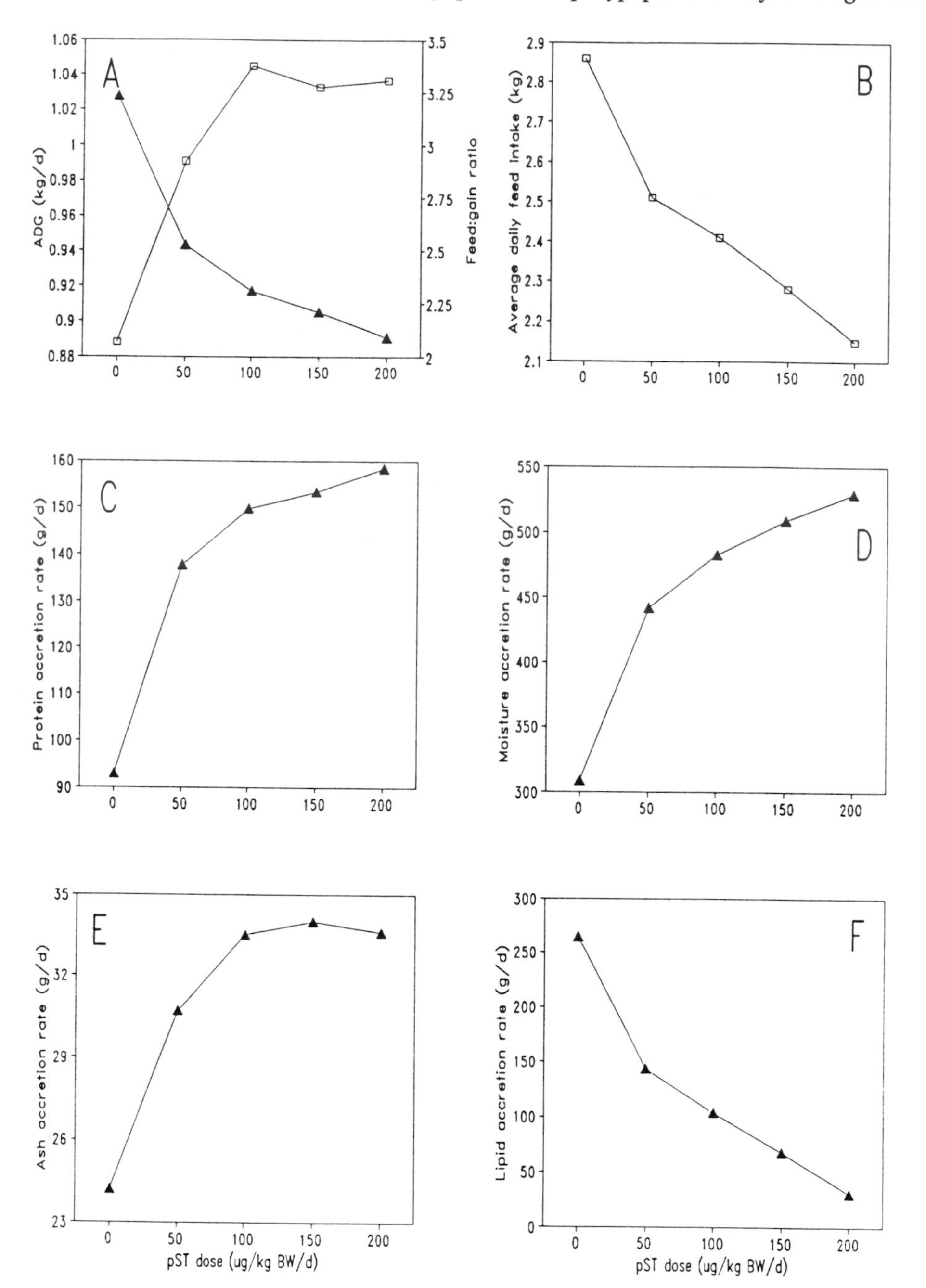

Figure 1 Growth performance and carcass accretion rate responses of barrows to daily administration of excipient or pST (50, 100, 150 or 200 μg/kg BW from 30 to 90 kg, Krick *et al*, 1991). Protein (C), moisture (D), ash (E) and lipid (F) accretion rates were determined by comparative slaughter.

phase of growth, gender and genotype. These issues are briefly discussed in this review because important considerations to peptide or polypeptide manipulated growth emerge. We will also review recent work which explores the potential with growth hormone releasing-factor (GRF), insulin-like growth factor(s) and associated analogues. Mechanisms(s) of ST action will not be addressed in detail, because a number of excellent reviews have been written recently (Isaksson, Eden, Jansson, Lindahl, Isgaard and Nilsson, 1986; Florini, 1987; Boyd and Bauman, 1989; Etherton, 1989; Vernon and Flint, 1989; Beermann and DeVol, 1991). Although IGF is considered a possible target for manipulation, much remains to be established regarding the importance of circulating concentrations versus *in situ* tissue production of growth factors (*ie* IGF-1), and the role of binding proteins in mediation of the biological effect(s) in growing animals.

Effects of exogenous somatotropin on growth and composition of gain

DOSE-RESPONSE RELATIONSHIPS IN GROWING PIGS

The nature and magnitude of growth in response to exogenous ST is best portrayed by the results of dose-response studies in growing pigs. Average daily gain (ADG) is increased with increments of porcine ST (pST) (up to 20% with 150 μg/kg body weight (BW) per day), and feed conversion efficiency is improved through an even greater dose range (Figure 1). The latter is explained in part by the continued reduction in intake to the highest dose. These relationships have been documented to some extent in other studies with growing pigs fed *ad libitum* (Boyd *et al*, 1986; Etherton, Wiggins, Evock, Chung, Rebhun, Walton and Steele, 1987; McLaren, Bechtel, Grebner, Novakofski, McKeith, Jones, Dalrymple and Easter, 1990). We are unaware of dose-response studies conducted with restricted feed intake.

The ST-induced shift in nutrient use toward greater rates of protein accretion and reduced lipid accretion also results in an improvement in feed efficiency. Although the energetic cost of lipid and protein accretion is relatively similar, protein deposited as skeletal muscle has associated with it approximately four units of water per unit of protein. This results in a decided advantage for muscle over adipose in tissue weight gain per unit of diet consumed. Carcass protein accretion rates can be increased up to 71%, coincident with an 88% decrease in lipid accretion rate when pST is administered from 30 kg to 90 kg BW (Krick, Roneker, Boyd, Beermann, David and Meisinger, 1992; Figure 1). Water accretion rates parallel changes in protein deposition. Ash accretion rate is increased 26% to 40%, thus stimulation of bone growth by ST is also dose-dependent with near maximal response being achieved at the 100 μg pST/kg BW dose.

The relative reductions in lipid accretion rate or mass of adipose tissue at constant slaughter weight are greater than the relative increases in protein deposition rate or muscle mass, respectively (Thiel, Beermann and Boyd, 1990;

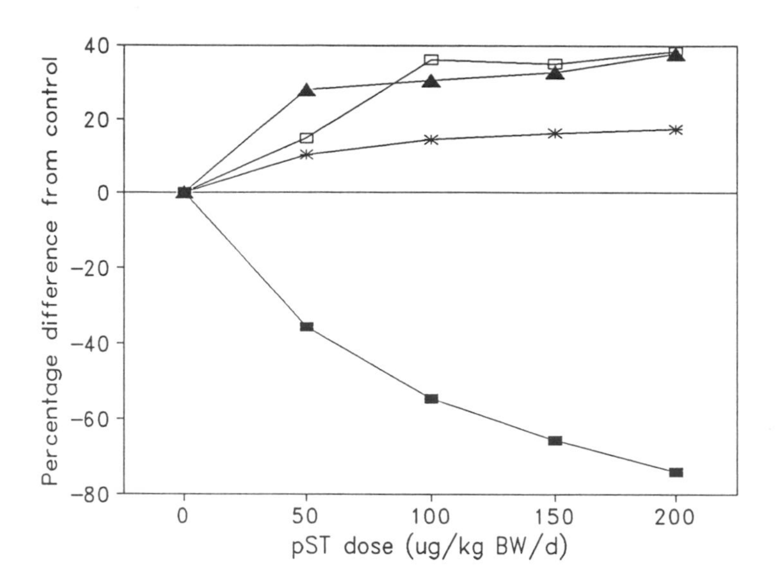

Figure 2 Dose-response changes in dissected weights of skeletal muscle (▲), adipose tissue (▪), skin (*) and bone (▫) from carcasses of barrows injected daily with excipient or pST from 30 to 90 kg BW (n=10 per pST dose). Control carcasses weighed 76.6% of live weight and contained 30.9 kg of muscle, 22.8 kg of adipose, 3.7 kg of skin and 7.3 kg of bone (Thiel *et al*, 1990).

Figure 2). Total carcass skeletal muscle mass is increased 28% to 38% (+8.5 kg to +11.5 kg) in a 90 kg pig, while adipose tissue mass is decreased 36% to 74% (-8.2 kg to -16.8 kg). These results clarify two important aspects of the relationships between growth performance and ST administration. First, the maximum increase in rate of body-weight gain may be constrained by a reduction in adipose weight which more than offsets the increase in muscle mass. Second, the diminishing relative response of protein accretion to pST doses above 75 to 100 μg/kg BW is not observed in lipid accretion rate. The reduction in lipid accretion rate appears linear from 50 to 200 μg pST/kg BW. These results suggest that the physiological effects of ST on composition of gain reflect independent effects on skeletal muscle and adipose tissue. Significant information is now available to verify this.

In addition to the changes in skeletal muscle and adipose tissue mass which are brought about with pST administration, significant dose-dependent changes in proximate composition of these tissues is also observed. Protein concentration is increased by approximately 4% to 5% in muscle with 50 to 200 μg pST/kg BW, and a 29% to 70% reduction in percentage intramuscular lipid is observed at these same doses (Beermann, Thiel and Prusa, 1990). Lipid concentration in adipose tissue is likewise reduced in a dose-dependent manner (15% to 50%), while the percentage protein and water is increased 100% or more at the 150 and 200 μg pST/kg doses. These changes represent significant improvement in nutrient composition of meat from pST treated pigs. However, cholesterol concentration of longissimus muscle is not changed, and minor increases in percentage of total unsaturated fatty acids are observed in intramuscular or subcutaneous fat from pigs administered a wide range of pST dose (Prusa, 1989; Beermann, Thiel and Prusa, 1990; Mourot, Bonneau, Charlotin and Lefaucher, 1992). Thus total energy in 100 g of longissimus muscle was reduced by 10% and percent derived from fat was reduced 50%.

The differential contributions of skeletal muscle and adipose tissue to total carcass protein, lipid, moisture and ash yield under the influence of exogenous pST are shown in Figure 3. Half-carcass lipid content exhibits a greater relative (-81%) and absolute (-3.8 kg) decrease at the highest pST dose than the respective 31% and 1.2 kg increases in half-carcass protein content. It is interesting to note that total adipose tissue protein, water and ash content declined with increasing pST dose, while all three constituents increased in skeletal muscle. Whether or not adipose cellularity was altered by pST was not determined.

DOSE-RESPONSE RELATIONSHIPS IN GROWING RUMINANTS

Growing ruminants also respond to exogenous ST, but improvements have not been of the magnitude observed with pigs. Until recently, it was unclear if this was the result of a biological difference between species, or whether the more complex digestive system of the ruminant imposed nutritional constraints. Although significant effects and dose-response relationships have been likewise demonstrated in lambs (Johnsson, Hathorn, Wilde, Treacher and Butler-Hogg, 1987; Zainur, Tassell, Kellaway and Dodemaide, 1989) and cattle (Crooker, McGuire, Cohick, Harkins, Bauman and Sejrsen, 1990; Moseley, Paulissen, Goodwin, Alaniz and Claflin, 1990; see review by Enright, 1989), these responses are impressive, but not equal to those observed in pigs.

Typical responses of growing lambs to exogenous ST are shown in Table 1. Average daily gain is increased 12 to 19%, and feed conversion efficiency is increased 20 to 22% when doses of 100 to 200 μg ST/kg BW are administered for 8 to 12 weeks. Feed intake has generally not changed with ST treatment in lambs, and a 13% increase was observed in lambs treated with 100 μg bovine somatotropin (bST)/kg BW for 11 weeks (Pell, Elcock, Harding, Morrell, Simmonds and Wallis, 1990). Carcass protein and moisture accretion rates were increased 36% and 33%, respectively, and lipid accretion rates were reduced 30% in the recent study by Beermann, Hogue, Fishell, Aronica, Dickson and Schricker (1990). These relative responses are approximately half those observed in pigs administered similar doses of ST for similar treatment periods. However, the

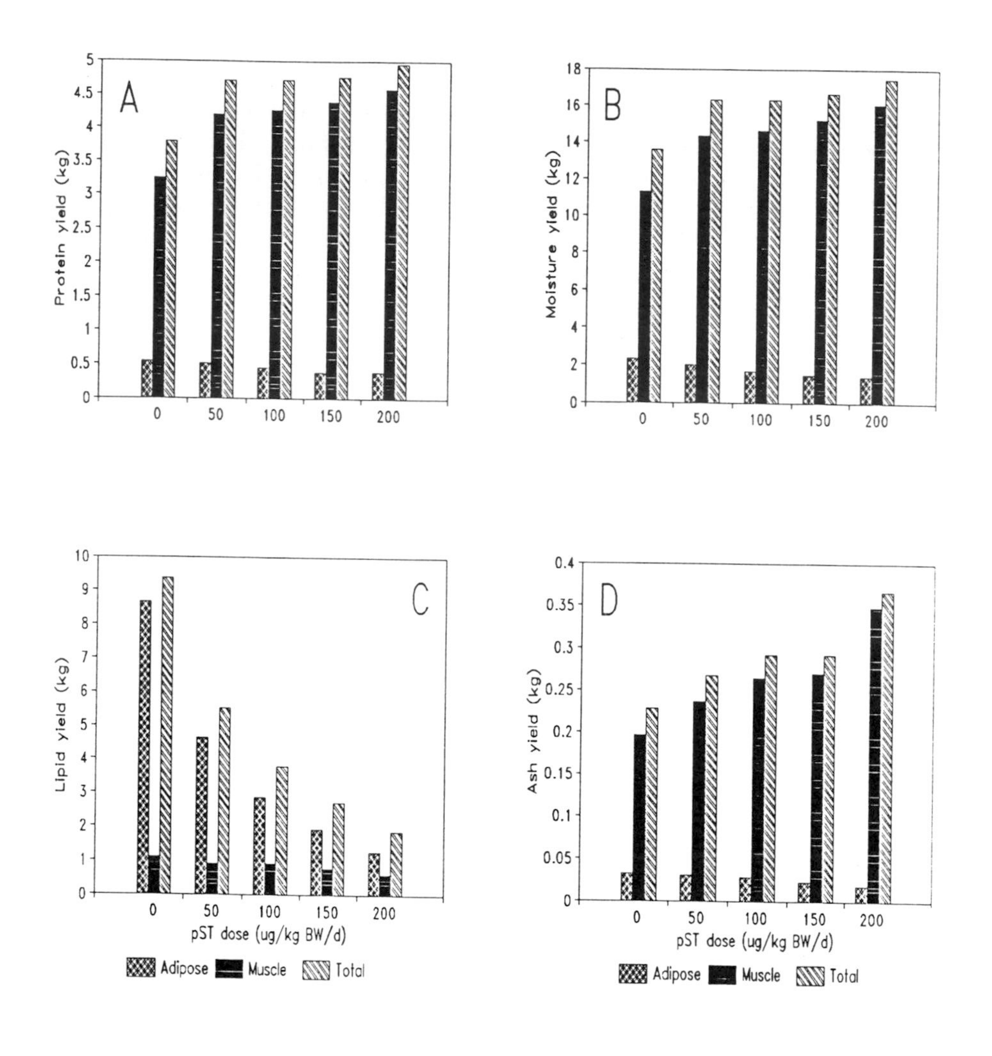

Figure 3 Protein (A), moisture (B), lipid (C) and ash (D) content of dissected skeletal muscle and adipose tissue from half-carcasses of barrows injected with excipient or increasing dose of pST from 30 to 90 kg BW (n=10 per pST dose; Thiel and Beermann, unpublished data).

18% increase in individual muscle weights (Beermann *et al*, 1990) and 24% increase in total dissected muscle (Johnsson *et al*, 1985) observed in lambs treated with ST are not markedly different from results observed in pigs.

Dose-response data for effects ST on nitrogen retention also exhibit smaller relative increases for growing cattle than for growing pigs. Crooker *et al* (1990) observed a graded increase in growing dairy heifers given 6.7, 33, 67 and 100 μg bST/kg BW. A maximal (+23%) response was observed with a 200 μg/kg dose (Figure 4). Nitrogen retention was increased approximately 30% in growing steers (Houseknecht, Bauman, Fox, Smith and Musso, 1990) and lambs (Beermann, Robinson, Byrem, Bell, Hogue and McLaughlin, 1991) administered twice-daily injections of 100 μg recombinant bovine somatotropin (rbST) for 14 days. In contrast, Wray-Cahen, Ross, Bauman and Boyd (1991) observed a 67% increase in nitrogen retention in barrows weighing 61 kg BW and administered 120 μg pST/kg BW for 19 days.

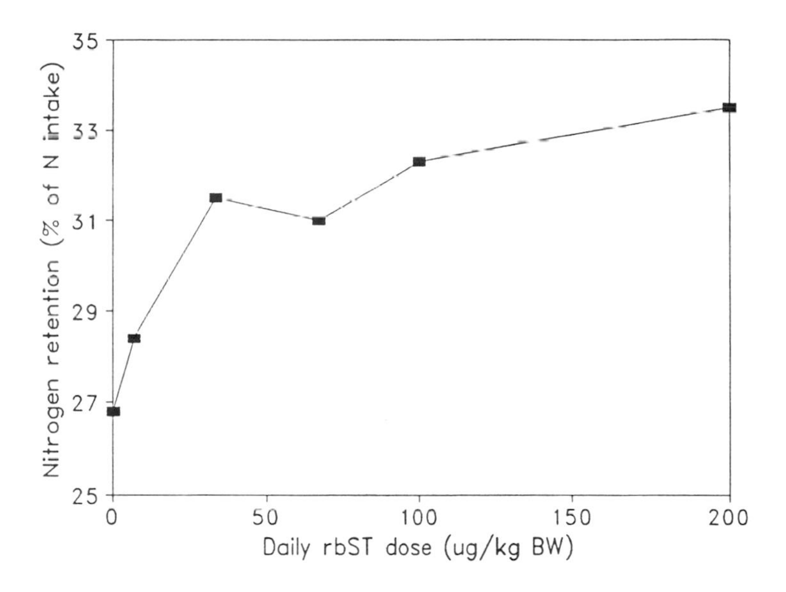

Figure 4 Dose-response curve for effects of daily administration of recombinant bST on nitrogen balance in growing dairy heifers (n=6; 104 kg initial BW). Treatments were administered for 14 days. Data were replotted from Crooker *et al* (1990).

Table 1 EFFECTS OF PITUITARY-DERIVED OVINE SOMATOTROPIN (oST) ON GROWTH PERFORMANCE AND COMPOSITION OF GAIN IN LAMBS[1]

		oST Treatment		
Response	Control	8 week	6 week + 2 week withdrawal	Sx
Number of animals	18	9	10	
Plasma hormone concentration, ng/ml				
oST	2.15	22.32*	1.98	.71
IGF-I	278.4	469.0*	284.3	35
		% Difference vs control		
Average daily gain, g	304	+14*	+10	12
Feed:Gain[2]	4.99	-22.4*	-15.0	.24
Carcass weight, kg	22.5	+6.6	0	.75
Carcass accretion rate, g/d				
Protein	17.2	+36**	+23.7*	.9
Moisture	55.6	+33.5**	+21.3*	2.7
Lipid	80.0	-30.4**	-22.7*	3.4
Ash	4.2	+18*	+14.3	.57
Semitendinosus wt, g	91.6	+20**	+16*	2.1
Semimembranosus wt, g	261.5	+15*	+10.5*	5.8

[1] Animals were housed in pairs and were treated with excipient or one-fourth of the total 160 μg oST/kg BW dose at 0100, 0700, 1300 and 1900 h. Half of the lambs were withdrawn from treatment after 42 days. Carcass composition of gain data were analyzed by analysis of variance using carcass weight as a covariate. Summarized from Beermann *et al* (1990).
[2] Feed intake data are averages for both non-withdrawal and 2-week withdrawal groups.

* (P<.05) ** (P<.01)

Although several studies have demonstrated that bST reduces fat content of the carcass in veal calves and finishing cattle, it appears that in only one study has significant effects of ST dose on skeletal muscle growth in cattle been observed (Moseley *et al*, 1990). Percentage protein in the 9th-to-11th rib section increased in linear manner with ST dose (+10.6%, +13.6% and +39.4% with 33, 100 and 300 μg ST/kg BW, respectively) in finishing steers treated from 393 kg to 540 kg (Table 2). The lack of skeletal muscle growth enhancement in other studies may be explained by the use of lower doses (Early, McBride and Ball, 1990), long withdrawal times (Enright, Quirke, Gluckman, Breier, Kennedy, Hart, Roche, Coert and Allen, 1990) and use of much younger cattle (Groenewegen, McBride, Burton and Elsasser, 1990).

Table 2 EFFECT OF RECOMBINANT bST ON GROWTH PERFORMANCE AND CARCASS COMPOSITION IN GROWING-FINISHING STEERS[1]

		rbST dose (μg/kg per day)			
Response	Control	33	100	300	P-value
		% Difference vs control			
Average gain, kg/d	1.14	+7.9	-9.2	-37.7	0.01
Feed intake, kg/d	8.4	-6.0	-13.1	-16.6	0.03
Feed:Gain, kg/kg	7.34	+11.6	+6.3	-35.4	0.03
Composition of 9th-11th rib section					
% Protein	13.2	+10.6	+13.6	+39.4	0.001
% Water	44.1	+10.2	+15.2	+39.9	0.001
% Fat	41.9	-13.4	-21.2	-54.4	0.001

[1] Summarized from Moseley *et al* (1990). N=96 steers in the study.

Moseley *et al* (1990) also observed a dose-dependent decrease in feed intake (to 16% at the highest dose). This could potentially compromise the ability of ST to enhance protein accretion because nutrient intake would be decreased accordingly. In other studies, feed intake was either not changed or was increased slightly in veal calves (Groenewegen *et al*, 1990; Maltin, Delday, Hay, Innes and Williams, 1990), growing steers (Peters, 1986; Enright *et al*, 1990; Early *et al*, 1990) or heifers (Sandles and Peel, 1987) fed *ad libitum* and administered bST.

It seems apparent, although not confirmed, that the effects of exogenous ST in cattle are more pronounced at the stage of physiological maturity where rates of lipid accretion are high relative to protein accretion rates (*ie* near market weight). The greater relative response of pigs during the 50 to 100 kg stage of growth, compared to the 20 to 50 kg BW range, is evident (Boyd, Bauman, Fox and Scanes, 1991). A marked decrease in responsiveness for each component of growth (lipid, protein and ash) was observed for the earlier phase of growth in pigs. Thus, responsiveness to exogenous ST may be a function of stage of maturity.

AGE, GENDER AND GENOTYPE INTERACTIONS

An animal's capacity for protein synthesis and accretion is dependent upon age, gender and genotype. Rates of nitrogen retention or protein deposition decline with increasing age in growing pigs (Carr, Boorman and Cole, 1977; Dunkin and Black, 1985) and ruminants (Black and Griffiths, 1975), reflecting the normal allometric pattern of tissue growth (Shields, Mahan and Graham, 1983). As an animal approaches market weight, or mature weight, nutrient requirements for lean tissue (muscle) growth decline and greater proportions of nutrients consumed are deposited as fat. These patterns are different among sexes and genotypes. Campbell, Taverner and Curic (1985) demonstrated that when protein intake is adequate, intact males exhibit superior rates of protein accretion and lesser rates of lipid accretion across a wider range of energy intake than their castrate male and normal female counterparts. Campbell and Taverner (1988) demonstrated similar differences for fast and slow growing strains of pigs at 80 kg BW.

Experiments were conducted at Cornell to compare responses of gender (boars and barrows) and genetic strain (barrows) to pST dose (Krick *et al*, 1992). Pigs received daily injections of excipient or 50, 100, 150 or 200 μg recombinant porcine somatotropin (rpST)/kg BW from 30 to 90 kg BW. Diets were formulated to provide sufficient protein intake (approximating an ideal amino acid pattern) to support whole-body protein accretion rates of 280 g/day. Relative increases in the rate of gain and protein accretion, as well as relative decreases in feed:gain ratios and lipid deposition were similar for boars and barrows at most doses. However, sex differences were not completely eliminated until the 200 μg/kg BW dose (Figure 5). Similar effects to reduce sex differences among boars, barrows and gilts administered one dose of pST have been reported (Campbell, Steele, Caperna, McMurtry, Solomon and Mitchell, 1989).

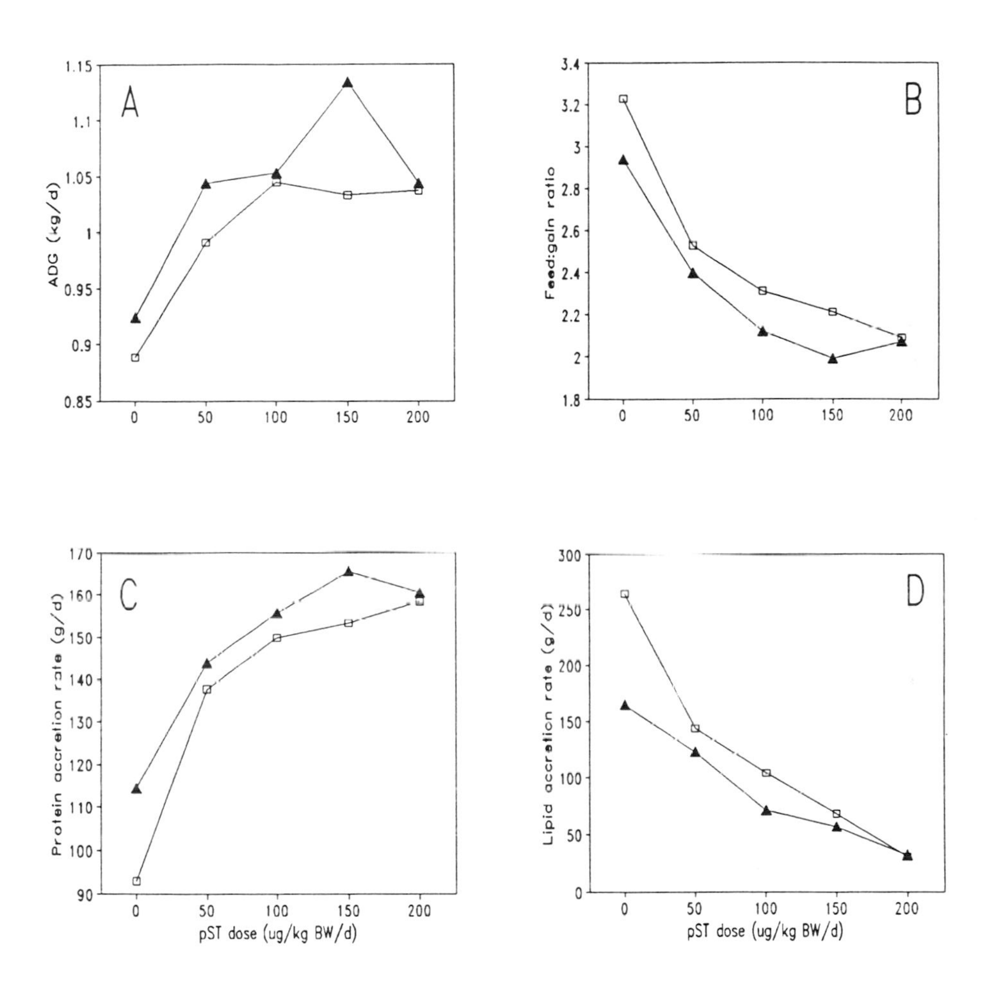

Figure 5 Comparisons of dose-response curves in barrows and boars for effects of daily pST administration on average daily gain (A), kg feed/kg gain (B), carcass protein accretion rate (C) and carcass lipid accretion rate (n=10 per subclass; Krick *et al*, 1992)

The dose-response curves for two genotypes are presented in Figure 6. Growth performance of the inferior genotype never equalled that of the superior genotype at any dose of pST. Carcass protein accretion rates were approximately 40% greater in barrows of the superior genotype without pST treatment, although lipid accretion rates were similar (260 g/d). Administration of pST brought the protein (and moisture) accretion rates closer together with increasing dose, but differences remained (21% and 11%, respectively) throughout the dose range.

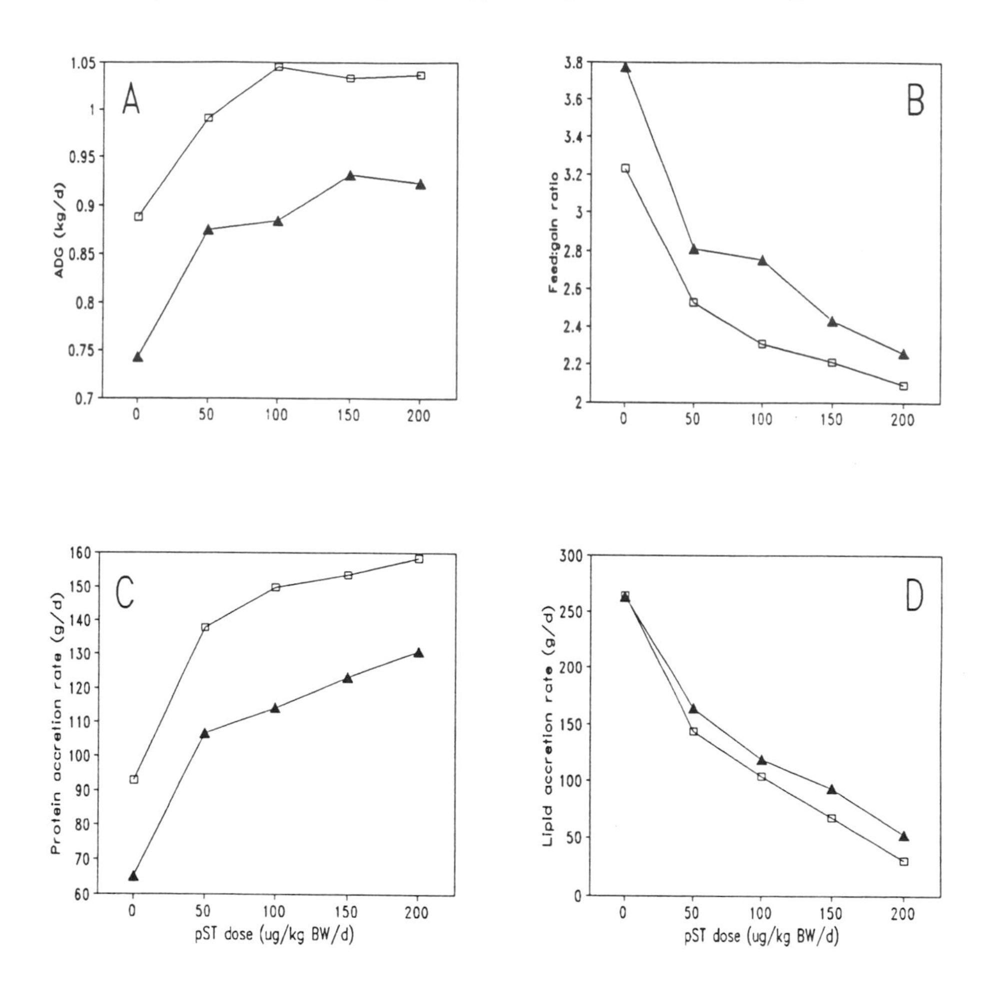

Figure 6 Comparisons of pST dose-response curves in barrows of two genotypes differing in growth performance and protein accretion rate (n=10 per subclass). Pigs were injected daily from 30 to 90 kg BW (Krick *et al*, 1992). Differences between genotypes remained across all doses of pST.

Studies with genotypes which exhibit more extreme differences have produced similar results. Pigs of three genotypes differing markedly in lipid accretion rates exhibited similar relative growth performance responses to 14 mg pST injected twice per week from 60 to 100 or 140 kg BW (Kanis, Nieuwhof, deGreef, van der Hel, Verstegen, Huisman and van der Wal, 1990). However, backfat was reduced least and lean tissue growth rate was increased least in the Pietrain genotype, compared with the intermediate and fattest genotypes. The fattest genotype exhibited the greatest relative composition changes, but never achieved the lean:fat ratio of the Pietrain pigs. Likewise, Beijing Black pigs exhibited larger relative improvements in growth performance and carcass merit than those observed in other breeds (McLaughlin, Baile, Shun-Zhang, Lian-Chun and Jin-Pu, 1989), but absolute growth rates, feed efficiency and carcass merit are superior in pST treated pigs from European and US breeds. The authors are unaware of reports in which systematic comparison of the effects of ST among sexes or breeds of growing cattle or lambs have been conducted at similar doses.

RELATIONSHIPS BETWEEN NUTRITIONAL ADEQUACY AND RESPONSE TO SOMATOTROPIN

The importance of accommodating nutrient requirements for lean tissue growth cannot be overemphasized. Protein accretion and skeletal muscle growth may be constrained by inadequate intake of either amino acids or energy. The interrelationships between these diet constituents and protein accretion have been systematically studied in several species, but the data are most extensive for growing pigs (see review by Campbell, 1988). The gains in lean tissue (skeletal muscle) growth and productive efficiency which are achievable with growth enhancers are critically dependent upon adherence to the fundamental concepts of protein and energy nutrition. Failure to recognize and adhere to these principles may compromise the biological response which is sought. Important aspects of these concepts have been discussed in recent reviews (Reeds and Mersmann, 1991; Boyd *et al*, 1991).

The ruminant presents some special challenges in accommodating tissue amino acid requirements because of rumen fermentation. As a result, the actual amino acid supply from microbial protein and ruminal escape protein cannot be predicted with a high degree of certainty. Furthermore, amino acid requirements have not been defined (NRC, 1985). Comparison of the amino acid pattern of microbial protein with that of body tissues suggests that it may have a high biological value (Boyd *et al*, 1991), but it provides only approximately 50 to 70% of the requirement (NRC, 1985). Accommodating residual needs for essential amino acids requires that protein which escapes the rumen be of sufficient quantity and appropriate amino acid composition to match tissue needs. This is difficult to determine, but nevertheless essential, particularly with ST treatment.

Two research groups at Cornell University adopted a similar approach to address the question of amino acid constraints on protein deposition response to ST administration. In one study N retention was measured in growing lambs receiving abomasal infusion of casein with and without ST administration (Beermann *et al*, 1991). Lambs weighing 23 kg were surgically fitted with abomasal cannulae and fed a total mixed diet at 85% of *ad libitum* intake. All lambs received continuous infusion of 2 litres of water or the amount of casein equalling 25% of *ad libitum* N intake and twice daily subcutaneous injections of 0 or 100 μg recombinant methionyl bovine somatotropin (Sometribove®) per kg BW for 15 days. Casein infusion increased N balance 43% (P<.001), and rbST increased N balance 34% (P<.001), without significant interaction (Figure 7). Combined effects of casein and rbST were additive, resulting in an 89% increment in N balance when compared with the water plus excipient control treatment.

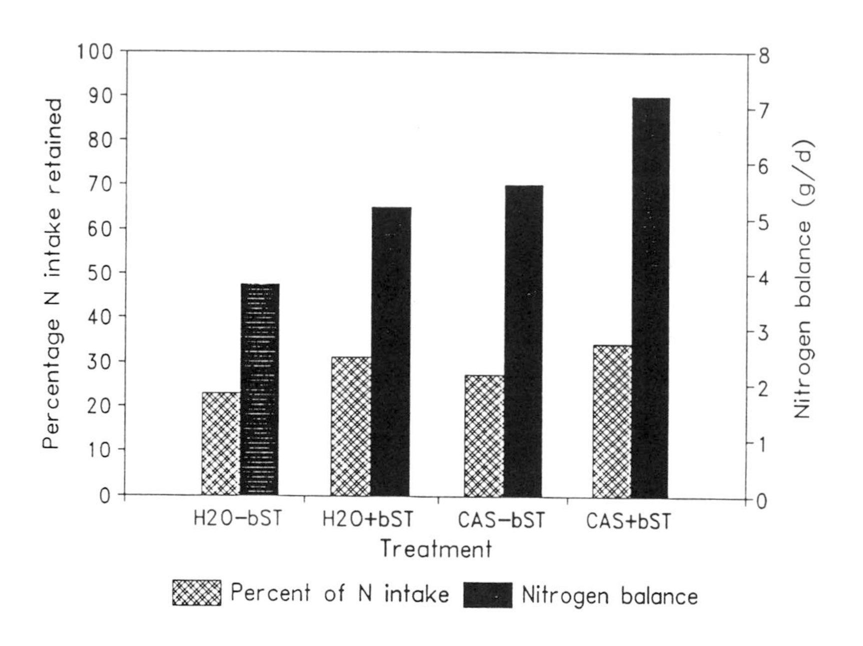

Figure 7 Effects of abomasal casein (CAS) infusion and recombinant methionyl bST on nitrogen utilization in growing wether lambs (n=7; 29 kg BW) fed a complete mixed dry diet at 85% of *ad libitum* intake. Lambs were injected subcutaneously twice daily with excipient or 100 μg rbST/kg BW for 14 days; N balance was determined on d 6 to d 14 (Beermann *et al*, 1991).

Similar effects of abomasal casein infusion and bST treatment on N balance have recently been reported in growing steers (230 kg) fed a conventional diet at 2.5 times maintenance and 115% of NRC requirements for metabolizable protein (Houseknecht *et al*, 1990). Infusion of casein and administration of bST individually increased N retention (21% and 33%, respectively), but the combination resulted in a 75% increase relative to the control group (water infusion). Results of the lamb and steer studies concur with those of several detailed studies with pigs (Krick, Roneker, Harrell, Boyd, Beermann and Kuntz, 1990; Campbell, Johnson, King, Taverner and Meisinger, 1990; Caperna, Steele, Komarek, McMurtry, Rosebrough, Solomon and Mitchell, 1990) in which the efficiency of utilization of absorbed amino acids for protein deposition was significantly increased (approximately 40%) with ST administration. Thus, this metabolic effect of ST would permit an increase in protein deposition in growing ruminants fed conventional diets, but the full biological potential would have been masked without amino acid (casein) infusion. It is not yet apparent how bST effects this change in the efficiency of N use.

Pell and Bates (1990) found that the increase in muscle mass in lambs administered bST could be explained by the magnitude of increase in non-collagen protein synthesis. Whole-body fractional rates of protein synthesis were increased, and rates of leucine oxidation were reduced in growing heifers administered bST and fed presumed adequate amounts of protein for near maintenance levels of energy (Eisemann, Tyrrell, Hammond, Reynolds, Bauman, Haaland, McMurtry and Varga, 1986). Similar effects on N and protein metabolism were observed in growing steers fed adequate protein and energy (Eisemann, Hammond, Rumsey and Bauman, 1989).

Another aspect of ST influence on nutrient requirement is the increase in bone growth which occurs in a dose-dependent manner, as evidenced by increased ash accretion rates in lambs (Beermann *et al*, 1990) and pigs (Boyd and Bauman, 1989; Campbell *et al*, 1989). Definitive information on this subject is needed before conclusions can be made.

BIOLOGICAL ACTIVITY OF SOMATOTROPIN ANALOGUES

A recent paper from our laboratory demonstrates that the potency of the ST molecule could be increased by structural modification. We compared the biological activity of a novel 21 kDa variant of the 22 kDa form of pST (Boyd, Beermann, Roneker, Bartley and Fagin, 1988). This novel variant was derived through genetic manipulation so that deletion of amino acids in positions 32 to 38 occurred. In preliminary experiments with rats, it was shown to exhibit approximately twice the activity of the 22 kDa form using the hepatic membrane receptor and hypophysectomized rat growth assay systems (K. D. Fagin, personal communication). The 21 kDa analogue exhibited greater nutrient partitioning activity in growing pigs with improvements in both composition and efficiency of gain. The biological activity was estimated through carcass protein and lipid accretion rates to be improved approximately 18% to 40%, depending on the component used as the criterion. This is the first report of the finding that alterations of the 22 kDa form of ST increases the effectiveness of the molecule for altering lean and fat deposition in meat animals.

Effects of exogenous growth hormone-releasing factor on growth performance and composition of gain

Intermittent exogenous administration of human growth hormone-releasing factor (hGRF) offers an alternative means by which chronic elevation of circulating concentrations of ST can be achieved in young bulls (Enright, Zinn, Chapin and Tucker, 1984), steers (Moseley, Krabill, Freedman and Olsen, 1985; Al-Raheem, Wheaton, Massri, Marcek, Goodrich, Vale and Rivier, 1986), calves (Della-Fera, Buonomo and Baile, 1986), and heifers (Petitclerc, Pelletier, Lapierre, Gaudreau, Couture, Dubreuil, Morisset and Brazeau, 1987). Similar efficacy exists for hGRF in sheep (Hart, Chadwick, Coert, James and Simmonds, 1985; Kensinger, McMunn, Stover, Schricker, Maccecchini, Harpster and Kavanaugh, 1987) and swine (Kraft, Baker, Ricks, Lance, Murphy and Coy, 1985; Della-Fera *et al*, 1986; Petitclerc *et al*, 1987). A major difference exists between GRF and ST administration, however, for the time course and duration of sustained elevation of ST in circulation. Single intravenous or subcutaneous injections of GRF stimulate ST secretion to peak concentrations in blood within 10 to 15 minutes, which subside within 30 to 60 minutes to near basal concentrations. Furthermore, the overall mean concentration of ST achieved with 4 times per day administration of hGRF [(1-44)NH_2] at doses of 5 or 10 μg/kg BW in growing lambs is only 2.5-fold that of controls, while 4 times per day subcutaneous injection of 40 μg oST/kg BW raised overall mean ST concentration to 10-fold that of excipient-treated lambs (Beermann, Hogue, Fishell, Aronica, Dickson and Schricker, 1990). The refractoriness to GRF administration observed in early studies in which doses of 0.016 to 0.065 nmol/kg were administered for 3 to 5 days was not present in subsequent studies using higher doses and chronic administration in lambs, calves and pigs (Della-Fera *et al*, 1986; Kensinger *et al*, 1987; Petitclerc *et al*, 1987; Beermann, Hogue, Fishell, Aronica, Dickson and Schricker, 1990), even after 6 or 8 weeks of administration.

Exogenous administration of GRF does indirectly elicit a protein anabolic response in growing animals. Continuous intravenous administration of hGRF at a dose of 24.3 μg/kg BW for 6 days increased N retention 18% in bull calves (Moseley, Huisman and VanWeerden, 1987). A similar magnitude of response was observed with 6 daily intravenous injections of hGRF at a dose of 0.4 μ/kg BW for 6 days, while a 35% increase in N retention was achieved in young bulls injected subcutaneously twice daily with 5 μg hGRF/kg BW (Lapierre, Petitclerc, Pelletier, Dubreuil, Gaudreau, Couture, Morisset and Brazeau, 1988). Multiple daily subcutaneous injcctions of hGRF [(1-44) NH_2] also reduce the rate of carcass lipid accretion, coincident with enhancing growth performance, skeletal muscle mass and carcass protein accretion rates in growing lambs (Beermann,

Table 3 EFFECTS OF HUMAN GROWTH HORMONE RELEASING FACTOR (hGRF) ON GROWTH AND COMPOSITION OF GAIN IN GROWING LAMBS[1]

		hGRF Treatment per kg body weight		
Response	Control	5 μg	10 μg	Sx
Plasma hormone concentration, ng/ml				
oST	2.15	4.74*	5.14*	.92
IGF-I	278.4	453.2*	444.1*	27
Growth performance		% Difference vs control		
Number of animals	18	20	20	--
Average daily gain	304	+13*	+1.6	12
Feed:Gain	4.99	-18*	-19*	.24
Composition of carcass gain				
Number of animals	9	10	10	--
Protein accretion, g/d	17.2	+30.8*	+34.9*	1.0
Water accretion, g/d	55.6	+19.6*	+28.8*	2.7
Lipid accretion, g/d	79.9	-21.2*	-28.4*	3.4
Ash accretion, g/d	5.0	+32**	+42**	.6
Semitendinosus weight, g	91.6	+10.5*	+15*	2.1
Semimembranosus weight, g	261.5	+10.7	+7.6*	5.8

[1] Lambs received saline or the indicated dose of hGRF four times per day (0100, 0700, 1300 and 1900 h) for 42 or 56 days. Half of the lambs were withdrawn from treatment after 42 days. Data shown are for lambs treated 56 days. Carcass composition data were analyzed by analysis of variance using carcass weight as the covariate. Data are summarized from Beermann *et al* (1990).

* (P<.05) vs control ** (P<.01) vs control

Hogue, Fishell, Aronica, Dickson and Schricker, 1990; see Table 3). In the latter study, hGRF administered at 5 μg/kg BW 4 times per day was nearly as effective as 40 μg oST/kg BW administered 4 times daily. Some evidence of dose-response relationships was apparent for hGRF. A 10 μg/kg BW dose of hGRF exhibited a trend toward greater increases in protein, moisture and ash accretion rates, as well as greater reduction of carcass lipid accretion rates. However, these differences between doses were not statistically significant. Twice-daily subcutaneous injections of 10 or 20 μg hGRF (1-44)NH_2/kg BW for 36 days in barrows weighing 78 kg improved feed conversion efficiency and lean content of the ham (Johnson, Coffey, Ebenshade, Schricker and Pilkington, 1990). However, treatment with hGRF was less effective than injection of 20 or 40 μg pST/kg BW at the same frequency.

Because continuous subcutaneous infusion of hGRF is less effective than intermittent injection for stimulating ST secretion (Kensinger *et al*, 1987), higher doses may be required for improving growth performance and composition of gain in growing animals. Dose-response effects of continuous subcutaneous infusion of hGRF on circulating concentrations of ST indicate that maximal responsiveness is achieved at doses below 100 μg/kg BW (D.H. Beermann, unpublished data). Continuous subcutaneous infusion of 80 μg hGRF/kg BW per day for 35 days in growing lambs elicited similar changes in circulating hormone and metabolite concentrations, and produced similar changes in carcass composition as achieved with intermittent subcutaneous injection of hGRF which totalled 40 μg/kg BW per day (Byrem, Dwyer, Aronica, Dickson, Schricker and Beermann, 1989; see Table 4). Feed intake was not altered, but daily gain was increased 32% and feed conversion efficiency was improved 30% (both $p<.05$) with hGRF infusion. Carcass weight was not affected, but carcasses from lambs infused with hGRF contained 0.36 kg (12%) more protein and 1.53 kg (21%) less lipid. Plasma urea nitrogen concentrations were chronically decreased 33% in lambs infused with hGRF, consistent with the reduced urinary N excretion observed in ST treated animals.

Potent analogues of hGRF [(1-29)NH_2] were found to exhibit greater *in vitro* and *in vivo* biological activity than native GRF [(1-44)NH_2] in stimulating ST release (Felix, Heimer, Mowles, Eisenbeis, Leung, Lambros, Ahmed, Wang and Brazeau, 1986; Mowles, Stricker, Eisenbeis, Heimer, Felix and Brazeau, 1987). A tri-substituted analogue of GRF (des-NH_2 Tyr_1, D-Ala_2, Ala_{15}) exhibited greater potency (approximately 10-fold) in swine than the 1-44 hGRF construct. The increased potency was attributed to enhanced stability of the NH_2-terminus to enzymatic degradation by a plasma diaminopeptidase. Intermittent (3 times per day) administration of this analogue produced dose-dependent increases in weight gain, carcass length, loin eye area and lean meat yield in pigs treated from 49.5 to 100 kg live weight (Dubreuil, Petitclerc, Pelletier, Gaudreau, Farmer, Mowles and Brazeau, 1990; Pommier, Dubreuil, Pelletier, Gaudreau, Mowles and Brazeau, 1990). Total feed consumed, feed:gain ratios, backfat thickness and total carcass adipose tissue yield were decreased in linear response to increasing hGRF analogue dose. Responses to the highest dose of hGRF were equivalent to those

Table 4 EFFECTS OF CONTINUOUS SUBCUTANEOUS hGRF INFUSION ON GROWTH PERFORMANCE AND CARCASS COMPOSITION IN GROWING WETHER LAMBS[1]

	Treatment			
Response	Control	hGRF Injection	hGRF Infusion	Sx
Number of lambs	6	7	7	
Average gain, g/d	172[a]	271[b]	227[b]	73
Feed:Gain, kg/kg	7.70[a]	4.82[b]	5.45[b]	.71
<u>Carcass Variables</u>				
Weight, kg	22.8	23.8	22.8	.40
Protein, %	12.7[a]	14.6[b]	14.3[b]	.44
Water, %	51.3[a]	53.8[ab]	56.2[b]	1.10
Lipid, %	32.1[a]	27.5[b]	25.4[b]	1.25
Ash, %	2.70	2.85	3.06	.15
<u>Plasma Variables</u>				
oST, ng/ml	2.31[a]	3.93[a]	9.79[b]	.99
IGF-I, ng/ml	199.2[a]	547.5[b]	578.0[b]	33
Urea N, mg/dl	24.3[a]	19.4[b]	16.2[b]	1.61
Glucose, mg/dl	69.1[a]	79.7[a]	100.6[b]	4.4
Insulin, μU/ml	22.5[a]	157.2[b]	179.4[b]	32.4

[1] Crossbred wether lambs with initial live weight of 35.9 kg were housed in metabolism crates and injected 4 times per day with saline (control) or 5 μg hGRF/kg BW or sub-cutaneously infused with 80 μg hGRF/kg BW per day

[a,b] Means within a row with different superscripts differ (P<.05)

observed with moderate dose of pST administration.

Thyrotropin-releasing hormone, an active hypothalamic peptide, has been shown to stimulate secretion of TSH, prolactin and ST in cattle (Kesner, Convey and Davis, 1977; Hodate, Johke and Ohashi, 1985). The non-specific action of TRF on ST secretion was also shown to act in synergy with hGRF [(1-29)NH_2] when both were administered at submaximal doses in dairy calves (Lapierre, Petitclerc, Pelletier, Debreuil, Morisset, Gaudreau, Couture and Brazeau, 1987). Efficacy of the hGRF analogue (1-29)NH_2 and thyrotropin-releasing factor (TRF), administered alone and in combination, for improving growth performance and (or) composition has been evaluated in grain-fed dairy calves (Lapierre, Pelletier, Petitclerc, Dubreuil, Morisset, Gaudreau, Couture and Brazeau, 1991). Twice-daily subcutaneous injections of 5 μg hGRF/kg BW, 1 μg TRF/kg BW or the combination of both from 70 to 223 kg live weight did not improve daily gain or efficiency of feed conversion. Treatment with hGRF increased DM, N and energy digestibility (all $P<.05$), and tended to increase N retention (+31%; $P<.19$). TRF treatment had no effect on these variables. The combination of GRF and TRF decreased lipid content ($P<.05$) and tended to increase protein content ($P<.10$) of the 9th-to-11th rib section. It would appear from these results that administration of peptides exhibiting hypothalamic activity on ST secretion offer little advantage over administration of hGRF alone for altering fat and lean deposition in young growing ruminants.

Potential for altering fat and lean deposition with exogenous administration of IGF-I or other growth factors

Because exogenous administration of ST at doses which promote enhanced rates, efficiency and composition of gain in growing animals also markedly increases circulating concentrations of IGF-I (see Tables 1, 3 and 4) the potential benefits of exogenous administration of IGF-I have also been considered. IGF-I is a single chain polypeptide that is believed to mediate many of the effects of ST (Daughaday, Hall, Raben, Salmon, Van den Brande and Van Wyk, 1972; Schoenle, Zapf, Hauri, Steiner and Froesch, 1985). Exogenous administration of ST produces a dose-dependent increase in circulating concentrations of IGF-I. Likewise, IGF-I administration restores growth in the hypophysectomized rat in a dose-dependent manner, nearly equivalent to exogenous ST (Schoenle *et al*, 1985). Subsequently, Hizuka, Naomi, Takano, Shizume, Asakawa, Miyakawa, Tanaka and Horikawa (1986) and Juskevich and Guyer (1990) reported that continuous exogenous administration of IGF-I stimulated growth and increased tibia epiphyseal plate width and weights of several organs in normal growing rats. Exogenous administration of IGF-I, but not ST, also restores growth in streptozotocin-induced diabetic rats (Scheiwiller, Guler, Merryweather, Scandella, Maerki, Zapf and Froesch, 1986).

Originally, IGF-I was thought to be produced only by the liver, and to exert its effect in an endocrine, receptor-mediated fashion. The relative importance of an endocrine role for IGF-I has recently come into question for two main reasons. First, essentially all of the IGF-I in circulation is associated with the 150 kDA IGF-I binding proteins, the abundance of which is also influenced by ST. Little or virtually no free IGF-I is available in circulation to interact with its receptor. Second, local cellular production of IGF-I has been demonstrated in many tissues and organs, where it may exert direct effects on target tissues in a paracrine or autocrine manner. Our understanding of the mechanisms by which anabolic actions of IGF-I on tissues of economic importance in growing animals are regulated is largely unknown. A brief discussion of the relationships between IGF-I mRNA expression in skeletal muscle and associated changes which occur with muscle growth manipulation will provide a useful model for discussing the role of IGF-I in growth regulation. A more detailed discussion of the influences and interactions of ST and IGF-I on skeletal muscle growth has recently been published (Beermann and DeVol, 1991). Regulation of IGF-I abundance or action may provide strategies for enhancing skeletal muscle growth in farm animals.

IGF-I is a potent mitogen for many cell types and stimulates proliferation and differentiation of myoblasts or satellite cells in developing and growing skeletal muscle (Florini, 1987; Allen and Rankin, 1990). IGF-I also stimulates protein synthesis in cultured myoblasts (Bagley, May, Szabo, McNamara, Ross, Francis, Ballard and Wallace, 1989) and may exert acute metabolic actions. Abundance of IGF-I mRNA has been measured in normal (Murphy, Bell, Duckworth and Friesen, 1987) and regenerating skeletal muscle (Jennische and Hansson, 1987). Expression of IGF-I in muscle is enhanced with exogenous administration of ST in rats (Turner, Rotwein, Novakofski and Bechtel, 1988), and is stimulated by work-induced hypertrophy of skeletal muscle (DeVol, Rotwein, Sadow, Novakofski and Bechtel, 1990). This latter phenomenon was found to be independent of ST. The response was observed in both pituitary-intact and hypophysectomized rats. Therefore, it is apparent that two mechanisms exist for the control of IGF-I gene expression in skeletal muscle: 1. a ST-dependent mechanism, and 2. a ST-independent mechanism which is apparently controlled by unknown local factors. Edwall, Schalling, Jennische and Norstedt (1989) also reported a ST-independent expression of the IGF-I gene in regenerating rat skeletal muscle. Further investigation into the role of circulating concentrations of IGF-I, function(s) of the various IGF-I binding proteins and local production of IGF-I in muscle are necessary to provide the basis for developing strategies for muscle growth enhancement via manipulation of IGF-I action. It is unclear whether IGF-I is capable of orchestrating the diverse effects on tissue metabolism now ascribed to ST.

Removing the amino-terminal amino acids of IGF-I results in a growth factor (des(1-3)IGF-I) which is more potent in stimulating protein synthesis in cultured myoblasts (Bagley *et al*, 1989; Gillespie, Read, Bagley and Ballard, 1990). It is also more potent in other cell culture assays than the full-length peptide (Ballard, Francis, Ross, Bagley, May and Wallace, 1987; Francis, Upton, Ballard, McNeil and Wallace, 1988). The increased potency of the truncated form of IGF-I is apparently the result of lesser binding by the IGF-I binding proteins secreted by cultured cells (Gillespie *et al*, 1990), which would increase free concentrations and make more of the des (1-3)IGF-I available for binding to cell receptors. Plasma clearance and tissue distribution of IGF-I, IGF-II and des (1-3)GRF-I was studied in rats, and the authors suggested that plasma binding proteins inhibit the transfer of the IGF growth factors to their tissue sites of action (Ballard, Knowles, Walton, Edson, Owens, Mohler and Ferraiolo, 1991). They also suggested that IGF analogues that are cleared more rapidly from circulation may have greater biological potencies *in vivo*.

In vivo assessment of des(1-3)IGF-I potency was evaluated in mice homozygous for the *lit* mutation. These *lit/lit* mice exhibit low circulating ST levels and low somatomedin activity, although hepatic ST receptors are normal. The des(1-3) IGF-I injected at a dose of 3 or 30 μg per day for 21 days increased total length and nose-rump length more than in controls or animals that received 3 μg IGF-I per day, and increased weights of liver, kidney, heart, lungs and stomach, relative to controls. Whether this truncated form of IGF-I can enhance growth of skeletal muscle or bone in normal mice or farm animals is of particular interest. Investigations of metabolic effects of des(1-3)IGF-I on muscle growth are currently being evaluated *in vivo* (P.C. Owens, personal communication).

Placental lactogen, another protein in the ST/prolactin gene family, is considered an important growth factor in the mammalian fetus, and has also been evaluated for its effects on postnatal growth. Similar significant increases in weight gain were achieved with exogenous subcutaneous administration of 4 doses (0.19 to 5 mg/day) of recombinant bST and recombinant bovine placental lactogen (bPL) in mature (200 g) female rats treated for 10 days (Byatt, Staten, Schmuke, Buonomo, Galosy, Curran, Krivi and Collier, in press). The slope of the response curves was different with bPL exhibiting a larger effect at the low dose and no differences between the two at the three higher doses. Feed consumption was increased more by bPL than bST at all doses, and the authors suggested that the somatogenic effects of bPL may be mediated through both ST and lactogenic receptors.

The effects of IGF-I, fibroblast growth factor, transforming growth factor-ß and others on myogenic cell proliferation and differentiation *in vitro* are rapidly being elucidated (Allen and Rankin, 1990). It is premature to speculate on whether an alteration of concentrations of these peptides or polypeptides in the circulation or in muscle tissue *in vivo* could be shown an effective strategy for altering lean and fat deposition in growing animals. Placental lactogen and other peptides of the ST gene family may warrant further study as possible growth regulators.

Acknowledgements

The authors express their gratitude to Todd Robinson for his assistance with preparation of figures and to Joanne Parsons for her assistance with preparation of the tables and manuscript.

References

Allen, R.E. and Rankin, L.L. (1990) *Proceedings of the Society for Experimental Biology and Medicine*, **194**, 81-86

Al-Raheem, S.N., Wheaton, J.E., Massri, Y.G., Marcek, J.M., Goodrich, R.D., Vale, W. and Rivier, J. (1986) *Domestic Animal Endocrinology*, **3**, 87-94

Bagley, C.J., May, B.L., Szabo, L., McNamara, P.J., Ross, M., Francis, G.L., Ballard, F.J. and Wallace, J.C. (1989) *Biochemical Journal*, **259**, 665-671

Ballard, F.J., Francis, G.L., Ross, M., Bagley, C.J., May, B. and Wallace, J.C. (1987) *Biochemical and Biophysical Research Communications* , **149**, 398-404

Ballard, F.J., Knowles, S.E., Walton, P.E., Edson, K., Owens, P.C., Mohler, M.A. and Ferraiolo, B.L. (1991) *Journal of Endocrinology*, **128**, 197-204

Bauman, D.E., Eisemann, J.H. and Currie, W.B. (1982) *Federation Proceedings*, **41**, 2538-2544

Beermann, D.H. and DeVol, D.L. (1991) In *Growth Regulation in Farm Animals* (Advances in Meat Research) Vol 7, Chapter 13 (ed A.M. Pearson and T.R. Dutson), Essex, England, Elsevier Publishing, pp 373-426

Beermann, D.H., Hogue, D.E., Fishell, V.K., Aronica, S., Dickson, H.W. and Schricker, B.R. (1990) *Journal of Animal Science*, **68**, 4122-4133

Beermann, D.H., Thiel, L.F. and Prusa, K. (1990) In *Biotechnology for Control of Growth and Product Quality in Meat Production Implications and Acceptability International Symposium Proceedings*, Washington, DC, pp 183-193

Beermann, D.H., Robinson, T.F., Byrem, T.M., Bell, A.W., Hogue, D.E. and McLaughlin, C.L. (1991) *Journal of Nutrition*, **121**, 2020-2028

Black, J.L. and Griffiths, D.A. (1975) *British Journal of Nutrition*, **33**, 399-413

Boyd, R.D. and Krick, B.J. (1989) *Proceedings Cornell Nutrition Conference*, Syracuse, NY, pp 149-161

Boyd, R.D. and Bauman, D.E. (1989) In *Animal Growth Regulation*, (ed D.R. Campion, G.J. Hausman and R.J. Martin), Chapter 12, New York, Plenum Publishing Corp, pp 257-293

Boyd, R.D., Bauman, D.E., Beermann, D.H., DeNeergard, A.F., Souza, L. and Butler, W.R. (1986) *Journal of Animal Science*, **63** (Suppl 1), 218 (abstr)

Boyd, R.D., Beermann, D.H., Roneker, K.R., Bartley, T.D. and Fagin, K.D. (1988) *Journal of Animal Science*, **66** (Suppl 1), 256 (abstr)

Boyd, R.D., Bauman, D.E., Fox, D.G. and Scanes, C.G. (1991) *Journal of Animal Science*, **69** (Suppl 2), 56-75

Byatt, J.C., Staten, N.R., Schmuke, J.J., Buonomo, F.C., Galosy, S.S., Curran, D.F., Krivi, G.G. and Collier, R.J. (1991) *Journal of Endocrinology*, **130** (1), 11-20

Byrem, T.M., Dwyer, D.A., Aronica, S.M., Dickson, H.W., Schricker, B.R. and Beermann, D.H. (1989) *FASEB Journal*, **3(4)**, Part II, A938

Campbell, R.G. (1988) *Nutrition Research Reviews*, **1**, 233-253

Campbell, R.G. and Taverner, M.R. (1988) *Journal of Animal Science*, **66**, 676-686

Campbell, R.G., Taverner, M.R. and Curic, D.M. (1985) *Animal Production*, **40**, 497-503

Campbell, R.G., Steele, N.C., Caperna, T.J., McMurtry, J.P., Solomon, M.B. and Mitchell, A.D. (1988) *Journal of Animal Science*, **66**, 1643-1655

Campbell, R.G., Steele, N.C., Caperna, T.J., McMurtry, J.P., Solomon, M.B. and Mitchell, A.D. (1989) *Journal of Animal Science*, **67**, 177-186

Campbell, R.G., Johnson, R.J., King, R.H., Taverner, M.R. and Meisinger, D.J. (1990) *Journal of Animal Science*, **68**, 3217-3225

Caperna, T.J., Steele, N.C., Komarek, D.R., McMurtry, R.W., Rosebrough, R.W., Solomon, M.B. and Mitchell, A.D. (1990) *Journal of Animal Science*, **68**, 4243-4252

Carr, J.R., Boorman, K.N. and Cole, D.J.A. (1977) *British Journal of Nutrition*, **37**, 143-155

Chung, C.S., Etherton, T.D. and Wiggins, J.P. (1985) *Journal of Animal Science*, **60**, 118-130

Crooker, B.A., McGuire, M.A., Cohick, W.S., Harkins, M., Bauman, D.E. and Sejrsen, K. (1990) *Journal of Nutrition*, **120**, 1256-1263

Daughaday, W.J., Hall, K., Raben, M.S., Salmon, W.D. Jr., Van den Brande, J.L. and Van Wyk, J.J. (1972) *Nature*, **235**, 107-108

Della-Fera, M.A., Buonomo, F.C. and Baile, C.A. (1986) *Domestic Animal Endocrinology*, **3**, 165-176

DeVol, D., Rotwein, P., Sadow, J.L., Novakofski, J. and Bechtel, P.J. (1990) *American Journal of Physiology*, **259** (Endocrinol Metab 22), E89-E95

Dubreuil, P., Petitclerc, D., Pelletier, G., Gaudreau, P., Farmer, C., Mowles, T.F. and Brazeau, P. (1990) *Journal of Animal Science*, **68**, 1254-1268

Dunkin, A.C. and Black, J.L. (1985) In *Energy Metabolism of Farm Animals*, (ed P.W. Moe, H.F. Tyrell and P.J. Reynolds), European Association of Animal Production Publication no 32, New Jersey, Rowman and Littlefield, pp 110-114

Early, R.J., McBride, B.W. and Ball, R.O. (1990) *Journal of Animal Science*, **68**, 4134-4143

Edwall, D., Schalling, M., Jennische, E. and Norstedt, G. (1989) *Endocrinology*, **124**, 820-825

Eisemann, J.H., Tyrrell, H.F., Hammond, A.C., Reynolds, P.J., Bauman, D.E., Haaland, G.L., McMurtry, J.P. and Varga, G.A. (1986) *Journal of Nutrition*, **116**, 157-163

Eisemann, J.H., Hammond, A.C., Rumsey, T.S. and Bauman, D.E. (1989) *Journal of Animal Science*, **67**, 105-115

Enright, W.J. (1989) In *Use of Somatotropin in Livestock Production* (ed K. Sejrsen, M. Vestergaard and A. Neimann-Sorensen), London, Elsevier, pp 132-156

Enright, W.J., Zinn, S.A., Chapin, L.T. and Tucker, H.A. (1984) *Journal of Animal Science*, **59** (Suppl 1), 224

Enright, W.J., Quirke, J.F., Gluckman, P.D., Breier, B.H., Kennedy, L.G., Hart, I.G., Roche, J.F., Coert, A. and Allen, P. (1990) *Journal of Animal Science*, **68**, 2345-2356

Etherton, T.D. (1989) In *Biotechnology in Growth Regulation* (ed R.B. Heap, C.G. Prosser and G.E. Lamming), London, Butterworths, pp 97-105

Etherton, T.D., Wiggins, J.P., Chung, C.S., Evock, C.M., Rebhun, J.F. and Walton, P.E. (1986) *Journal of Animal Science*, **63**, 1389-1399

Etherton, T.D., Wiggins, J.P., Evock, C.M., Chung, C.S., Rebhun, J.F., Walton, P.E. and Steele, N.C. (1987) *Journal of Animal Science*, **64**, 433-443

Felix, A.M., Heimer, E.P., Mowles, T.F., Eisenbeis, H., Leung, T., Lambros, T.J., Ahmed, M., Wang, C.T. and Brazeau, P. (1986) *Proceedings 19th European Peptide Symposium*, Chalkidiki, Greece, pp 481-487

Florini, J.R. (1987) *Muscle and Nerve*, **10**, 577-598

Francis, G.L., Upton, F.M., Ballard, F.J., McNeil, K.A. and Wallace, J.C. (1988) *Biochemical Journal*, **151**, 95-103

Gillespie, C., Read, L.C., Bagley, C.J. and Ballard, F.J. (1990) *Journal of Endocrinology*, **127**, 401-405

Groenewegen, P.P., McBride, B.W., Burton, J.H. and Elsasser, T.H. (1990) *Domestic Animal Endocrinology*, **7**, 43-54

Hart, I.C., Chadwick, P.M.E., Coert, A., James, S. and Simmonds, A.D. (1985) *Journal of Endocrinology*, **105**, 113-119

Hizuka, Naomi, Takano, K., Shizume, K., Asakawa, K., Miyakawa, M., Tanaka, I. and Horikawa, R. (1986) *European Journal of Pharmacology*, **125**, 143-146

Hodate, K., Johke, T. and Ohashi, S. (1985) *Endocrinology Japan*, **32**, 375-383

Houseknecht, K.L., Bauman, D.E., Fox, D.G., Smith, D.F. and Musso, T.M. (1990) *Journal of Animal Science*, **68** (Suppl 1), 272 (abstr)

Isaksson, O.G.P., Eden, S., Jansson, J., Lindahl, A., Isgaard, J. and Nilsson, A. (1986) *Journal of Animal Science*, **63** (Suppl 2), 48

Jennische, E. and Hansson, H.-A. (1987) *Acta Physiologica Scandinavica*, **130**, 327-332

Johnson, J.L., Coffey, M.T., Ebenshade, K.L., Schricker, B.R. and Pilkington, D.H. (1990) *Journal of Animal Science*, **68**, 3204-3211

Johnsson, I.D., Hart, I.C. and Butler-Hogg, B.W. (1985) *Animal Production*, **41**, 207-217

Johnsson, I.D., Hathorn, D.J., Wilde, R.M., Treacher, T.T. and Butler-Hogg, B.W. (1987) *Animal Production*, **44**, 405-414

Juskevich, J.C. and Guyer, C.G. (1990) *Science*, **249**, 875-884

Kanis, E., Nieuwhof, G.J., deGreef, K.H., van der Hel, W., Verstegen, M.W.A., Huisman, J. and van der Wal, P. (1990) *Journal of Animal Science*, **68**, 1193-1200

Kensinger, R.S., McMunn, L.M., Stover, R.K., Schricker, B.R., Maccecchini, M.L., Harpster, H.W. and Kavanaugh, J.F. (1987) *Journal of Animal Science*, **64**, 1002-1009

Kesner, J.S., Convey, E.M. and Davis, S.L. (1977) *Journal of Animal Science*, **44**, 784-790

Kraft, L.A., Baker, P.K., Ricks, C.A., Lance, V.A., Murphy, W.A. and Coy, D.H. (1985) *Domestic Animal Endocrinology*, **2**, 133-139

Krick, B., Roneker, K.R., Boyd, R.D., Beermann, D.H. and Ross, D.A. (1990) *Journal of Animal Science*, **68** (Suppl 1), 384 (abstr)

Krick, B.J., Roneker, K.R., Harrell, R.J., Boyd, R.D., Beermann, D.H. and Kuntz, H.T. (1990) *Journal of Animal Science*, **68** (Suppl 1), 383 (abstr)

Krick, B., Roneker, K.R., Boyd, R.D., Beermann, D.H., David, P. and Meisinger, D.J. (1992) *Journal of Animal Science* (in press)

Lapierre, H., Petitclerc, D., Pelletier, G., Dubreuil, P., Morisset, J., Gaudreau, P., Couture, Y. and Brazeau, P. (1987) *Domestic Animal Endocrinology*, **4**, 207-214

Lapierre, H., Petitclerc, D., Pelletier, G., Dubreuil, P., Gaudreau, P., Couture, Y., Morisset, J. and Brazeau, P. (1988) *Journal of Animal Science*, **66** (Suppl 1), 438 (abstr)

Lapierre, H., Pelletier, G., Petitclerc, D., Dubreuil, P., Morisset, J., Gaudreau, P., Couture, Y. and Brazeau, P. (1991) *Journal of Animal Science*, **69**, 587-598

Maltin, C.A., Delday, M.I., Hay, S.M., Innes, G.M. and Williams, P.E.V. (1990) *British Journal of Nutrition*, **63**, 535-545

McLaren, D.G., Bechtel, P.J., Grebner, G.L., Novakofski, J., McKeith, F.K., Jones, R.W., Dalrymple, R.H. and Easter, R.A. (1990) *Journal of Animal Science*, **68**, 640-651

McLaughlin, C.L., Baile, C.A., Shung-Zhang, Q., Lian-Chun, W. and Jin-Pu, X. (1989) *Journal of Animal Science*, **67**, 116-127

Moseley, W.M., Huisman, J. and VanWeerden, E.J. (1987) *Domestic Animal Endocrinology*, **4**, 51-59

Moseley, W.M., Krabill, L.F., Freedman, A.R. and Olsen, R.F. (1985) *Journal of Endocrinology*, **104**, 433-439

Moseley, W.M., Paulissen, J.B., Goodwin, M.C., Alaniz, G.R. and Claflin, W.H. (1990) *Journal of Animal Science*, **68** (Suppl 1), 273 (abstr)

Mourot, J., Bonneau, M., Charlotin, P. and Lefaucheur, L. (1992) *Meat Science*, **31**, 219-227

Mowles, T., Stricker, P., Eisenbeis, H., Heimer, E., Felix, A. and Brazeau, P. (1987) *Endocrinology Japan*, **34** (Suppl), 148

Murphy, L.J., Bell, G.I., Duckworth, M.L. and Friesen, H.G. (1987) *Endocrinology*, **121**, 684-691

NRC (1985) *Ruminant Nitrogen Useage*, Washington, DC, National Academy Press

Pell, J.M. and Bates, P.C. (1990) *Nutrition Research Reviews*, **3**, 163-192

Pell, J.M., Elcock, C., Harding, R.L., Morrell, D.J., Simmonds, A.D. and Wallis, M. (1990) *British Journal of Nutrition*, **63**, 431-445

Peters, J.P. (1986) *Journal of Nutrition*, **116**, 2490-2503

Petitclerc, D., Pelletier, G., Lapierre, H., Gaudreau, P., Couture, Y., Dubreuil, P., Morisset, J. and Brazeau, P. (1987) *Journal of Animal Science*, **65**, 996-1005

Pommier, S.A., Dubreuil, P., Pelletier, G., Gaudreau, P., Mowles, T.F. and Brazeau, P. (1990) *Journal of Animal Science*, **68**, 1291-1298

Prusa, K.J. (1989) In *Biotechnology for Control of Growth and Product Quality in Swine* (ed P. van der Wal, G.J. Nieuwhof and R.D. Politiek), Wageningen, Netherlands, Pudoc Wageningen, pp 183-189

Reeds, P.J. and Mersmann, H.J. (1991) *Journal of Animal Science*, **69**, 1532-1550

Sandles, L.D. and Peel, C.J. (1987) *Animal Production*, **44**, 21-27

Scheiwiller, E., Guler, H.-P, Merryweather, J., Scandella, C., Maerki, W., Zapf, J. and Froesch, E.R. (1986) *Nature*, **323**, 169-171

Schoenle, E., Zapf, J., Hauri, C., Steiner, T. and Froesch, E.R. (1985) *Acta Endocrinologica*, **108**, 167-174

Shields, R.G. Jr., Mahan, D.C. and Graham, P.L. (1983) *Journal of Animal Science*, **57**, 43-65

Thiel, L.F., Beermann, D.H. and Boyd, R.D. (1990) *Journal of Animal Science*, **68** (Suppl 1), 340 (abstr)

Turner, J.D., Rotwein, P., Novakofski, J. and Bechtel, P.J. (1988) *American Journal of Physiology*, **255**, E513-E517

Vernon, R.G. and Flint, D.J. (1989) In *Biotechnology in Growth Regulation* (ed R.B. Heap, C.G. Prosser and G.E. Lamming), London, Butterworths, pp 57-72

Wray-Cahen, D., Ross, D.A., Bauman, D.E. and Boyd, R.D. (1991) *Journal of Animal Science*, **69**, 1501-1514

Zainur, A.S., Tassell, R., Kellaway, R.C. and Dodemaide, W.R. (1989) *Australian Journal of Agricultural Research*, **40**, 195-206

13

THE USE OF NON-PEPTIDE HORMONES AND ANALOGUES TO MANIPULATE ANIMAL PERFORMANCE

D.B. LINDSAY, R.A. HUNTER and M.N. SILLENCE
CSIRO Division of Tropical Animal Production, Rockhampton, Queensland 4702, Australia

For many hundreds of years there has been implicit appreciation of the performance-enhancing value of some anabolic agents through the recognition of the differences between males and castrates. Over the last 15-20 years there has been a much fuller pragmatic understanding of the effects of steroids on the growth rate of animals and the ability to modify body composition. At a previous Easter School (Buttery, Haynes and Lindsay, 1986) an excellent summary of some of these effects was provided by Roche and Quirke (1986). In the same volume was also offered an early assessment of a newer class of anabolic agents - the ß-agonists (Stock and Rothwell, 1986). Outside the peptides and proteins there have been no significant developments of agents which fall outside the general classes of steroid hormones (and analogues) and the cationic amines (and analogues). There have been occasional claims of growth-promoting properties of compounds not in these classes (*eg* ferulic acid) but such claims seem to be anecdotal with little evidence from properly designed trials. Thus it is proposed to confine comment to these main classes. Moreover there is extensive evidence of the value of such compounds in conventional husbandry practices. We propose to concentrate attention on somewhat different conditions and then consider later the implications of our findings for more conventional husbandry.

The Beef Cattle industry of tropical Australia is pastoral and thus availability of food is largely determined by rainfall. Growth of cattle occurs generally in the summer (wet) season (see Figure 1, from Norman and Stewart, 1964). In the dry (winter) period there is frequently stasis or negative growth (weight loss). Economic analysis has suggested that the time taken to reach market weight, which is determined by overall growth rate is perhaps the major limiting factor for the beef industry. In considering ways by which the overall growth rate of cattle might be improved in tropical conditions we felt there might be advantages in minimising the weight loss that occurs in the dry season, rather than enhancing the growth rate of the summer season. With this approach, the mean (overall) growth rate would be enhanced although there would be no improvement in maximal growth rate. Since nutrient availability is restricted, the strategy we have taken is to improve the efficiency of utilisation of food, by reducing basal metabolic rate.

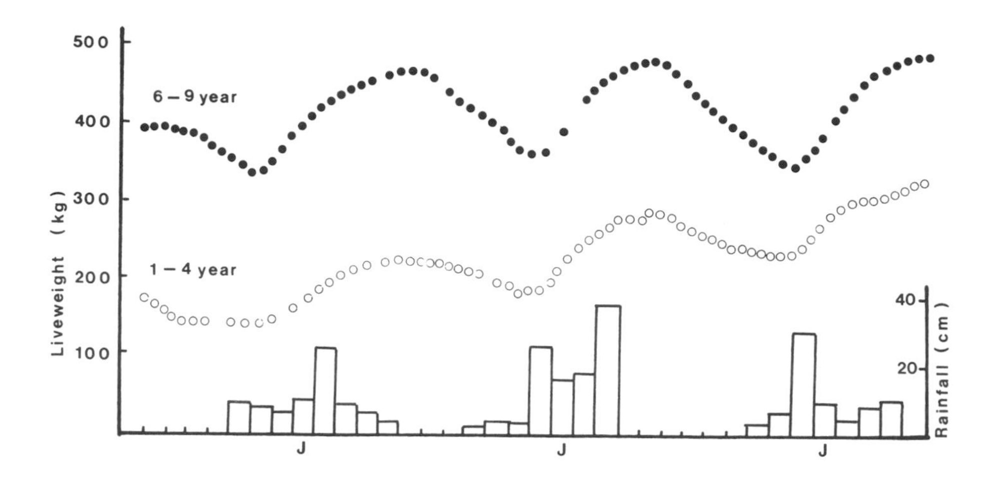

Figure 1 Liveweight gains and losses over three years in cattle grazing in tropical Australia : J = January

The value of the xenobiotic growth promoter trenbolone was examined. It was found (Hunter and Vercoe, 1987) that trenbolone was effective in reducing fasting metabolic rate of steers by about 11% (83.3 to 74.5 kJ/kgd). However this did not result in the expected saving in liveweight loss when steers were given free access to poor quality hay because voluntary food intake was reduced. The blood urea concentration was significantly reduced in trenbolone-treated animals (Figure 2) and this probably resulted in significantly lesser amounts of urea being transferred to the rumen from the blood (Hunter and Magner, 1990). On the low-nitrogen hay provided, the already low rumen NH_3 fell even further and the low microbial population supported would be expected to limit voluntary feed intake. When a supplement of urea (and sulphur) was provided, voluntary food intake was equalised for control and trenbolone-treated animals but the latter maintained a lower fasting metabolic rate (and also a substantial advantage in liveweight gain). It has also been shown that in animals maintained on the same (restricted) diet, for trenbolone-implanted animals there is a significant attenuation of liveweight loss over 50-60 days (Figure 3). Although only limited studies have been made of

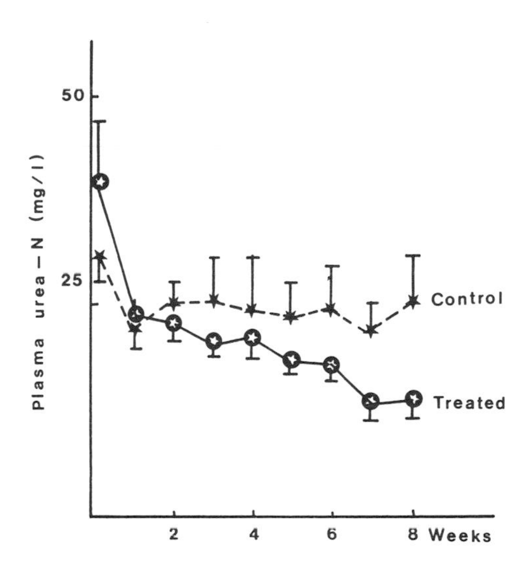

Figure 2 Effect of a 300 mg implant of trenbolone acetate in cattle on plasma urea concentrations

the effect of dose, it appears that the standard 300 mg implant (Finaplix) developed by Roussel-Uclaf for use of trenbolone acetate as a growth promoter is not far from optimal for minimising liveweight loss.

In contrast to trenbolone, oestradiol-17ß was ineffective in lowering fasting metabolic rate of steers and showed no ability to attenuate liveweight loss of steers given a poor-quality hay insufficient for maintenance (Hunter and Vercoe, 1988) (Table 1). There is implicit support for this finding from the work of Rumsey and Hammond (1990). They showed for the synthetic oestrogen analogue diethylstilboestrol and for a mixture of oestrogen and progesterone (Synovex) that effectiveness (increasing bodyweight or nitrogen gain) depended markedly on energy intake. When bodyweight or nitrogen gain was low or negative, oestrogen had no or even negative effect on performance.

Testosterone has also been found ineffective in lowering fasting metabolic rate (Table 1) or in attenuating liveweight loss in steers fed below maintenance (O'Kelly, 1985; Hunter, 1989). The metabolic rate was not affected by testosterone whether measured before or after a period of liveweight loss. Testosterone was administered at a fairly large dose (0.5 mg/kg body weight, three times weekly). Testosterone concentrations in blood were maintained at about the **peak** levels found in bulls, that is, substantially greater than the **mean** concentrations found in bulls. In underfed animals treated with testosterone at the same dose, the mean concentrations were markedly higher even than this, perhaps reflecting a decreased capacity to metabolise testosterone. Growth stimulation when food was freely available was large - an increase in growth rate from 0.95 to 1.48 kg/day. The rate was particularly striking since the food available was lucerne hay, with no concentrate. After a period of undernutrition, growth rate was high in untreated animals (1.47 kg/day) and in these circumstances (compensatory growth) testosterone had no effect on growth rate. Even eight weeks after the period of underfeeding, when growth rate had declined to about 1.2 kg/day, testosterone was unable to stimulate growth.

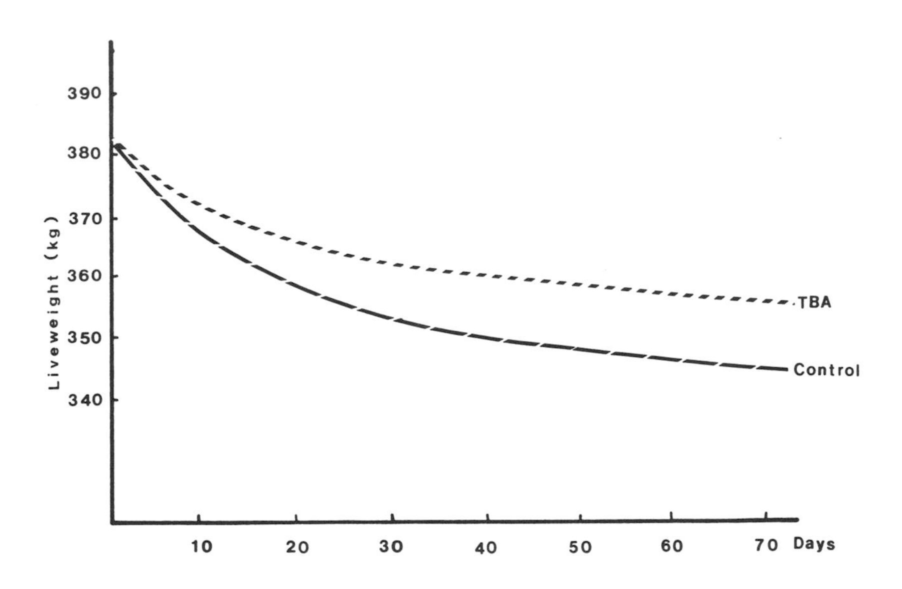

Figure 3 Liveweight loss of steers given a fixed sub-maintenance diet. Each curve shows the best-fit line for the mean values from 7 animals. TBA: animals initially implanted in the ear with 300 mg trenbolone acetate

Table 1 EFFECT OF SEX-STEROIDS ON FASTING METABOLIC RATE AND DAILY LIVEWEIGHT LOSS OF STEERS FED A BELOW-MAINTENANCE ROUGHAGE DIET

	Fasting metabolic rate (kJ/kg/day)			Liveweight change (kg/d)		
	control	treated	(SE)	control	treated	(SE)
oestradiol-17ß[a]	78.3	77.9	(0.20)	-0.38	-0.41	(0.203)
testosterone[b]	65.9	63.5	(2.04)	-0.65	-0.71	(0.020)

[a] Brahman-Hereford steers n = 6/group
[b] Africander-Hereford steers n = 4/group

Other metabolic effects

In undernourished steers, synthesis of the whole body, hide and muscle protein is greatly reduced (Hunter and Magner, 1990) compared to the values in fed cattle (Lobley, Milne, Lovie, Reeds and Pennie, 1980). Trenbolone treatment did not significantly affect the values, although because of the inevitably limited precision a modest change in rates cannot be excluded. However there was a reduced nitrogen loss and a significant reduction in 3-methyl histidine excretion as a result of trenbolone treatment. This probably indicates a decrease in rate of muscle protein degradation, as found by Vernon and Buttery (1978) for rats; by Sinnett-Smith, Dumelow and Buttery (1983) for sheep and by Lobley, Connel, Mollinson, Brewer, Harris and Buchan (1985) for well-nourished cattle. Trenbolone was originally selected for use in lowering metabolic rate because of its known capacity to reduce rate of protein degradation from which it seemed plausible that energy utilisation would be minimised. However, one may calculate that the observed energy savings are substantially greater than might be expected on the basis of the reduction in rate of protein degradation alone.

In contrast, testosterone has been supposed to act as a growth promoter by increasing the rate of protein synthesis in sheep (Martinez, Buttery and Pearson, 1984) and humans (Griggs, Kingston, Josefowicz, Herr, Forbes and Halliday, 1989) with an assumed increase in the rate of protein degradation. Thus testosterone might be expected to stimulate energy use, and have no capacity to lower metabolic rate. Lobley, Connell, Buchan, Skene and Fletcher (1987) showed that testosterone actually increased the metabolic rate of wethers, but since amino acid oxidation was reduced, these workers suggested that the rate of protein degradation was reduced. This was further supported (Lobley, Connell, Milne, Buchan, Calder, Anderson and Vint, 1990) when they were unable to demonstrate an effect of testosterone on muscle protein synthesis.

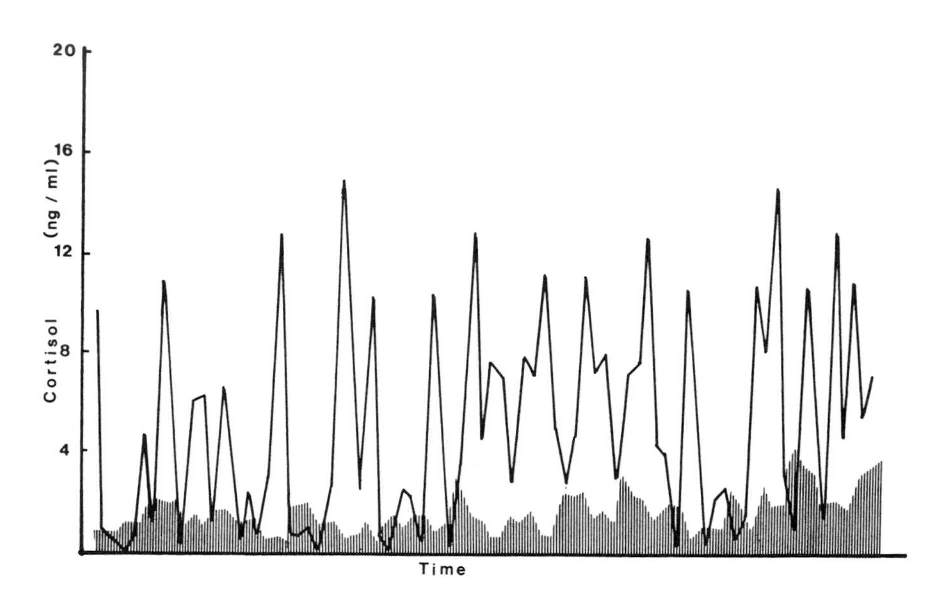

Figure 4 Comparison of plasma cortisol concentration at 20 min intervals over 24 hours in a control (unhatched)and ACTH-immunised (hatched) steer. Treated animal was immunised against synthetic ACTH (1-24) coupled to human serum albumin. Primary and booster vaccinations were given 3 months and two weeks before treatment. The **mean** cortisol concentrations (for 6 animals) was about 30% of the control values

Although the effects of androgens on protein turnover are debated, the contrast in action of trenbolone and testosterone with respect to metabolic rate is particularly striking. There is some evidence that glucocorticoids can have effects on metabolic rate (Coyer, Cox, Rivers and Millward, 1985) and energy balance (Freedman, Horwitz and Stern, 1986). For trenbolone (Sharpe, Buttery and Haynes, 1986) and to a lesser degree testosterone (Lobley *et al*, 1990) it has been suggested that their mode of action involves glucocorticoids. However evidence in rats (Sillence and Rodway, 1990) shows that while the growth-promoting action of trenbolone can at least in part be explained by decrease in corticosterone, this explanation is not plausible for testosterone where glucocorticoid concentrations are actually increased.

An attempt was made to mimic the action of trenbolone by immunising cattle against ACTH (Jones, Hunter, Magner, Hoskinson and Wynn, 1990). In about half the animals there was a significant antibody titre and in such animals the plasma cortisol concentration was significantly reduced (Figure 4). However there was no effect on metabolic rate, nor any ability to attenuate the body weight loss in undernourished animals, as was shown for trenbolone. Sharpe *et al* (1986) suggested that trenbolone action involved a reduction in corticosteroid receptor numbers. Receptor numbers were not measured in ACTH-immunised animals. With female rats, Sillence, Jones, Lowry and Bassett (unpublished), were able to enhance growth by passive immunisation against ACTH (Figure 5), although there was no advantage in feed-restricted animals. Thus while there is increasing evidence that corticosteroids restrict normal growth, it is not clear whether this is any way related to the effect on metabolic rate.

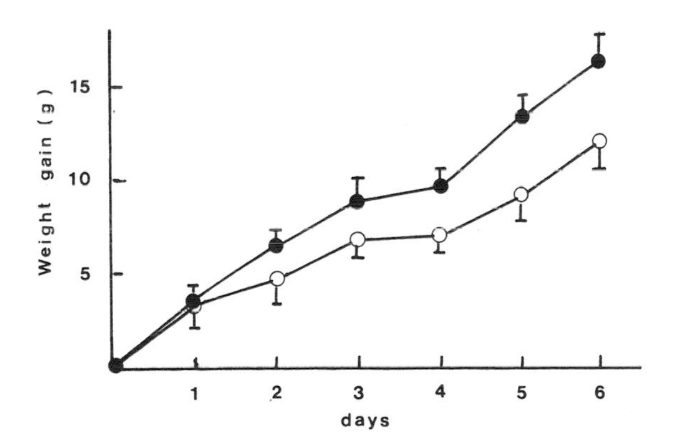

Figure 5 Mean (± SEM) cumulative weight gain of female rats given daily injections of saline (-o-) or sheep anti-ACTH (1-24) antiserum (-•-). Weight gains after 6 days were significantly different ($P < 0.05$). Mean initial weight 189g ($n = 12$)

CATIONIC AMINES

In searching for additional ways to lower basal metabolic rate, it was attractive to consider possible manipulation of the catecholamine system. There are several indications that suggest such manipulation could improve animal production.

1. All three of the natural catecholamines have been known for many years to stimulate metabolic rate. Particularly striking was a study by Staten, Matthews, Cryer and Bier (1987) in which very small doses of adrenaline (0.1 - 1 μg/min) were infused into humans. There was an increase in metabolic rate even with the lowest dose, when plasma concentrations overlapped the normal range. The estimated threshold concentration for effect on metabolic rate was similar to that for increase in free fatty acids.
2. Several new sympathomimetics have shown promise in **increasing** energy utilisation. For example Connacher, Jung and Mitchell (1988) showed for obese humans on a restricted diet that weight loss was enhanced by administration of the drug BRL26830A.
3. The well-known action of several sympathomimetic agonists in stimulating oxygen consumption, reducing the accretion of fat and increasing accretion of protein (*eg* Stock and Rothwell, 1986).
4. Adrenaline infusion into humans reduces protein turnover without change in plasma insulin (Castellino, Luzi, Del Prato and De Fronzo, 1990).

While a method of inhibiting catecholamine action might well result in lower metabolic rate it was not clear if this would result in attenuation of weight loss in undernourished animals.

CATECHOLAMINE RECEPTORS

Catecholamine action was probably the first to which the concept of receptor initiation was applied (Dale, 1906). Ahlquist (1948) showed that there were two kinds of catecholamine receptor - α and ß. The basis for this classification was examination of a number of catecholamine-responsive actions, and ranking of the potency of adrenaline, noradrenaline and three or four known synthetic sympathomimetics. This ranking order fell into two patterns, from which it was deduced there were two kinds of receptor. Much the same sort of argument has been used for further subdividing these major classes. Lands, Arnold, McAuliff, Luduena and Brown (1967) used 19 different compounds and measured their effectiveness on several actions considered to be ß-mediated. They found that there was strong correlation between cardiac and lipolytic actions, and between bronchodilator and vasodepressor actions, but very weak correlation between cardiac and bronchodilator. Thus they sub-classified $ß_1$ and $ß_2$-receptors. Subdivision of α receptors is historically more complicated because it was originally developed from the concept that α_2-receptors acted before a synaptic junction and α_1 after. It is now recognised that this is over-simplistic and subsequent development has depended much more on the selectivity of a range of drugs. More recently a further subdivision of α_1- and α_2-receptors has been proposed (see *eg* Docherty, 1989), as well as possible additional ß-receptors (*eg* Arch, 1989) - again generally on the basis of selective action of drugs. With the ability to prepare membranes which are greatly enriched with respect to amount of receptor, much recent analysis has been concentrated on the kinetics of drug/receptor combination. It is important to recognise that while permitting much more systematic analysis, this method has important limitations.

Functional activity

It is possible to link some receptor types to differences in biochemical mode of action. ß-receptor stimulation results in a rise in cyclic AMP; α_2-action generally results in a fall in cyclic AMP; while α_1-action is linked to a change in intracellular calcium concentration. In each case membrane transduction involves different guanine nucleotide binding proteins, (G-proteins) - G_s (ß); G_i (α_2); and G_x (α_1). This too may be over-simplified. Some α_2-receptors at least appear to have actions which do not involve cyclic AMP; in other cases a decrease in cyclic AMP may occur but is not responsible for the physiological action (Bylund, 1988). While these features appear remote from our particular interest, they can be relevant. In the isolation of receptor-enriched membranes, the G-proteins are sometimes separated from the receptor protein. Analysis of receptor/drug combination can tell us about the affinity of a compound for the receptor, but this says nothing about biological activity - one cannot distinguish between blocker and agonist just on this basis - one needs to know whether there is a biological response, or at least a change in cyclic AMP in response to receptor-agonist combination. Further complicating the situation, the affinity of some compounds for the receptor may be greatly increased by the presence of the G-protein.

ACTION OF α_2 -AGONISTS

α_2-agonists seemed promising candidates for lowering metabolic rate since it had been suggested that among other actions, they exerted negative feedback on the release of noradrenaline from synapses. Initial studies were made with guanfacine using mice. As Figure 6 shows, there is a marked dose-dependent decrease in metabolic rate following injection of guanfacine (Sillence, Matthews, Spiers and Lindsay, 1990).

However, when growth rate was measured in animals treated chronically with the agonist, it was found that there was a parallel decrease with dose (Figure 6). While in part this was due to a decrease in food intake, there was also a fall in the efficiency of utilisation of food.

It seemed very surprising that a fall in metabolic rate was associated with a fall in efficiency of utilisation of food. Further analysis, made with rats (which also respond to guanfacine with a fall in metabolic rate and efficiency of food utilisation) showed that there was an elevation of plasma glucose, and a striking glycosuria (although attenuated, this was still significant even after 6 days) (Spiers, Sillence and Lindsay, 1990). There is evidence of α_2-receptors in the pancreas which could inhibit insulin secretion (Nakaki, Nakadate and Kato, 1980) and thereby produce a mild diabetic state although it seems unlikely the mild hyperglycaemia observed would result in the marked glycosuria. However, α_2-receptors are also found in the kidney (Sanchez and Pettinger, 1981) and this could conceivably enhance any glycosuria. There was also significant inhibition of the growth of a skeletal muscle bundle (soleus, gastrocnemius and plantaris). A possible explanation for this and indeed the overall catabolic action stems from the finding that plasma corticosterone was significantly increased. The extent of enhancement observed was sufficient to produce significant inhibition of muscle growth (Sillence and Etherton, 1991).

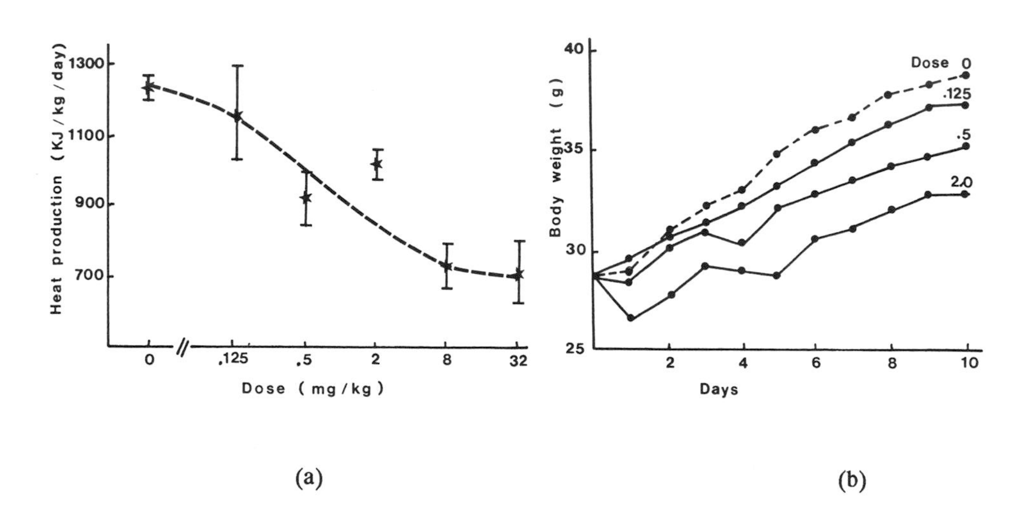

Figure 6 (a) Heat production measured (in pairs of mice) over a 4-hour period commencing 1 hour after administration of guanfacin, (b) body weight gain in control mice and animals given guanfacin at 3 dose levels. The drug was given orally each day (12 animals per group)

Table 2 EFFECT OF A CONTINUOUS INFUSION OF GUANFACINE ON WEIGHT CHANGE, FASTING METABOLIC RATE AND FEED INTAKE OF STEERS RECEIVING A POOR-QUALITY ROUGHAGE DIET

	guanfacine μg/kg				
	0	30	40	80	
	control	low	medium	high	SE
Feed intake (kg/d)	16.1	14.3	15.1	14.5	1.32
Fasting metabolic rate (kJ/kg/day)	86.0	73.7	65.3	67.4	1.94
Liveweight loss (kg/d)	-0.12	-0.04	-0.02	-0.02	0.013

In view of these findings, guanfacine would have seemed an unpromising candidate for study in cattle. As it happened however, such a study began before these findings had been completed. It was shown (Table 2) that guanfacine infusion in cattle could induce a significant fall in metabolic rate (Hunter, 1991). Cattle appear to be quite sensitive to the drug, a significant effect being seen with 20 μg/kg, a dose about 20-fold lower than that used in rats. Moreover there was no significant effect on food intake. Finally, when given by constant infusion over about 6 weeks, in animals losing weight on a sub-maintenance diet, there was a progressive attenuation in weight loss, just as has been shown earlier for trenbolone. In contrast with studies with rats, there was no significant glycosuria (detectable glucose only in 1/12 animals), although surprisingly there was a small but significant hyperglycaemia (increase of about 5%). Yet there was not a detectable increase in plasma insulin. It is suggested that α_2-agonists stimulate the production of growth hormone. An increase was observed in one, but not a second experiment. As referred to earlier, the α_2-receptor is supposed to be located, among other sites, at a pre-synaptic site in neurons. It is considered to function by negative feedback, so that a proportion of noradrenaline, formed at the junction stimulates the α_2-receptor and this reduces release of noradrenaline from the synapse. The possibility was tested by determining 24-hour excretion of noradrenaline in the urine of treated animals. There was no difference between treated and control animals (nor of excretion of the other catecholamines, adrenaline and dopamine). In fact only about 2% of noradrenaline formed appears to be excreted as such in cattle urine. However by using ^{3}H-noradrenaline, it was found that there was no difference in excretion between control and treated animals, nor was there a difference in rate of production (net production/%recovery). There was also no appreciable difference between the groups in the concentration of plasma noradrenaline, nor in the plasma turnover rate as measured by changing specific activity over time. Thus for neither a short nor long-time basis could guanfacine be seen to change the rate of noradrenaline production. A further check on the likelihood of central nervous action was made by calculating the probable concentration within the brain and infusing guanfacine directly into the brain to match this concentration. No response was seen using this route of administration. It thus seems highly probable that the action of guanfacine in reducing metabolic rate involves direct stimulation of a peripheral receptor.

There have been few studies of α_2-agonists in agricultural animals. Gorewit (1981) administered clonidine (one of the earliest α_2-agonists developed) to dairy cows, in the hope of using it to stimulate growth hormone production. While there was some increase in plasma growth hormone, there was also an increase in plasma cortisol. Kennedy and Belluk (1987) administered clonidine to sheep and suggested there could be some modest redistribution of fat as a result of the drug. Schaefer, Jones, Kennedy, Tong and Onishuk (1990) administered clonidine to beef cattle with little effect except a trend to increased growth hormone and insulin after a meal; and some reduction in body cavity fat. It should be noted in these experiments, clonidine was given in the feed and the doses used were 0.25 to 1 gm/kg body weight. This range is lower than that used clinically but is more than 1000-fold greater than the dose of guanfacine as described here.

Clonidine undoubtedly acts at least in part in the central nervous system. Schaefer *et al* (1990) comment on the sedative effect - which was not evident in our studies with guanfacine. However, clonidine does reduce metabolic rate, although it is much less effective in this respect than guanfacine.

We have also conducted studies with another α_2-agonist, tizanidine. This compound also reduced metabolic rate in mice. However it did not appear to have the deleterious effects seen with guanfacine. In a growth trial there was a significant stimulation of growth, although the effect appeared mediated mainly by a stimulation of food intake. In some initial studies in cattle, although it did reduce oxygen consumption, it was immediately clear tizanidine was not a suitable agonist - there was a very marked sedative effect, and very profuse salivation. Schaefer *et al* (1990) also comment on excessive salivation in the early stages of treatment with clonidine, in contrast to studies in humans where clonidine is reported to inhibit salivation and a 'dry mouth' is a common side effect.

ACTION OF ß-AGONISTS

Perhaps one of the most striking findings in recent years has been the evidence that ß-adrenergic agents have anabolic actions. Clenbuterol was the agent most studied originally, but more recently studies have been made more with cimaterol (also a Cyanamide product) and ractopamine (Lilley), perhaps because less side effects have been observed with these products. In this section attention will largely be concentrated on clenbuterol. Three distinct metabolic actions are reported (1) stimulation of metabolic rate (2) reduction in deposition of fat (3) increase in deposition of protein. In connection with studies of undernourished cattle, there was particular interest in (1). Any method of manipulating catecholamine activity to reduce metabolic rate should not also reduce protein accretion. Thus a crucial question is whether these responses are linked. Clenbuterol was originally developed as an agonist for bronchodilation, which indicates it is $ß_2$ selective. Reeds, Hay, Dorwood and Palmer (1987) showed that the non-selective ß-blocker, propranolol could markedly attenuate the effect of clenbuterol in rats in increasing energy utilisation and reducing fat deposition, while no effect was seen on muscle protein deposition. Maltin, Delday, Hay, Smith and Reeds (1987) showed even more strikingly that in muscle propranolol attenuated the increase in fibre size induced by clenbuterol but there was no effect on muscle protein accretion. Sillence, Spiers and Lindsay (1990) confirmed these findings, showing that propranolol attenuated both the fall in fat deposition and the increase in muscle water, but had no effect on muscle protein deposition. It was thus suggested that the protein-accreting action of clenbuterol was not ß-receptor induced. However, MacLennan and Edwards (1989) have shown that these apparently clear results are not so definite. In the experiments above propranolol (and clenbuterol) were given orally and in these conditions, the inability of propranolol to attenuate the protein-accreting effect of clenbuterol was confirmed. However, when propranolol was given intraperitoneally, it could block the effect of clenbuterol on protein accretion. The key factor is that in the rat propranolol is short-acting, because it is very rapidly metabolised. Differences in response at different sites might be because the relative affinity of propranolol and clenbuterol differs between sites. It is also striking that MacLennan and Edwards (1989) were able to demonstrate a rise in cyclic AMP, glycogen and lactate in muscle within 30 minutes of injection of clenbuterol; this effect too was inhibited by intraperitoneal injection of propranolol one hour previously. This strongly suggests there is a direct ß-agonist action on muscle.

Another characteristic of ß-receptors is their down-regulation (decrease in receptor numbers) by continuing treatment with a ß-agonist. Kim and Sainz (1990) showed in rats that the increase in weight of plantaris muscle following clenbuterol treatment was greatest over the first seven days; subsequent growth was little greater than controls. In parallel with this response, they found that ß-receptor numbers showed a maximum difference over controls at about 7 days after clenbuterol injection (Figure 7). It is possible that this accounts for the sort of response seen in cattle. In results reported by Schiavetta, Miller, Lundt, Davis and

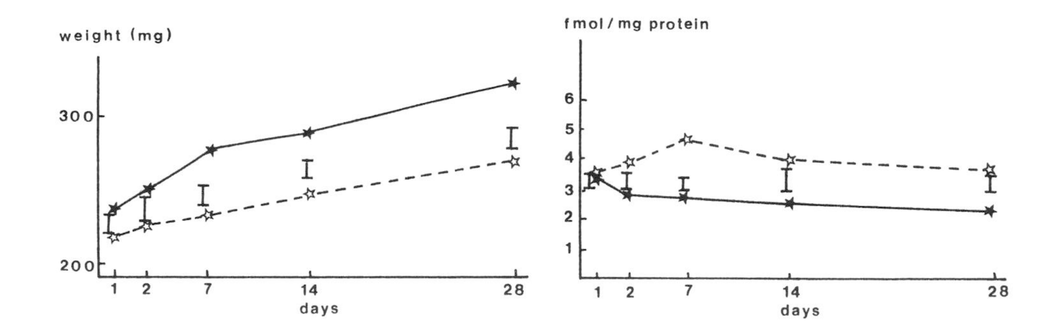

Figure 7 Effect of a ß-agonist (cimaterol) on (a) weight gain (b) ß-2 receptor density of plantaris muscle of rats (from Kim & Sainz, 1990). I = S.E. difference

Smith (1990) for growth of beef steers following treatment with clenbuterol, the stimulating effect is clearly declining by 50 days. After withdrawal of the drug, there would appear to be an initial 'rebound' effect, but later the difference between liveweight of control and treated animals declined.

The ß-receptor present in adipose tissue was generally assumed to be a $ß_1$ subtype (Lands *et al*, 1967), that in skeletal muscle was identified as $ß_2$ (Elfellah and Reid, 1987). It is thus understandable that a $ß_2$ selective compound such as clenbuterol might be more readily displaced in adipose tissue than in muscle by a non-selective competitor.

A more selective blocker should give further support to this concept. Sillence, Matthews, Spiers, Pegg and Lindsay (1991) carried out further experiments on blocking the action of clenbuterol in muscle comparing a long-acting non-specific blocker (sotalol) and the $ß_2$-selective blocker ICI118551 (Figure 8). There was a very clear-cut effect of both blockers in attenuating the growth-stimulating effect of clenbuterol on a hind-limb muscle bundle (gastrocnemius, soleus and plantaris). Even more striking, the blockers, and especially ICI118551 actually inhibited normal muscle growth. In addition there was a marked fall in receptor density in response to clenbuterol, and this effect was attenuated by sotalol. (It was not possible to demonstrate this effect with ICI1118551 because this compound has such high affinity for the $ß_2$-receptor that following injection it is not possible to sufficiently remove the blocker from the membrane preparation to permit estimation of receptor density; even 10^{-4}% remaining is sufficient to interfere significantly). It was also found that the blockers attenuated the stimulating effect

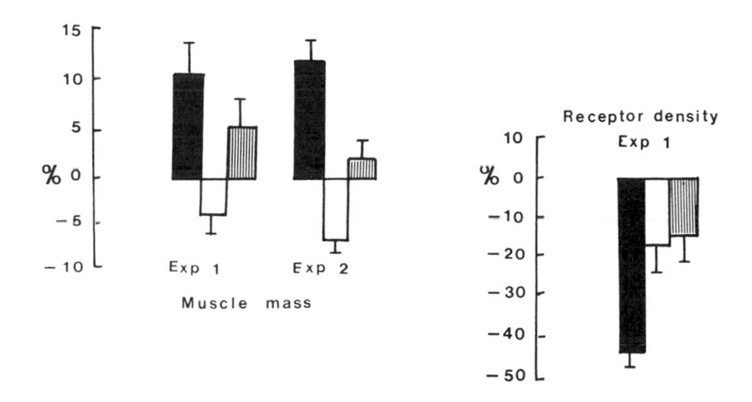

Figure 8 Effect of clenbuterol (solid), a ß-receptor antagonist (unhatched), or combination (hatched), on the mass and ß-receptor density of a muscle bundle (gastrocnemius, plantaris and soleus) of the hind limb of rats. Animals were treated 7 days (exp 1) or 10 days (exp 2). Results were expressed as mean % (+ s.e.m) of control rats (for each experiment) treated with saline. ß-antagonists used were sotalol (exp 1) and ICI 118551 (exp 2)

of clenbuterol on growth of the heart. In the case of ICI118551, this was sufficiently powerful to actually restrict the normal growth of the heart. These experiments suggest then not only that clenbuterol acts in muscle via a ß-adrenergic receptor, but that this is a $ß_2$-receptor. The inhibitory action in the absence of clenbuterol also suggests that this receptor is active in the normal control of growth.

In experiments made about the same time to analyse the action of clenbuterol in muscle, Pegg, Lindsay, Matthews and Sillence (1990) prepared a series of compounds chemically fairly closely related to clenbuterol. They prepared receptor-enriched membrane fragments from bovine (*L. dorsi*) muscle and obtained the dissociation constants for these compounds with respect to muscle receptors by measuring the extent of displacement of ^{125}I-iodocyanopindolol (ICYP) with increasing concentrations of the ligands. A test of biological potency was also developed by determining the effect of increasing concentration in reducing urinary nitrogen loss. ED_{50} was defined as the concentration of drug that would reduce loss by 50% of the maximum achievable effect. It was found (see Table 3) that there was no relation between K_D and ED_{50}. It was argued that this threw doubt on whether the clenbuterol action in muscle is mediated by stimulation of $ß_2$-adrenoceptors. However this argument does depend on there being only one receptor in skeletal muscle. In further experiments (Moore, Sillence, Pegg and Lindsay, 1990) careful analysis was made of the kinetics of ICYP binding to membrane fragments from rat skeletal muscle. Initially the binding appeared to be indicative of a single site or receptor, as published

Table 3 COMPARISON BETWEEN ß$_2$ RECEPTOR AFFINITY IN BOVINE SKELETAL MUSCLE AND RELATIVE POTENCY IN LOWERING URINARY N OUTPUT (ED_{50} $\mu g/kg$) OVER 48 HOURS IN FEMALE RATS, USING CLENBUTEROL AND 5 ANALOGUES

compound	K_D	ED_{50}
clenbuterol	62	270
mono-iodo	310	3960
methoxy	647	3900
ethoxy	1480	1330
keto	1640	6
precursor (no chlorine)	5160	1970

information had suggested; and this was supported by competition studies with several ligands. However careful study at low ICYP concentrations suggested there was heterogeneity in the receptor population and this has been confirmed by studies with ICI118551 and even more effectively by a novel ligand (the subject of a patent application) developed by the group. Study with a ß$_1$ selective agent (RO363) showed weak binding to a single site. Thus there appears to be an additional receptor (accounting for about 20% of the ß-receptor population in skeletal muscle). Further studies will be needed to see if this plays a significant part in the action of clenbuterol. It is striking that in adipose tissue which was supposed to contain first only ß$_1$-receptors, then a mixture of ß$_1$- and ß$_2$-receptors, the receptor is now described as atypical ß (Arch, 1989). Emorine, Marullo, Briend-Sutren, Patey, Tate, Delavier, Klutchko and Strosberg (1989) have been able to clone three ß-adrenergic receptors into E.Coli. They claim to have identified the atypical ß of human adipose tissue as ß$_3$. While we have identified a sequence corresponding to ß$_2$ in the bovine genome, there is some indication that the atypical ß in skeletal muscle differs to significant degree from the ß$_3$ of human adipose tissue. Perhaps even more confusing, Smith, Lee and Coutinho (1990) have concluded that the ß-agonist growth promoter ractopamine developed by Lilley (which has many similar actions to clenbuterol and cimaterol) is probably ß$_1$-selective on the basis of affinity studies with rat cell glioma membranes.

Mechanism of action

The mode of action of ß-agonists in stimulating protein accretion has proved curiously difficult to define. Evidence that clenbuterol increases protein synthesis (as indicated by incorporation of labelled amino acid) in rats has been provided by Emery, Rothwell, Stock and Winter (1984); in mice by Rothwell and Stock (1985); Pell, Bates, Elcock, Lane and Simmonds (1987); in sheep (Cleays, Mulvaney, McCarthy, Gore, Marple and Sartin (1989); in pigs by Helferich, Jump, Anderson, Skjaerlund, Merkel and Bergen (1990). In contrast, in rats (Reeds, Hay, Dorwood and Palmer, 1986) and sheep (Bohorov, Buttery, Correia and Soar, 1987) no change or even a fall in rate of protein synthesis was observed in response to clenbuterol. These authors therefore suggested that clenbuterol must affect the rate of protein degradation. In further support of this, clenbuterol treatment has been shown to result in a decrease in urinary 3-methyl histidine excretion in rats (Kim and Lee, 1990) and in calves (Williams, Pagliani, Innes, Pennie, Harris and Garthwaite, 1987); with a similar effect reported in rabbits treated with cimaterol (Forsberg, Ilian, Ali-Bar, Cheeke and Wehr, 1989). While it is now recognised that excretion of 3-methyl histidine is not necessarily a quantitative measure of skeletal muscle protein degradation, the decrease observed in the above species is at least consistent with a decrease in rate of protein breakdown. In addition, Higgins, Lasslett, Bardsley and Buttery (1988) showed that changes in calcium-dependent proteinases and calpastatin in sheep muscle following clenbuterol treatment were consistent with a decrease in rate of protein degradation; similar findings were reported by Wang and Beermann (1988) for sheep; by Forsberg *et al* (1989) for rabbits treated with cimaterol; and by Kretchmar, Hathaway, Epley and Dayton (1990) for lambs treated with yet another ß-agonist (L-644,969; Merck, Sharpe and Dohm).

It is difficult to account for these marked variations. One possible explanation is variability in the dose of agonist used. In Figure 9 are shown results by Eadara, Dalrymple, DeLay, Ricks and Romsos (1989) in rats. These workers used the accretion of 3-methyl histidine in tissues (plus that lost by excretion) as an index of protein synthesis, and urinary 3-methyl histidine excretion as an index of protein degradation. Two doses of clenbuterol were used. As may be seen, while the lower dose produced a marked effect on 3-methyl histidine excretion, there was only a small further increase with the higher dose of clenbuterol. However, while the lower dose resulted in a small increase in rate of synthesis, there was a much larger increase with the higher dose. There were no significant effects by days 15-29 after treatment.

Nevertheless, the time course could also be important. The change in urinary 3-methyl histidine excretion was maximal after about 24 hours, and had largely disappeared by 6-7 days after treatment began. Perhaps changes in rate of protein synthesis developed much more slowly. Thus the primary action could be a change in rate of protein degradation.

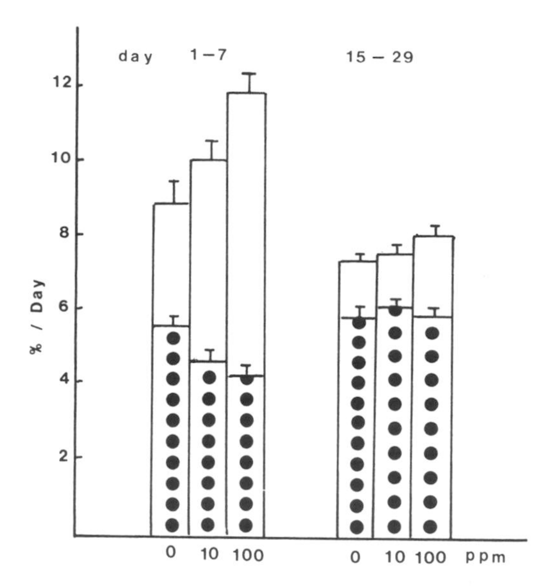

Figure 9 The effect of changes in the dose of dietary cimaterol, given over 30 days to female rats, on overall fractional rates of protein synthesis, degradation and net accretion. Degradation ●, Accretion □, Sum = synthesis. Results of Eadara *et al* (1989)

The role of protein degradation is of some practical importance since to it has been attributed a reduced tenderness of meat induced by ß-agonists (Hamby, Stouffer and Smith, 1986; Miller, Garcia, Coleman, Ekeren, Lunt, Wagner, Procknor, Welsh and Smith, 1988; Kretchmar *et al*, 1990; Fabry, Deronne, Buts and Demeyer, 1990).

Conclusions

In searching for means of controlling metabolic rate, trenbolone was originally selected because the action probably involved a reduction in the rate of protein degradation. Yet reduction in metabolic rate is greater than can reasonably be attributed to the decreased protein turnover. Studies with the catecholamines emphasise there is no necessary link. The α_2-agonist guanfacine lowers metabolic rate but does not affect protein degradation, at least in skeletal muscle. In contrast ß-agonists such as clenbuterol probably reduce the rate of protein degradation and increase energy utilisation; but the two effects can be dissociated and must be assumed to be determined by different ß-receptors. How then may metabolic rate best be decreased? It is tempting to suppose that metabolic rate is defined by the balance between an α-lowering and a ß-elevating action. The release of fatty acids from adipose tissue represents an analogous model. Indeed, the release of fatty acids itself affects oxygen consumption. However adipose tissue cannot be the only important site since in rats, adipose tissue contains very few α_2-receptors but guanfacine can readily lower metabolic rate.

In studies described here the prime aim was to develop a means of minimising liveweight loss, through lowering metabolic rate. Although we have some distance still to travel, we remain optimistic that a specific agent can be found without side effects, which can produce this result. Because of consumer attitudes, it is likely that in the medium term any successful technique developed will need to be immunological. In the longer term the technique required will need to involve genetic manipulation. For all three approaches, the only way to go is to understand more thoroughly the processes that control growth.

This study was specifically aimed at a problem in tropical agriculture. Does it have any significance for conventional agriculture? We believe it may have. In enhancing growth promotion an improvement in the utilisation of energy is a prime factor and it probably is unimportant whether this is obtained by reducing basal metabolic rate or by improving the efficiency of use of above-maintenance energy. At least in assessing whether this concept can be rebutted we may further clarify the control of growth.

Acknowledgment

We are grateful to the Australian Meat and Livestock Research and Development Corporation for financial support for most of the work described here.

References

Ahlquist, R.P. (1948) *American Journal of Physiology*, **154**, 586-600

Arch, J.R.S. (1989) *Proceedings of the Nutrition Society*, **48**, 215-223

Bohorov, O., Buttery, P.J., Correia, J.H.R.D. and Soar, J.B. (1987) *British Journal of Nutrition*, **57**, 99-107

Buttery, P.J., Haynes, N.B. and Lindsay, D.B. (1986) (Editors) *Control and Manipulation of Animal Growth*, London, Butterworths

Bylund, D.B. (1988) In *The α_2-adrenergic receptors* (ed L.E. Limbird), Clifton, New Jersey, Humana Press, pp 1-13

Castellino, P., Luzi, L., Del Prato, S. and De Fronzo, R.A. (1990) *American Journal of Physiology*, **258**, E117-E125

Cleays, M.C., Mulvaney, D.R., McCarthy, F.D., Gore, M.T., Marple, D.N. and Sartin, J.L. (1989) *Journal of Animal Science*, **67**, 2245-2254

Connacher, A.A., Jung, R.T. and Mitchell, P.E.G. (1988) *British Medical Journal*, **296**, 1217-1220

Coyer, P., Cox, M., Rivers, J.P.W. and Millward, D.J. (1985) *British Journal of Nutrition*, **53**, 491-499

Dale, H.H. (1906) *Journal of Physiology*, **34**, 163-206

Docherty, J.R. (1989) *Pharmacology and Therapeutics*, **44**, 241-284

Eadara, J.K., Dalrymple, R.H., DeLay, R.L., Ricks, C.A. and Romsos, D.R. (1989) *Metabolism*, **38**, 883-890

Elfellah, M.S. and Reid, J.L. (1987) *European Journal of Pharmacology*, **139**, 67-72

Emery, P.W., Rothwell, N.J., Stock, M.J. and Winter, P.D. (1984) *Bioscience Reports*, **4**, 83-91

Emorine, L.J., Marullo, S., Briend-Sutren, M., Patey, G., Tate, K., Delavier, K., Klutchko, C. and Strosberg, A.D. (1989) *Science*, **245**, 1118-1121

Fabry, J., Deronne, C., Buts, B. and Demeyer, D. (1990) *Journal of Animal Science*, **68** Suppl 1, 332

Forsberg, N.E., Ilian, M.A., Ali-bar, A., Cheeke, P.R. and Wehr, N.B. (1989) *Journal of Animal Science*, **67**, 3313-3321

Freedman, M.R., Horwitz, B.A. and Stern, J.S. (1986) *American Journal of Physiology*, **250**, R595-R607

Gorewit, R.C. (1981) *Journal of Endocrinological Investigation*, **4**, 135-139

Griggs, R.C., Kingston, W., Josefowicz, R.F., Herr, B.A., Forbes, G. and Halliday, D. (1989) *Journal of Applied Physiology*, **66**, 498-503

Hamby, P.L., Stouffer, J.R. and Smith, S.B. (1986) *Journal of Animal Science*, **63**, 1410-1417

Helferich, W.G., Jump, D.B., Anderson, D.B., Skjaerlund, D.M., Merkel, R.A. and Bergen, W.G. (1990) *Endocrinology*, **126**, 3096-3100

Higgins, J.A., Lasslett, Y.W., Bardsley, R.G. and Buttery, P.J. (1988) *British Journal of Nutrition*, **60**, 645-652

Hunter, R.A. (1989) *Journal of Agricultural Science*, **112**, 257-263

Hunter, R.A. (1991) *British Journal of Nutrition*, (In Press)

Hunter, R.A. and Magner, T. (1990) *Journal of Agricultural Science*, **115**, 121-127

Hunter, R.A. and Vercoe, J.E. (1987) *British Journal of Nutrition*, **58**, 477-483

Hunter, R.A. and Vercoe, J.E. (1988) *Journal of Agricultural Science*, **111**, 187-190

Jones, M.R., Hunter, R.A., Magner, T., Hoskinson, R.M. and Wynn, P.C. (1990) *Proceedings of the Society of Animal Production*, **18**, 500

Kennedy, A.D. and Belluk, B.M. (1987) *Canadian Journal of Animal Science*, **67**, 417-425

Kim, Y.S. and Lee, Y.B. (1990) *Journal of Animal Science*, **68** Suppl 1, 317

Kim, Y.S. and Sainz, R.D. (1990) *Journal of Animal Science*, **68** Suppl 1, 318

Kretchmar, D.H., Hathaway, M.R., Epley. R.J. and Dayton, W.R. (1990) *Journal of Animal Science*, **68**, 1760-1772

Lands, A.M., Arnold, A., McAuliff, J.P., Luduena, F.P. and Brown, T.G.jr (1967) *Nature*, **214**, 597-598

Lobley, G.E., Connell, A., Milne, E., Buchan, V., Calder, A.G., Anderson, S.E. and Vint, H. (1990) *British Journal of Nutrition*, **64**, 691-704

Lobley, G.E., Connell, A., Mollinson, G.S., Brewer, A., Harris, C.I. and Buchan, V. (1985) *British Journal of Nutrition*, **54**, 681-694

Lobley, G.E., Milne, V., Lovie, J.M., Reeds, P.J. and Pennie, K. (1980) *British Journal of Nutrition*, **43**, 491-502

Lobley, G.E., Connell, A., Buchan, V., Skene, P.A. and Fletcher, J.M. (1987) *Journal of Endocrinology*, **115**, 439-445

MacLennan, P.A. and Edwards, R.H.T. (1989) *Biochemical Journal*, **264**, 573-579

Maltin, C.A., Delday, M.I., Hay, S.M., Smith, F.G. and Reeds, P.J. (1987) *Bioscience Reports*, **7**, 51-57

Martinez, J.A., Buttery, P.J. and Pearson, J.T. (1984) *British Journal of Nutrition*, **52**, 515-521

Miller, M.F., Garcia, D.K., Coleman, M.E., Ekeren, P.A., Lunt, D.K., Wagner, K.A., Procknor, M., Welsh, Jr T.H. and Smith, S.B. (1988) *Journal of Animal Science*, **66**, 12-20

Moore, N.G., Sillence, M.N., Pegg, G.G. and Lindsay, D.B. (1990) *Proceedings of the Nutrition Society of Australia*, **15**, 164

Nakaki, T., Nakadate, T. and Kato, R. (1980) *Naunyn Schmiedebergs Archives of Pharmacology*, **313**, 151-153

Norman, M.I.T. and Stewart, G.H. (1964) *Journal of the Australian Institute of Agricultural Science*, **30**, 39-46

O'Kelly, J.C. (1985) *Nutrition Reports International*, **32**, 935-942

Pegg, G.G., Lindsay, D.B., Matthews, M.L. and Sillence, M.N. (1990) *Proceedings of the Australian Society of Animal Production*, **18**, 537

Pell, J.M., Bates, P.C., Elcock, C., Lane, S.E. and Simmonds, A.D. (1987) *Journal of Endocrinology*, **115** Suppl, 68

Reeds, P.J., Hay, S.M., Dorwood, P.M. and Palmer, R.M. (1986) *British Journal of Nutrition*, **56**, 249-258

Reeds, P.J., Hay, S.M., Dorwood, P.M. and Palmer, R.M. (1987) *Comparative Biochemistry and Physiology*, **89C**, 337-341

Roche, J.F. and Quirke, J.F. (1986) In *Control and Manipulation of Animal Growth*, (eds P.J. Buttery, N.B. Haynes and D.B.Lindsay), London, Butterworths, pp 36-53

Rothwell, N.J. and Stock, M.J. (1985) *Bioscience Reports*, **5**, 755-760

Rumsey, T.S. and Hammond, A.C. (1990) *Journal of Animal Science*, **68**, 4310-4318

Sanchez, A. and Pettinger, W.A. (1981) *Life Sciences*, **29**, 2795-2802

Schaefer, A.L., Jones, S.D.M., Kennedy, A.D., Tong, A.K.W. and Onischuk, L.A. (1990) *Canadian Journal of Animal Science*, **70**, 857-866

Schiavetta, A.M., Miller, M.F., Lunt, D.K., Davis, S.K. and Smith, S.B. (1990) *Journal of Animal Science*, **68**, 3614-3623

Sharpe, P.M., Buttery, P.J. and Haynes, N.B. (1986) *British Journal of Nutrition*, **56**, 289-304

Sillence, M.N. and Etherton, T.G. (1991) *Journal of Animal Science*, **69**, 2815-2821

Sillence, M.N., Matthews, M.L., Spiers, W.G. and Lindsay, D.B. (1990) *Proceedings of the Nutrition Society of Australia*, **15**, 170

Sillence, M.N., Matthews, M.L., Spiers, W.G., Pegg, G.G. and Lindsay, D.B. (1991) *Naunyn-Schmiederbergs Archives of Pharmacology*, **344**, 449-453

Sillence, M.N. and Rodway, R.G. (1990) *Journal of Endocrinology*, **126**, 461-466

Sillence, M.N., Spiers, W.G. and Lindsay, D.B. (1990) *Proceedings of the Australian Society of Animal Production*, **18**, 550

Sinnett-Smith, P.A., Dumelow, N.W. and Buttery, P.J. (1983) *British Journal of Nutrition*, **50**, 225-234
Smith II, C.K., Lee, D.E. and Coutinho, L.L. (1990) *Journal of Animal Science*, **68** Supp 1, 284
Spiers, W.G., Sillence, M.N. and Lindsay, D.B. (1990) *Proceedings of the Nutrition Society of Australia*, **15**, 172
Staten, M.A., Matthews, D.E., Cryer, P.E. and Bier, D.M. (1987) *American Journal of Physiology*, **253**, E322-E330
Stock, M.J. and Rothwell, N.J. (1986) In *Control and Manipulation of Animal Growth*, (eds P.J. Buttery, N.B. Haynes and D.B.Lindsay), London, Butterworths, pp 249-258
Vernon, B.G. and Buttery, P.J. (1978) *Animal Production*, **26**, 1-9
Wang, S-Y. and Beerman, D.H. (1988) *Journal of Animal Science*, **66**, 2545-2550
Williams, P.E.V., Pagliani, L., Innes, G.M., Pennie, K., Harris, C.I. and Garthwaite, P. (1987) *British Journal of Nutrition*, **57**, 417-428

14

REGULATION OF FAT AND LEAN DEPOSITION BY THE IMMUNE SYSTEM

D. J. FLINT
Hannah Research Institute, Ayr, Scotland, KA6 5HL, UK

Introduction

The increasingly sedentary lifestyle adopted in the developed world has led to considerable concern regarding energy intake and has progressed to an almost obsessive desire to drastically reduce fat consumption. Along with the increasing demand for greater choice of foods has developed a desire therefore to consume leaner cuts of meat. For the producer, too, the production of excess fat is a costly business, so there is an equivalent desire to improve lean:fat ratios in livestock from a purely commercial standpoint.

There is no doubt that we already have the wherewithal to greatly improve carcass composition. Indeed the use of anabolic steroids was widespread in Europe until the EC banned their use, although these compounds are still used in the United States. Although pressure was brought to bear on the basis of concern over their safety, they were in fact banned shortly before a scientific committee, brought together to consider their use, found no grounds for such a ban. Partly in response to the loss of such steroid growth promoters, two alternative hormonal strategies have been developed, perhaps more rapidly than they would otherwise have been. The ß-agonists, based upon ß adrenergic compounds such as noradrenaline, have marked effects in increasing muscle mass and decreasing carcass fat content. These compounds however suffer the same drawback as anabolic steroids in being orally active in humans. Much greater promise has however been shown by the development of recombinant forms of both bovine and porcine growth hormones (GH). Being peptide in nature, they require injection as they are orally inactive and indeed they are considered to be inactive in humans even if injected. GH not only improves carcass composition, but also dramatically increases milk yield in dairy cattle, making it of considerable interest to the dairy industry. Despite its apparent "clean bill of health", however, GH has many vociferous opponents and attempts to obtain a licence for its use have resulted in producing far and away the most exhaustively-tested compound proposed for animal use. Concern now exists that GH has already passed the three normal hurdles of safety, efficacy and quality but that it faces the introduction, some would say existence, of the so-called fourth hurdle;

that of need.

It is with these problems in mind that the remainder of my discussions turns to yet another series of alternative strategies, involving immune intervention in biological processes. It should be stressed, however, that even if these approaches are perceived as more "consumer friendly", safer and more cost-effective, they will probably have to jump just as high as growth hormone to avoid falling at the final hurdle.

Techniques for immunomanipulation

For the purposes of this discourse, I have divided the various approaches into groups which I have classified as immunoneutralization, where antibodies are effective simply by means of binding to and preventing the biological function of hormones; immunoenhancement, where the converse is true; true cytotoxic antibodies, in which the effects are dependent upon the activation of immune effector systems; and immuno-mimics, antibodies which look like and function like hormones.

Finally, although these techniques attempt to produce a more natural and acceptable approach than hormones, I will consider the possibility that there already exist normal physiological relationships between adipose tissue, muscle and the immune system which may provide us with a completely novel series of approaches to occupy us well into the 21st century.

Immunoneutralization

GH release from the pituitary is stimulated by growth hormone releasing factor (GRF), whilst somatostatin inhibits GH release. Immunoneutralization of somatostatin release should result in increased GH concentrations in serum and improved growth rates and/or carcass composition and Spencer, Garssen and Hart (1983a,b) described a dramatic increase in body weight gain of St Kilda sheep treated in this way. Similar findings have been demonstrated in sheep (Bass, Gluckman, Fairclough, Peterson, Davis and Carter, 1987) and in cattle (Lawrence, Schelling, Byers and Greene, 1986) although others have failed to find such an effect (Fitzsimons and Hanrahan, 1984; Galbraith, Wigzell, Scaife and Henderson, 1985; Trout and Schanbacher, 1990) and at least one study produced inhibition of growth (Varner, Davis and Reeves, 1980).

Although passive immunization against somatostatin resulted in increased concentrations of GH in serum of rats (Arimura, Smith and Schally, 1976; Terry and Martin, 1981) as did some studies involving active immunization in sheep (Varner *et al*, 1980; Spencer *et al*, 1983a,b), pigs (Dubreuil, Pelletier, Peticlerc, Lapierre, Gaudreau and Brazeau, 1989) and cattle (Peticlerc, Pelletier, Dubreuil, Lapierre, Farmer and Brazeau, 1988) other studies have reported no such change in sheep (Laarveld, Chaplin and Kerr, 1986) or cattle (Lawrence *et al*, 1986).

The fact that growth effects were produced in some cases without convincing changes in GH concentrations in blood and indeed without any change in the protein:fat ratio within the carcass (which is normally increased by GH) suggested that the hypothesis was too simplistic, since somatostatin also regulates thyrotrophin (TSH) (Reichlin, 1983), adrenocorticotrophin (ACTH) (Reisine, 1985), glucagon and insulin secretion (Bozikov, 1980). Somatostatin is also present in the gastrointestinal tract and passive immunization against somatostatin significantly reduced digesta flow (Fadlalla, Spencer and Lister, 1985) which might explain the increased efficiency of food utilization described by Spencer *et al*, (1983b).

Immunomodulation of hormone binding proteins

Although antibodies can be used to neutralize or even enhance the actions of hormones by a direct interaction between antibody and hormone, antibodies can, under appropriate circumstances, be used to influence hormone action in a more subtle, indirect fashion. This possibility arises because of the somewhat surprising finding that in an analagous fashion to insoluble steroid hormones, certain peptide hormones, including GH and insulin-like growth factors (IGF), interact in the blood stream with specific binding proteins which prolong their half-life in blood. The GH binding protein (Baumann, Stolar, Amburn, Barsano and De Vries, 1986) appears to be a truncated version of the GH receptor (Leung, Spencer, Cachianes, Hammonds, Collins, Henzel, Barnard, Waters, and Wood, 1987), whilst a complex array of large and small molecular weight binding proteins (which are unrelated to IGF receptors) interact with IGFs both in serum and extracellular fluid. Although the biological functions of these binding proteins are still the subject of much debate, it is likely that the large molecular weight binding protein in serum serves to prolong the half-life of IGFs in blood, whilst the smaller binding proteins produced by numerous cell types probably inhibit IGF action. This concept is supported by the observation that a truncated form of IGF-I which binds very poorly to certain IGF binding proteins, has increased biological activity (Szabo, Mottershead, Ballard and Wallace, 1988).

If these binding proteins do inhibit the effects of GH and IGF-I, then antibodies produced against such binding proteins, which are capable of reducing their IGF or GH binding capacity, might serve to enhance the anabolic actions of these hormones.

Immunoenhancement

The classical functions of antibodies are to identify and bind their target and produce immunoneutralization via a number of mechanisms, so it is perhaps surprising to find that antibodies complexed to hormones are actually capable of enhancing the actions of hormones such as GH.

The difficulties of understanding the mechanism of enhancement are illustrated by studies such as those of Aston, Holder, Preece and Ivanyi, (1986) which demonstrated monoclonal antibodies to human GH which were inhibitory *in vitro* and enhancing *in vivo*, with no apparent relationship between enhancement and receptor binding properties in various *in vitro* systems. The mechanism of enhancement does appear to be by improvement of GH-action since circulating IGF-I levels are increased by such approaches, a classical response to GH (Wallis, Daniels, Ray, Cottingham and Aston, 1987).

Despite the fact that GH bound to monoclonal antibodies may be 10-fold more active, for example in Snell Dwarf mice (Aston *et al*, 1986), this is not likely to be the most effective approach in animal production systems. True potential lies in the ability of such enhancing antibodies to complex with endogenous GH. Such an effect has been achieved in sheep when examining the "diabetogenic effect" of GH (its ability to antagonize the effects of insulin on blood glucose concentrations). Insulin given to sheep intravenously induced a decrease in blood glucose of approximately 2 mmol/l. GH antagonized the effect of insulin by reducing this decrease in blood glucose to around 1 mmol/l, but when GH was given pre-complexed to monoclonal antibody it was even more effective since the decline in blood glucose was abolished. Perhaps of greatest significance was the effect of a monoclonal antibody to GH when administered alone, since it appeared to be capable of complexing with and enhancing the effectiveness of endogenous GH (Pell, Johnsson, Pullar, Morrell, Hart, Holder and Aston, 1989).

An even more promising practical approach for enhancement of hormone action in this manner involves identification of the epitope(s) on GH involved in binding enhancing antibodies, then synthesizing the peptide and using it to evoke a polyclonal antibody response restricted to that region of GH, which should allow antibody to complex with and enhance GH action.

Recently such studies have been described in cattle where effects have been demonstrated in lactating animals. Rather than using monoclonal antibodies, polyclonal antibodies were produced against a small peptide derived from the sequence of GH. This produces an antiserum of "restricted specificity" probably reacting with a single epitope and thereby resembling a monoclonal antibody. When bovine GH was given to lactating dairy cows it prevented the decline in milk yield which normally occurs. When bovine GH was precomplexed to an antiserum to the peptide, it actually stimulated milk yield and serum IGF-I concentrations over and above those of GH alone (Pell, Elcock, Walsh, Trigg and Aston, 1989). Milk yield was also enhanced in animals given the antiserum alone, indicating its ability to complex with endogenous GH and enhance its actions. The production of enhancing antibodies by the use of synthetic linear peptide sequences looks promising but perhaps the production of synthetic peptides which simulate epitopes which involve discontinuous amino acid residues could yield even more potent agents for enhancement of hormone action.

Understanding the mechanism of this enhancement phenomenon has been made even more puzzling by the fact that conjugation of GH to ovalbumin or bovine serum albumin using cross-linking reagents significantly increased ^{35}S uptake into costal cartilage over that of growth hormone alone (Holder and Aston, 1989). When growth hormone was conjugated to itself, similar enhancement was obtained. Such findings suggest that enhancement may not be limited to particular epitopes on GH but may involve the increase in molecular size induced by binding to antibody. The determination of the mechanism(s) of enhancement should allow more precise control of this phenomenon and may provide new options for manipulation of hormone activity.

Amongst the possible mechanisms of action of enhancing antibodies are receptor cross-linking, which seems unlikely since Fab fragments are capable of enhancement (Aston *et al*, 1986) or antibody-induced conformational changes in the hormone structure. It is difficult to imagine how antibody binding to a number of different epitopes of the hormone would all lead to improved conformational changes. Alternatively the antibody could prevent hormone internalization, leading to increased duration of interaction of hormone with its receptor, increase its half-life in blood or lead to targetting of hormone to particular subsets of receptors. Subtypes of GH receptors do appear to exist (Barnard, Bundesen, Rylatt and Waters, 1985) and monoclonal antibodies to GH possess the ability to inhibit GH binding to some receptors but not others (Thomas, Green, Wallis and Aston, 1987).

Antibodies as mimics of hormones

The production of antibodies which mimic the structure and function of hormones has been of considerable interest to a number of research areas, not least because they are thought to play a role in certain autoimmune diseases. When antibodies are produced to hormones, these antibodies themselves can be used as an immunogen to produce anti-antibodies, a proportion of which will be anti-idiotypic antibodies or true "internal" images of the original hormone. Such anti-idiotypes have been produced for acetylcholine (Wasserman, Penn, Freimuth, Treptow, Wentzel, Cleveland and Erlanger, 1982), insulin (Sege and Peterson, 1978), ß-adrenergic compounds (Schreiber, Courgud, Andre, Vray and Strosberg, 1980), TSH (Farid, Pepper, Urbina-Briones and Islam, 1982) and GH (Gardner, Morrison, Stevenson and Flint, 1990). These antibodies may act as hormone antagonists whilst others may be capable of inducing the biological response of the hormone. Anti-idiotypes to ß-adrenergic compounds and GH are particularly interesting in view of the considerable interest in the use of both classes of hormone in animal growth and both types of hormone agonist have been successfully produced (Schreiber *et al*, 1980; Gardner *et al*, 1990). In studies involving the production of polyclonal anti-idiotypic antibodies to rat GH, we have shown that antibodies can mimic GH in a very specific fashion, inhibiting GH binding to receptors from sheep and rat liver and rat adipose tissue, whilst showing no ability to inhibit either prolactin or insulin binding. When these antibodies were administered to hypophysectomized rats, they were able to increase body weight gain in a fashion analagous to GH (Figure 1).

The elegant studies of Thomas *et al* (1987) demonstrated that certain monoclonal antibodies could inhibit binding of human GH to certain GH receptors but not to others and that some of those antibodies inhibited GH binding in livers from lean but

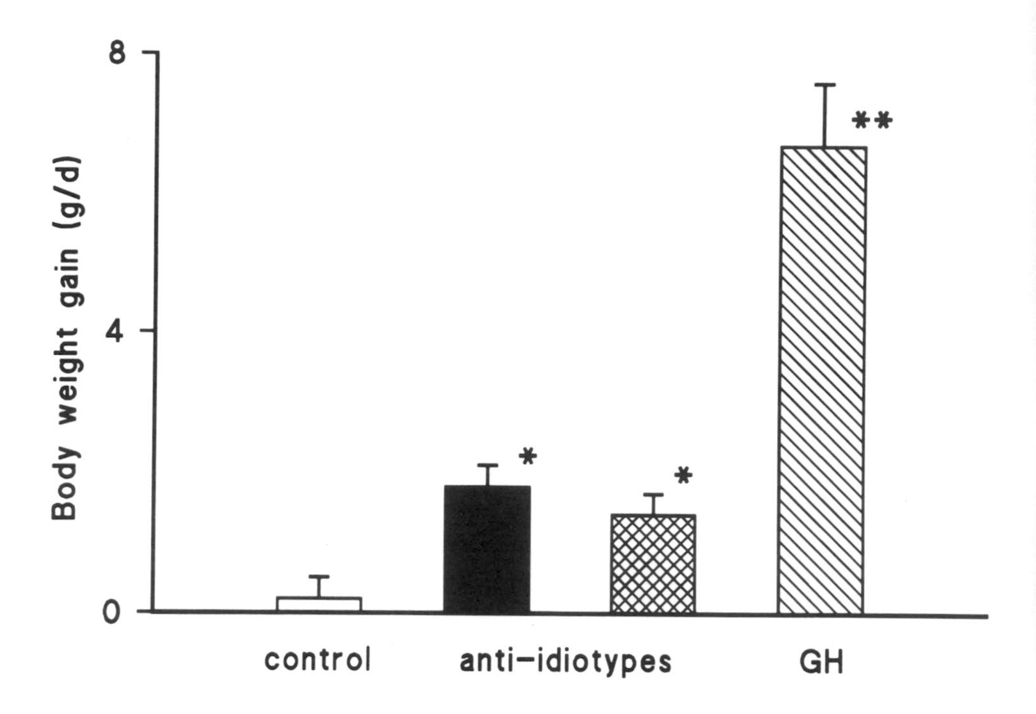

Figure 1 Effects of GH and anti-idiotypic antibodies to rGH on body weight gain in hypophysectomized rats.

Values are means ± SE. *P<0.05, **P<0.01 compared with control. For further details see Gardner *et al* (1990)

not obese mice. Such findings clearly indicate that there may be distinct GH receptors which recognize distinct epitopes on the GH molecule and that the proportions of these receptors may change with physiological state. Recent studies by Elbashir, Brodin, Akerstrom and Donner (1990) showed that monoclonal anti-idiotypic antibodies to human GH exhibited different effects dependent upon the specific receptors for GH to which they bound. Some antibodies inhibited GH binding to liver, increased it to the human GH serum binding protein and had no effect on GH binding to rabbit liver. This suggests that different epitopes of GH may be involved in binding to different receptor populations and to serum binding proteins.

Anti-idiotypic antibodies which mimic single epitopes on the GH molecule may thus provide GH mimics which are more specific than GH itself - perhaps enhancing only growth, or milk production or its diabetogenic effects. Anti-idiotypic antibodies can also be mimics of epitopes which involve discontinuous amino acid residues derived from different parts of the polypeptide chain, which offers a considerable advantage over linear peptide fragments of GH as mimics of GH.

From a practical point of view, anti-idiotypes have another advantage, since after identifying an idiotypic antibody to a hormone, (Ab_1), this could be used to immunize animals in order to induce endogenous production of Ab_2, the hormone image. This is an amplification process, requiring only small amounts of Ab_1 and since the half life of antibody is typically weeks compared with thc half lifc of GII (approximately 20 minutes), it would preclude the need for frequent injections. This approach does have limitations since no mechanisms exist for regulating either the size or the duration of the immune response. This is less of a drawback for GH however, since excess GH in the blood of growing animals has not been reported to produce problems for animal health.

The ability to mimic the three-dimensional structure of parts of the GH molecule using antibodies has considerable advantages over the attempts to produce fragments of GH by a variety of cleavage techniques. Although such techniques have demonstrated peptides which mimic certain functions of GH (see for example Wallis, 1982; Retegui, Milne, Cambiaso and Masson, 1982) the biological activity of the molecules is typically dramatically reduced by such modification, demonstrating the probable importance of the tertiary structure of the molecule. This fact is emphasized by the three-dimensional structure described for porcine GH by Abdel-Meguid, Shieh, Smith, Dayringer, Violand and Bentle (1987).

Cytotoxic antibodies

The control of abnormal tissues, such as tumours, has been a problem facing the medical world for centuries and in recent times increasingly sophisticated techniques have been attempted. One of these, the so-called "magic bullet" approach involves the destruction of tumours with antibodies produced against them. Unfortunatcly in many cases even considerable success in destroying cancerous cells serves only to delay the advance of rapidly-growing tumours. We reasoned, however, that in adipose tissue, which has a relatively low mitotic rate post-natally, such an approach may have a good chance of successfully reducing adipocyte numbers and, with it, the ability to store fat.

Antibodies were raised in sheep against plasma membranes from rat adipocytes. Such antibodies induced lysis of rat adipocytes *in vitro* and when they were injected into young growing rats they produced a number of effects. Initially, after injecting rats for four days with these antibodies, destruction of adipocytes was apparent and was accompanied by cellular infiltration of the tissue by lymphocytes and polymorphs presumably involved in the killing procedure (Futter and Flint, 1987). The relative importance of such cell-mediated cytotoxicity as opposcd to a complement-mediated mechanism, is as yet uncertain.

After 7 days the major fat depot in the rat, the parametrial depot, was reduced by 50% in weight due to a 50% reduction in the number of cells in the depot. Despite the fact that animals were subsequently untreated, by 8 weeks the weight of this depot was reduced to 25% of control values, indicating that lost adipocytes were not being replaced. In fact those adipocytes which were not lysed by antibodies were apparently less able to store triacylglycerol compared with adipocytes from control rats. At this stage, total body fat was reduced by 30% with a replacement of lost fat by protein (Futter and Flint, 1987).

In longer-term studies effects of treatment were still evident for at least 6 months (Panton, Futter, Kestin and Flint, 1990). Surprisingly, treated animals actually became heavier than controls exhibiting for a period of about 4 weeks hyperphagia, during which they ate 20% more than controls and gained 40% more body weight, resulting in an increase in food conversion efficiency of about 15%. Despite increased body weight 7 weeks after injection, there was a reduction in the amount of fat in several major depots (Table 1). Subsequently food intake and rate of body weight gain normalized although treated rats remained heavier than controls until the the experiment ended at 6 months.

The ability to destroy adipocytes *in vitro* has also been shown for a number of commercially-important species, including sheep, pig and chicken. Using local subcutaneous injections of antibodies into pigs (see Futter and Flint, 1990) and sheep, we have demonstrated that adipocytes can be destroyed *in vivo*. Reductions in fat deposition in sheep treated in similar fashion have also been reported by others (Moloney and Allen, 1988; Moloney, 1990; Nassar and Hu, 1991) and in studies administering anti-adipocyte antibodies intraperitoneally to young pigs, we have reduced back-fat thickness by 30% and fat content of limb joints by 25% with complete replacement by lean tissue 14 weeks later (Table 2).

This approach, involving a short treatment period with antibodies, followed by a long withdrawal period, using antibodies which are peptide in nature and therefore inactive orally, offers a very safe approach to reducing body fat deposition. The surprising beneficial effect upon lean body growth suggests that lean body mass is accumulating at submaximal rates in the presence of adipose tissue. It also questions the need for ß adrenergic compounds to have direct effects upon muscle and suggests that purely lipolytic (*ie*, not thermogenic) agents may also produce increases in lean body mass.

Immunological ablation of specific endocrine cells

Combining the approach of using antibodies to neutralize hormones with the concept of using antibodies as cytotoxic agents to destroy tissues leads naturally to the prospect of using antibodies to destroy hormone-producing cells within endocrine organs. Such a technique could follow that of tumour immunology, by identifying specific antigens on the surface of the cell and using these to target antibodies to them in a specific fashion. This can be a difficult and time-consuming approach because many cells share antigens such as histocompatibility antigens. It appears however that cells which secrete particular hormones also express those same hormones on the cell

Table 1 EFFECT OF PASSIVE IMMUNIZATION OF RATS ON BODY LENGTH, WEIGHT AND COMPOSITION 7 WEEKS POST-TREATMENT

		Control	Treated
Weight	(g)	247.0 ± 7.0	269.0 ± 17**
Length	(cm)	18.9 ± 0.3	19.8 ± 0.3
Protein	(g)	50.8 ± 2.2	56.5 ± 2.3*
Ash	(g)	13.1 ± 0.6	15.2 ± 0.8*
Fat	(g)	41.5 ± 1.8	34.1 ± 5.1*
Parametrial fat pad	(g)	7.1 ± 0.6	3.3 ± 0.8*
Subcutaneous fat pad	(g)	4.3 ± 0.1	2.9 ± 0.5*

Values are means ± SE. *$P<0.05$, **$P<0.01$ compared with control
Data adapted from Panton *et al* (1990)

Table 2 EFFECTS OF PASSIVE IMMUNIZATION (IP) WITH ANTIBODIES TO PIG ADIPOCYTE PLASMA MEMBRANES ON CARCASS QUALITY IN PIGS AT 20 WEEKS OF AGE (KESTIN, TONNER, KENNEDY AND FLINT, UNPUBLISHED OBSERVATIONS)

	Control	Treated
Backfat thickness at P2 site (mm)	12.9 ± 0.7	9.0 ± 0.7
Forelimb joint weights (kg)		
Fat	1.01 ± 0.10	0.76 ± 0.11
Lean	2.00 ± 0.07	2.23 ± 0.14

surface, possibly during secretion, suggesting that anti-hormone antibodies might be capable of destroying the cells producing them. In studies where antiserum to rat GH was administered *in vivo* a long-term reduction in growth and body weight gain of rats was evident. Such rats had a poor response to GH releasing factor, GRF (Flint and Gardner, 1989; Gardner and Flint, 1990), and their pituitary glands had reduced contents of GH and decreased numbers of somatotrophs indicative of destruction of GH producing cells (Figure 2).

Others have opted to deliver toxins to pituitary cells, not by means of antibodies, but by the releasing factors which bind to the pituitary and normally elicit hormone secretion. Schwartz, Penke, Rivier and Vale (1987) used corticotrophin releasing factor (CRF) linked to the plant toxin gelonin and showed that it could destroy pituitary ACTH producing cells, at least based on the inability of pituitary cells to produce ACTH in response to a subsequent challenge with CRF. Subsequently, however, Schwartz and Vale, (1988) showed a potential limitation of this approach since, after treatment with the CRF-gelonin conjugate, ACTH responses to CRF were severely reduced but the ACTH response to vasopressin was unaffected. This suggests that a population of ACTH-producing cells exists which have no CRF receptors. It is imperative, therefore, if total ablation is required, that receptors for the releasing factor exist on all hormone-producing cells. If total ablation is not required, however, then such a technique actually has considerable advantages. A second limitation is that not only should the receptors for the releasing factor be present on the target cells, but they should be absent on all other cell types. This is demonstrably not the case for thyrotrophin releasing hormone (TRF) receptors, since it has a potent stimulatory effect, not only on TSH release but also on prolactin release from the pituitary. This would imply that TRF-toxin conjugates would decrease prolactin secretion and this has been shown to be the case (Reichlin, Bacha and Murphy 1983).

Although it might seem likely that techniques involving immunodetection of the target hormone to be ablated might be preferable to the use of releasing factor-toxin conjugates, it should be borne in mind that even for hormones like GH which are considered classical endocrine hormones of pituitary origin, it is now evident that the placenta and lymphocytes also produce GH-like molecules. Indeed the immune system now appears to produce an array of hormones resembling those of the endocrine system including GH, prolactin and ACTH.

Interactions of adipose tissue and the immune system

The preceding discussions have concentrated on the use of the immune system, as a tool for manipulation of the endocrine system or directly against adipose tissue, but increasing evidence suggests that a two-way physiological relationship may exist between the two tissues. In cachexia, for example, where tumour necrosis factor (TNF, a product of macrophages) increases, the TNF inhibits both lipoprotein lipase and acetyl CoA carboxylase activity in adipose tissue although the physiological

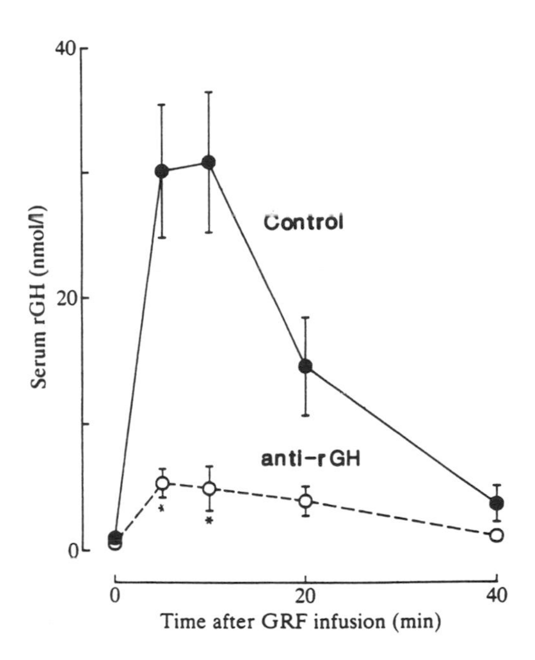

Figure 2 GH release in response to GRF treatment in rats treated neonatally with antiserum to rGH.

Values are means ± SEM. *P<0.01 compared with controls.

significance of such a finding is still a little perplexing. Even more recently, however, the identification of adipsin, produced only by adipocytes and the CNS, as a serine protease similar to factor D of the alternative pathway of complement makes for the intriguing possibility that immune activation within adipose tissue itself may serve to regulate adiposity (Cook, Min, Johnson, Chaplinsky, Flier, Hunt and Spiegelman, 1987; Rosen, Cook, Yaglom, Groves, Volanakis, Damm, White and Spiegelman, 1989). This is borne out to some extent by the fact that adipsin levels in adipose tissue are negatively correlated with some, though not all, forms of obesity (Spiegelman, Lowell, Napolitano, Dubuc, Barton, Francke, Groves, Cook and Flier, 1989; Platt, Min, Ross and Spiegelman, 1989; Dugail, Le Leipvre, Quignard-Boulangé, Pairault and Lavau, 1989).

Several other observations are worthy of mention in this context including the "glutamate cycle" which apparently operates between lymphocytes and muscle and the detection of both factors B and C3 which are required for activation of the alternative pathway of complement in adipose tissue (Choy, Rosen and Spiegelman, 1991) as well as the bidirectional communication which exists between the endocrine and immune systems, including lymphokine regulation of pituitary hormone secretion (Brown, Smith and Blalock, 1987; McCann, Rettori, Milenkovic, Jurcovicova, Synder and Beutler, 1989), mitogenic responses of lymphocytes to various pituitary hormones, and the production of pituitary-like hormones by the immune system (see Blalock, 1989 for review). Although many of these studies are in their infancy, they may provide an as yet untapped source of manipulative approaches for the regulation of body lean and fat deposition.

Summary

The limitations imposed upon the use of certain hormones and the increasing concern in general about their use in animal production systems has led to the pursuit of a variety of immunological approaches to achieve the same desirable effects on body composition. Although the use of antibodies to neutralize the hormone somatostatin produced dramatic and encouraging results, this particular approach has subsequently brought more controversy to the study of immunomodulation of growth than probably any other. It serves to highlight a number of problems when evoking immune responses within animals, which include control of the size and duration of the response, its specificity and, last but not least, the fact that although the target hormone is well characterized, its site of production and its functions within the body may not be. The induction of a number of complex changes within the body may result from such immunological intervention, making interpretation of the study difficult, if not impossible.

At the opposite end of the spectrum are the immuno-enhancement and hormone-imaging techniques which, as well as suffering the potential drawbacks described above, rely on the fact that antibodies induced in this way do not evoke activation of the destructive elements of the immune system, at least not to a major extent, which would result in inappropriate cytotoxic events.

Even for those approaches which deliberately use classical cytotoxic responses for their effects, there are potential problems. Precise targetting of antibodies, which is highly desirable, has been a considerable problem for immunotherapy in tumour biology, but may be less so for animal production systems because of the desire to remove specific tissues rather than a proportion of cells within a tissue which are exhibiting cancerous behaviour, whilst leaving the remaining "normal cells" undamaged. Cytotoxic approaches clearly require antibodies to activate the effector elements of the immune system, *ie* complement and cell-mediated immune killing, but once again the body possesses mechanisms designed to reduce, in particular, autoimmune responses of this nature, in at least two ways, decay accelerating factor, which clears complement rapidly, and homologous restriction factor, which restricts the ability of complement to lyse the host's own cells.

Notwithstanding these potential limitations, the approaches outlined in this article have already demonstrated the power of many of these techniques. Control of the immune response is currently the subject of considerable study and with an inevitable improvement in our knowledge of these processes it should be possible to design immunological approaches which are at least as safe as hormones, potentially simpler and more economic, and in some cases may lead to "novel" hormones of greater specificity.

Whether these approaches will be any more "consumer acceptable" than hormones themselves, however, and where they will fit in relation to the powerful potential of transgenic animals, remains to be seen.

References

Abdel-Meguid, S.S., Shieh, H-S., Smith, W.W., Dayringer, H.E., Violand, B.N. and Bentle, L.A. (1987) *Proceedings of the National Academy of Sciences, USA*, **84**, 6434-6437

Arimura, A., Smith, W.D. and Schally, A.V. (1976) *Endocrinology*, **98**, 540-543

Aston, R., Holder, A.T., Preece, M.A. and Ivanyi, J. (1986) *Journal of Endocrinology*, **110**, 381-388

Barnard, R., Bundesen, P.G., Rylatt, D.B. and Waters, M.J. (1985) *Biochemical Journal*, **231**, 459-468

Bass, J.J., Gluckman, P.D., Fairclough, R.J., Peterson, A., Davis, S.R. and Carter, W.D. (1987) *Journal of Endocrinology*, **112**, 27-31

Baumann, G., Stolar, M.W., Amburn, K., Barsano, C.P. and De Vries, B.C. (1986) *Journal of Clinical Endocrinology and Metabolism*, **62**, 134-141

Blalock, J.E. (1989) *Physiological Reviews*, **69**, 1-32

Bozikov, V. (1980) *Diabetologia Croatica*, **IX-1**, 8-58

Brown, S.L., Smith, L.R. and Blalock, J.E. (1987) *Journal of Immunology*, **139**, 3181-3183

Choy, L., Rosen, B.S. and Spiegelman, B.M. (1991) *Journal of Cell Biology Supplement*, **15B**

Cook, K.S., Min, H.Y., Johnson, D., Chaplinsky, R.J., Flier, J.S., Hunt, C.R. and Spiegelman, B.M. (1987) *Science*, **237**, 402-408

Dubreuil, P., Pelletier, G., Peticlerc, D., Lapierre, H., Gaudreau, P. and Brazeau, P. (1989) *Endocrinology*, **125**, 1378-1384

Dugail, I., Le Liepvre, X., Quignard-Boulangé, A., Pairault, J. and Lavau, M. (1989) *Biochemical Journal*, **257**, 917-919

Elbashir, M.I., Brodin, T., Akerstrom, B. and Donner, J. (1990) *Biochemical Journal*, **266**, 467-474

Fadlalla, A.M., Spencer, G.S.G. and Lister, D. (1985) *Domestic Animal Endocrinology*, **5**, 35-41

Farid, N.R., Pepper, B., Urbina-Briones, R. and Islam, N.R. (1982) *Journal of Cellular Biochemistry*, **19**, 305-313

Fitzsimons, J.M. and Hanrahan, J.P. (1984) *An Foras Taluntais Animal Production Report*, pp 80-81

Flint, D.J. and Gardner, M.J. (1989) *Journal of Endocrinology*, **122**, 79-86

Futter, C.E. and Flint, D.J. (1987) In *Recent Advances in Obesity Research*, (ed E.M. Berry), London, John Libbey and Co, pp 181-185

Futter, C.E. and Flint, D.J. (1990) In *The control of body fat*, (eds J.M. Forbes and G.R. Hervey), Smith-Gordon and Co Ltd, pp 87-99

Galbraith, H., Wigzell, S., Scaife, J.R. and Henderson, G.D. (1985) *Animal Production*, **40**, 523

Gardner, M.J. and Flint, D.J. (1990) *Journal of Endocrinology*, **124**, 381-386

Gardner, M.J., Morrison, C.A., Stevenson, L.W. and Flint, D.J. (1990) *Journal of Endocrinology*, **125**, 53-59

Holder, A.T. and Aston, R. (1989) In *Biotechnology in Growth Regulation*, (eds R.B. Heap, C.G. Prosser and G.E. Lamming), London, Butterworths, pp 167-177

Laarveld, B., Chaplin, R.K. and Kerr, D.E. (1986) *Canadian Journal of Animal Science*, **66**, 77-83

Lawrence, M.E., Schelling, G.T., Byers, F.M. and Greene, L.W. (1986) *Journal of Animal Science*, **63**, 215 (supplement)

Leung, D.W., Spencer, S.A., Cachianes, G., Hammonds, R.G., Collins, C., Henzel, W.J., Barnard, R., Waters, M.J. and Wood, W.I. (1987) *Nature*, **330**, 537-543

McCann, S.M., Rettori, V., Milenkovic, L., Jurocovicova, J., Synder, G. and Beutler, B.(1989) *Neurology and Neurobiology*, **50**, 333-349

Moloney, A.P. (1990) *Biochemical Society Transactions*, **18**, 336-337

Moloney, A.P. and Allen, P. (1988) *Proceedings of the Nutrition Society*

Nassar, A.H. and Hu, C.Y. (1991) *Journal of Animal Science*, **69**, 578-586

Panton, D., Futter, C., Kestin, S. and Flint, D. (1990) *American Journal of Physiology*, **258**, E958-E989

Pell, J.M., Elcock, C., Walsh, A., Trigg, T., and Aston, R. (1989) In *Biotechnology in Growth Regulation* (eds R.B. Heap, C. Prosser and G.E. Lamming), London, Butterworths, p 259

Pell, J.M., Johnsson, I.D., Pullar, R.A., Morrell, D.J., Hart, I.C., Holder, A.T. and Aston, R. (1989) *Journal of Endocrinology*, **120**, R15-R18

Peticlerc, D., Pelletier, G., Dubreuil, P., Lapierre, H., Farmer, C. and Brazeau, P. (1988) *Journal of Animal Science*, **66**, 389 (supplement). Abstract 408

Platt, K.A., Min, H.Y., Ross, S.R. and Spiegelman, B.M. (1989) *Proceedings of the National Academy of Sciences, USA*, **86**, 7490-7494

Reichlin, S (1983) *New England Journal of Medicine*, **309**, 1495-1501

Reichlin, S., Bacha, P. and Murphy, J.R. (1983) *Proceedings of the 8th American Peptide Symposium* (eds V.J. Hruby and D.H. Rich) p 837

Reisine, T. (1985) *Endocrinology*, **116**, 2259-2266 Rockford:Pierce Chemical Co

Retegui, L.A., Milne, R.W., Cambiaso, C.L. and Masson, P.L. (1982) *Molecular Immunology*, **19**, 865-875

Rosen, B.S., Cook, K.S., Yaglom, J., Groves, D.L., Volanakis, J.E., Damm, D., White, T. and Spiegelman, B.M. (1989) *Science*, **244**, 1483-1487

Schreiber, A.B., Courgud, P.D., Andre, C.L., Vray, B. and Strosberg, A.D. (1980) *Proceedings of the National Academy of Sciences, USA*, **77**, 7385-7389

Schwartz, J., Penke, B., Rivier, J. and Vale, W. (1987) *Endocrinology*, **121**, 1454-1460

Schwartz, J. and Vale, W. (1988) *Endocrinology*, **122**, 1695-1700
Sege, K. and Peterson, P.A. (1978) *Proceedings of the National Academy of Sciences, USA*, **75**, 2443-2447
Spiegelman, B.M., Lowell, B., Napolitano, A., Dubuc, P., Barton, D., Francke, U., Groves, D.L., Cook, K.S. and Flier, J.S. (1989) *Journal of Biological Chemistry*, **264**, 1811-1815
Spencer, G.S.G., Garssen, G.J. and Hart, I.C. (1983a) *Livestock Production Science*, **10**, 25-37
Spencer, G.S.G., Garssen, G.J. and Hart, I.C. (1983b) *Livestock Production Science*, **10**, 469-477
Szabo, L., Mottershead, D.G., Ballard, F.J. and Wallace, J.C. (1988) *Biochemical and Biophysical Research Communications*, **151**, 207-214
Terry, L.C. and Martin, J.B. (1981) *Endocrinology*, **109**, 622-627
Thomas, H., Green, I.C., Wallis, M. and Aston, R. (1987) *Biochemical Journal*, **243**, 365-372
Trout, W.E. and Schanbacher, B.D. (1990) *Journal of Endocrinology*, **125**, 123-129
Varner, M.A., Davis, S.L. and Reeves, J.L. (1980) *Endocrinology*, **106**, 1027-1032
Wallis, M. (1982) *Nature*, **296**, 112-113
Wallis, M., Daniels, M., Ray, K.P., Cottingham, J.D. and Aston, R. (1987) *Biochemical and Biophysical Research Communications*, **149**, 187-193
Wassermann, N.H., Penn, A.S., Freimuth, P.I., Treptow, N., Wentzel, S., Cleveland, W.L. and Erlanger, B.F. (1982) *Proceedings of the National Academy of Sciences, USA*, **79**, 4810-4814

15
GENETIC MANIPULATION OF ANIMALS WITH SPECIAL REFERENCE TO EXPRESSION OF GROWTH HORMONE AND GROWTH RELEASING HORMONE GENE CONSTRUCTS

B. BRENIG and G. BREM
Department of Molecular Animal Breeding, Veterinary Faculty, Ludwig-Maximilians-University of Munich, D-8000 Munich 22, Germany

Introduction

According to the classical laws of thermodynamics energy can neither be gained nor lost. Different forms are interconvertible. This is one of the fundamental problems in animal production, because a substantial portion of the dietary energy is converted into heat. Normally this is interpreted as 'loss' of energy, which is incorrect strictly speaking. To minimise this wastage efforts have been concentrated on the possibility of improving the efficiency of food conversion, meat quality, and quantity, culminating in the direct genetic manipulation of genes controlling growth and disease resistance. First attempts in genetic manipulation have focused on reproductive engineering, which is a prerequisite for genetic engineering technologies. Several techniques have been used subsequently to manipulate the genome of animals. Currently the only available and most efficient technique for genetic manipulation is the direct microinjection of DNA into the pronuclei of fertilized oocytes. Other techniques applicable to genetic manipulation, *eg* liposome-mediated gene transfer, transfection of pluripotent embryonal stem cells and the production of chimeras, or the use of retroviral vectors, have not yet been applied satisfactorily to most livestock. However, avian gene transfer has been limited to the use of retroviral vectors because of the unique physiology of the chicken (Briskin, Hsu, Boggs, Schultz, and Rishell and Bosselman, 1991).

Genetic engineering of livestock is not only complicated by techniques that facilitate gene transfer *per se*. An additional problem for molecular biologists stems from the number of genes involved in, or controlling, growth directly which have not been isolated or characterized biochemically. Furthermore the control of eukaryotic gene expression is very complex and thus transgene expression may be modified or even abolished after integration into the genome. Experiments in different species with ectopically expressed transgenes, not only those from the growth hormone cascade, have once again thrown light on the complexity of gene expression and endogenous hormonal feedback mechanisms. So far, using DNA microinjection, genes have only been added to the genome of livestock. As transgenes potentially integrate at any site in the genome, they may introduce undesirable mutations into essential genes. However, advances in the establishment of pluripotent embryonal stem cells will make it possible to use homologous recombination for the targeted replacement of defective genes. Many genetic disorders result from mutations of a single gene. A genetic disorder which has a great influence on carcass composition in pigs is the porcine stress syndrome or malignant hyperthermia. This disorder has been studied in great detail for about 30 years, but only recently has a candidate gene, the ryanodine receptor, been characterized.

Manipulation of growth by gene transfer

Gene transfer by microinjection of DNA (Gordon, Scangos, Plotkin, Barbosa and Ruddle, 1980) into fertilized oocytes has provided a powerful tool for the manipulation of complex endocrine regulatory systems, *eg* the growth hormone cascade (Figure 1).

Genes of the growth hormone cascade have been isolated from a variety of species and have been used in gene transfer experiments. Early studies in mice have shown that ectopic expression of growth hormone genes under the control of the murine metallothionein-I promoter results in an increased growth (Palmiter, Brinster, Hammer, Trumbauer, Rosenfeld, Birnberg and Evans, 1982; Palmiter, Norstedt, Gelinas, Hammer and Brinster, 1983). These and other experiments, using different promoters and fusion gene constructs, were readily repeated in rabbits, sheep, pigs,

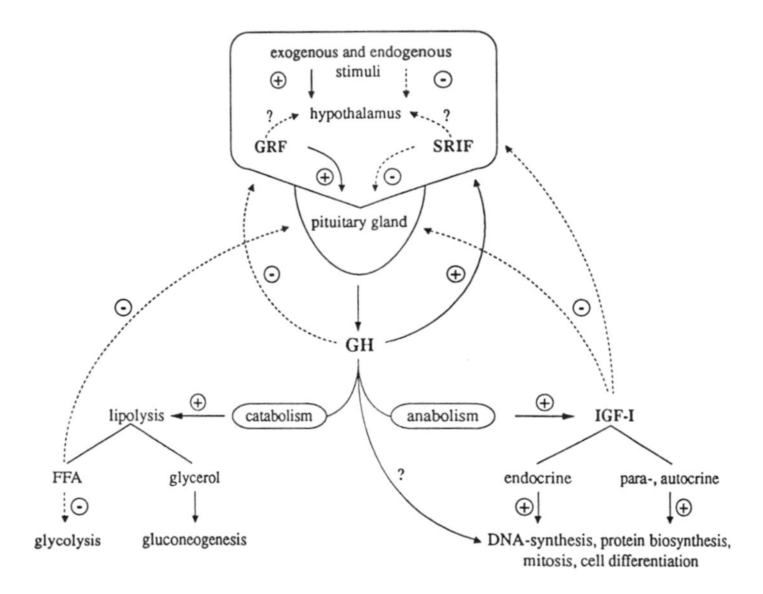

Figure 1 The growth hormone cascade: hormones, growth factors, feedback mechanisms, and pathways.
GRF: growth hormone releasing hormone; SRIF: growth hormone inhibiting hormone; GH: growth hormone; IGF-I: insulin-like growth factor I; FFA: free fatty acids; ±: positive, negative feedback mechanism. Potential interactions or feedback mechanisms that have not yet been identified are indicated with question marks.

and fish (Brem, Brenig, Goodman, Selden, Graf, Kruff, Springmann, Hondele, Meyer, Winnacker and Kräußlich, 1985; Michalska, Vize, Ashmann, Stone, Quinn, Wells and Seamark, 1986; Ebert, Low, Overstrom, Buonomo, Baile, Roberts, Lee, Mandel and Goodman, 1988; Brem, Brenig, Hörstgen-Schwark and Winnacker, 1988). However, in these species elevated levels of circulating heterologous or homologous growth hormones had aberrant effects. Transgenic pigs harbouring a metallothionein-I promoter/bovine growth hormone gene construct grew only 11-15% faster than control litter mates and only when they were placed on a diet with elevated crude protein and additional lysine (Pursel, Pinkert, Miller, Bolt, Campbell, Palmiter, Brinster and Hammer, 1989). As a consequence of reduced feed intake (17-20 %) pigs with GH transgenes did not reach their genetic potential for growth rate. However, high levels of GH in the plasma of the transgenic pigs prevented subcutaneous fat accretion and enhanced the utilization of nutrients for other carcass components. Lipolytic effects of excess GH may be attributed to inhibition of phosphatidylinositol phospholipase C in adipose plasma membranes (Roupas, Chou, Towns and Kostyo, 1991). Although growth promoting effects and alterations in carcass composition have been achieved in livestock, it was soon evident that high levels of ectopically expressed growth hormone induce a number of severe and detrimental side effects. Detailed patho-physiological, histopathological and morphological analyses in GH transgenic pigs and mice revealed many symptoms, *eg* gastric ulcers, synovitis, cardiac myocyte nuclear hypertrophy, dermatitis, nephritis, pneumonia and reduced life span (Doi, Striker, Quaife, Conti, Palmiter, Behringer, Brinster and Striker, 1988; Brem, Wanke, Wolf, Buchmüller, Müller, Brenig and Hermanns, 1989; Pursel *et al*, 1989). So far transgenic animals with other genes of the growth hormone cascade, *ie* growth hormone releasing factor gene or insulin-like growth factor I gene, have not shown such symptoms, but these animals await further and detailed analysis.

In conclusion, studies with transgenic animals have once again shown the complexity of eukaryotic gene expression. However, there are still a number of questions unanswered regarding tissue-specificity of gene expression, regulation of gene expression and influence of chromatin structure (DNA methylation, DNase I hypersensitivity, nucleosome arrangement) on transcription. In the following section some of the factors influencing transgene expression will be examined more closely.

Regulation of transgene expression

The expression of transgenes in vertebrates is not only subject to regulation by a number of *cis*- and *trans*-regulatory elements but also is influenced by the chromosomal position of the insertion *locus* (Al-Shawi, Kinnaird, Burke and Bishop, 1990). Integration occurs randomly either into regions that are transcriptionally active (active chromatin) or silent (inactive chromatin). These terms generally refer to specific states of DNA methylation, nuclease sensitivity and nucleosome distribution (Doerfler, Langner, Knebel, Hoeveler, Müller, Lichtenberg, Weisshaar and Renz, 1988; Gross and Garrard, 1988). By analogy with studies of a variety of endogenous genes, for example *D. melanogaster* major heat shock genes (see Cartwright and Elgin, 1988), it is assumed that non-expressed transgenes are located in inactive chromatin. So far however no reports are available concerning methylation and its effect on the formation of DNase I hypersensitive sites and nucleosome positioning of transgenes. Since copy numbers *per se* are not related to expression levels it is unclear why differences exist in the degree of expression of a transgene among lines of transgenic animals (see Al-Shawi *et al*, 1990). To date little is known about the mechanisms governing transgene integration into host genomes. It appears that after integration, methylation and the specific chromatin structure of the flanking regions are conferred upon the foreign gene *locus*, propagating from the 5′ and 3′ ends of the insertion (see Doerfler *et al*, 1988). After a specific configuration has been established the foreign gene complex is usually transmitted to progeny without any significant changes. DNA methylation is known to play an important role in the regulation of gene expression and the formation of active chromatin. Hypo- or demethylation seems to be necessary, but not sufficient, for the establishment of nuclease hypersensitive sites (Gross and Garrard, 1988). The methylation of a transgene which is microinjected in an unmethylated form is influenced by the chromosomal integration *locus* and/or *cis*-acting sequences present within the gene (Kolsto, Kollias, Giguere, Isobe, Prydz and Grosveld, 1986). Recently it has been shown that heterologous introns can enhance expression of mMT-I growth hormone transgene constructs by a number of different mechanisms (Palmiter, Sandgren, Avarbock, Allen and Brinster, 1991). Generally a transgene that becomes methylated is transcriptionally inactive and this is stably transmitted. The methylation pattern can vary within a particular cell type resulting in a mosaic phenotype when the transgenes have integrated at two separate chromosomal sites (McGowan, Campbell, Peterson and Sapienza, 1989).

Nucleosome positioning influences the function of *cis*-acting DNA elements by inhibiting the transcriptional initiation of RNA polymerase II (Lorch, LaPointe and Kornberg, 1987). DNase I hypersensitivity near the 5′ part of a gene can be used as a marker for the accessibility of specific DNA regions to proteins, for example transcription factors. But DNase I hypersensitivity only reflects the predisposition of chromatin to be transcribed, as there are examples of transcription of sequences which are DNase I resistant (Graessmann, Graessmann, Wagner, Werner and Simon, 1983; Jove, Sperber and Manley, 1984; Lois, Freeman, Villeponteau and Martinson, 1990).

STRUCTURE OF THE INTEGRATION LOCUS, INHERITANCE, AND EXPRESSION OF THE MTGHRH-1 TRANSGENES

To study the regulation of transgenes as a consequence of a specific chromatin structure at the integration *locus*, we have used two lines of transgenic mice produced in our laboratory harbouring multiple copies of a differentially expressed transgene, consisting of the mouse metallothionein-I promoter (1.8 kb *Eco*RI/*Bgl*II fragment), isolated from plasmid HamerMT (Hamer and Walling, 1982) and fused (*Bgl*II/*Eco*RI) to a 150 bp intronless human growth hormone releasing factor gene, designated

GHRH Leu27. The 3′ region (1.1 kb *Hae*III/*Bam*HI fragment) of the fusion gene construct was isolated from pSV2-cat^S (Gorman, Moffat and Howard, 1982) containing parts of the chloramphenicol acetyltransferase gene (CAT) and sequences of SV40 small t-antigen (splice donor-sites, splice acceptor-sites, polyadenylation signal). The gene construct was used as a linear 3.1 kb *Eco*RI/*Bam*HI fragment (Figure 2) for pronuclear microinjection as described previously (Brenig and Brem, 1989). Transgenic founder mice were bred to cstablish cxpressing (621-4) and non-expressing (621-2) lines harbouring approximately 50 (621-2) or 30 (621-4) unrearranged copies of the construct respectively. Copy numbers were determined by a dilution slot blot hybridization procedure (Brenig, Müller and Brem, 1989) and confirmed by Southern blot hybridization of *Hin*dIII digested DNA after pulsed field gel electrophoresis. In both lines integration occurred at a single chromosomal site with all copies of the transgenes oriented in a head-to-tail tandem array. Transmission of the transgenes was followed over 4 generations and no changes in copy numbers, restriction patterns, or expression (see below) were observed.

Figure 3a shows a Northern blot analysis of total RNA isolated from several tissues and probed with a 1.14 kb 3′ *Sal*I/*Bam*HI fragment (1.1 SB) of MTGHRH-1. A single band is detectable in the liver of mice from line 621-4 at approximately 1.3 kb, corresponding to the expected size of the processed MTGHRH-1 transcript. In 4 generations of line 621-2 MTGHRH-1 mRNA was not detectable in any tissucs

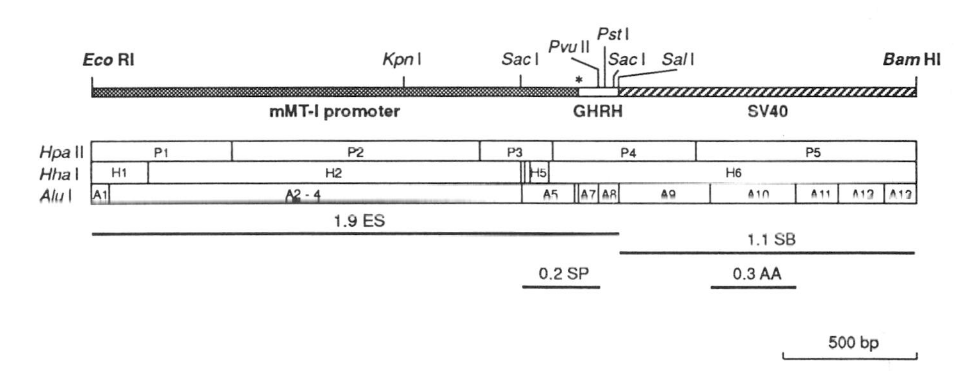

Figure 2 Structure and partial restriction map of MTGIIRII-1.
The 3.1 kb *Eco*RI/*Bam*HI MTGHRH-1 transgene was constructed by fusion of the mouse metallothionein-I promoter to a human GHRH cDNA fragment (the *Bgl*II/*Eco*RI fusion site is indicated with an asterisk). The 3′ part consists of a 1.14 kb *Hae*III/*Bam*HI fragment from pSV2-cat^S. The *Hae*III site has been converted to a *Sal*I site. Shaded box: mMT-I promoter, white box: GHRH Leu27, hatched box: SV40. The positions of restriction enzymes with one or two cleavage sites examined in this study are shown. Approximate fragment sizes resulting from digestion with *Hpa*II (or *Msp*I): P1 (540 bp), P2 (900 bp), P3 (260 bp), P4 (531 bp), P5 (855 bp), *Hha*I: H1 (210bp), H2 (1.37 kb), H3 (35 bp), H4 (24 bp), H5 (50 bp), H6 (1.39 kb), and *Alu*I: A1 (20 bp), A2-4 (1560 bp), A5 (199 bp), A6 (6 bp), A7 (68 bp), A8 (76 bp), A9 (357 bp), A10 (329 bp), A11 (155 bp), A12 (180 bp), A13 (120 bp) are indicated. The small fragments H3, H4, and A6 are not marked in the diagram. *Alu*I sites A2-4 in the promoter have not been mapped precisely, but were deduced from restriction analysis. DNA fragments used as probes are indicated beneath.

usually permitting transcription from the metallothionein-I promoter. In line 621-4 the transgenes were exclusively and abundantly expressed in liver tissue. Expression of MTGHRH-1 was not inducible in any other tissues after treatment with Zn^{2+}.

For S1 nuclease mapping two MTGHRH-1 fragments were used, *ie* a 1.95 kb *Eco*RI/*Sal*I promoter fragment (1.9 ES) and a 329 bp *Alu*I 3′ fragment (0.3 AA) containing the splice sites of the SV40 small t-antigen (Figure 3b, c). S1 nuclease digestion yields a 221 nt promoter fragment and two mRNA forms at the 3′ end of 256 nt (spliced) and 329 nt (unspliced) respectively. This demonstrates that the mRNA is correctly initiated at the 5′ end and processed at the splice of the small t-antigen.

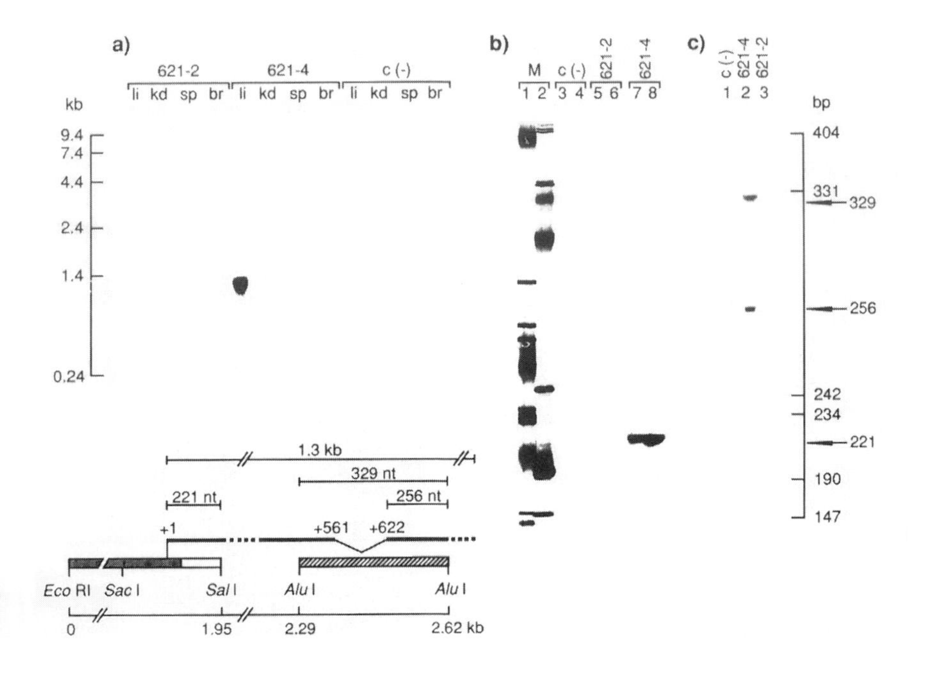

Figure 3 Transcription analysis of transgenic lines 621-2 and 621-4.
(a) Total RNA was isolated from liver, spleen, kidney, and brain, electrophoresed on a 1.2 % formaldehyde/agarose gel, and blotted onto a nylon membrane. The 1.1 SB fragment was used as probe. Sizes are indicated in kilobases (kb). (b) For S1 nuclease mapping total RNA was hybridized to the ^{32}P-5′ end-labeled 1.9 ES or (c) 0.3 AA fragment and electrophoresed on 7 M urea/8 % polyacrylamide gels after digestion with S1 nuclease. Sizes are indicated in basepairs (bp).

The diagram above outlines the experimental strategy of mapping the 5′ and 3′ ends of the construct and the expected effects of S1 digestions on the lengths of the terminally labeled fragments. Shaded box: mMT-I promoter, white box: GHRH Leu27, hatched box: SV40 A10 fragment. The broken thick line indicates the putative GHRH mRNA with its cap site at +1 and the splice sites at positions +561 and +622. Arrows mark the positions of resulting fragments after S1 digestion. Li: liver, kd: kidney, sp: spleen, br: brain; c (-): negative control (total liver RNA isolated from a non-transgenic mouse) M: marker (lane 1: ΦX174 *Hae*III; lane 2: pUC12 *Hpa*II); nt: nucleotides; (b) lanes 3-8: total RNA from two individuals respectively.

METHYLATION PATTERNS OF TRANSGENES

To analyse whether the lack of transcription in line 621-2 is due to complete methylation of the transgenes, DNAs from different animals and tissues (liver, kidney, spleen, brain) were digested with *Hpa*II, *Hha*I and *Msp*I. As depicted in Figure 4 (bottom) the mMT-I promoter contains three *Hpa*II or *Msp*I sites (for fragment lengths see Figure 2), and five *Hha*I sites. One *Hpa*II or *Msp*I site is located in the proximal 3′ region of the construct. The distal restriction sites in the mMT-I promoter are 20 bp upstream of the TATA-box and approximately 50 bp upstream of the transcription start site. Figure 4 shows the hybridization patterns of MTGHRH-1 in DNA isolated from (a) liver or (b) brain from animals of the two lines after digestion with *Pst*I and *Pst*I/*Hpa*II, *Pst*I/*Hha*I, or *Pst*I/*Msp*I, using the 1.1 SB fragment as a probe.

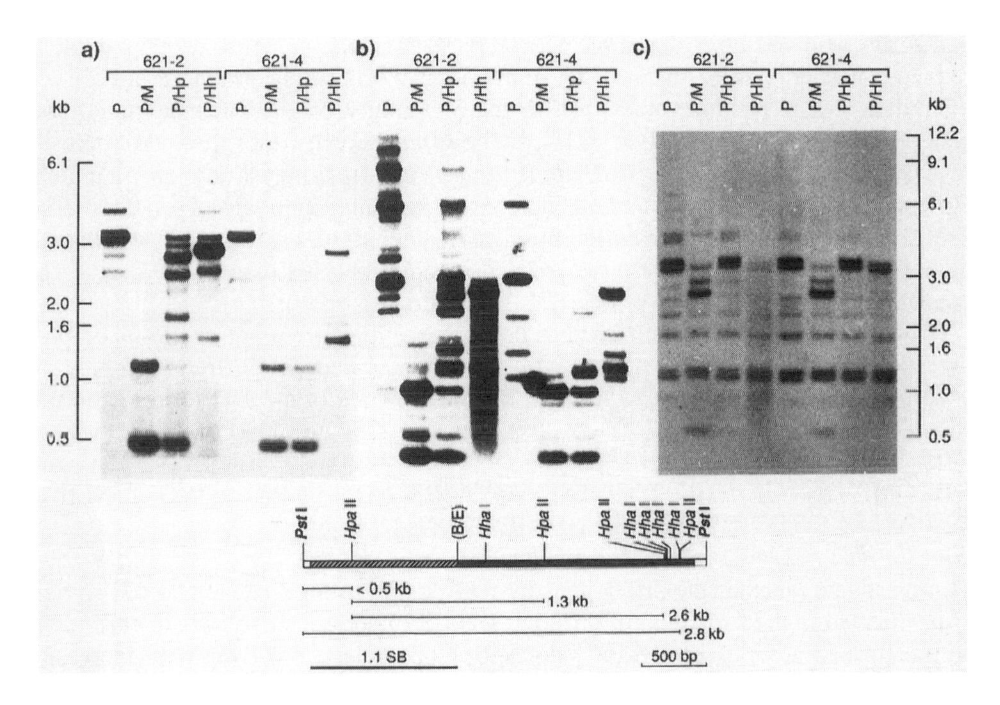

Figure 4 Analysis of methylation at *Hpa*II and *Hha*I sites of expressing and non-expressing transgenic mice.

DNA from (a) liver or (b) brain was isolated from animals of lines 621-2 and 621-4 and restricted with *Pst*I, *Pst*I/*Hpa*II, *Pst*I/*Msp*I and *Pst*I/*Hha*I. After electrophoresis on a 1.0 % agarose gel, DNA was transferred to a nylon membrane and hybridized to the 1.1 SB fragment. Sizes are indicated in kilobases (kb). (c) The nylon membrane (brain DNA) was reprobed with a murine 1.97 kb *Bam*HI/*Hin*dIII MxcDNA fragment (Staeheli, Haller, Boll, Lindenmann and Weissmann, 1986). (b) and (c) are aligned to the size standard on the right-hand side. The diagram above shows the fragment junction region of two adjacent copies after digestion with *Pst*I, *Hpa*II, *Hha*I sites and observed fragments are indicated. Shaded box: mMT-I promoter, white boxes: GHRH Leu27, hatched box: SV40. P: *Pst*I, M: *Msp*I, Hp: *Hpa*II, Hh: *Hha*I, (B/E): fragment junction site.

Although most of the transgenes in line 621-2 are methylated at *Hpa*II or *Hha*I sites in liver tissue several copies are hypo- or unmethylated at the restriction sites analysed. As estimated from the intensity of the signals approximately one-fifth of all copies are unmethylated at the *Hpa*II site in the proximal 3′ region (< 0.5 kb *Pst*I/*Hpa*II fragment) and one-tenth at the *Hpa*II site in the upstream promoter region (1.3 kb *Pst*I/*Hpa*II fragment). Interestingly a large fraction of the copies are not methylated at the *Hpa*II site upstream of the transcription start site (2.6 kb *Pst*I/*Hpa*II fragment). These data are consistent with the results obtained by digestion with *Pst*I/*Hha*I.

The prominent band at 2.8 kb which comprises at least 90 % of all copies results from the unmethylated *Hha*I sites upstream the transcription start site. In brain (Figure 4b) and other tissues of line 621-2 differences in the restriction patterns of methylation-sensitive and -insensitive restriction enzymes are detected. These differences can also be seen comparing different tissues of both lines. Additional bands are resolved after digestion with *Pst*I, *Sac*I, *Sal*I, or *Msp*I which cannot be reduced further. We assume that these partial digestion products are due to methylation at C residues other than in the 5′ vicinity of G. No methylation was observed in liver tissue of line 621-4 at the *Hpa*II sites in the promoter since the restriction patterns of the *Pst*I/*Hpa*II and *Pst*I/*Msp*I digestions are identical. However, approximately 50% of all copies are methylated at the distal promoter *Hha*I site (2.8 kb *Pst*I/*Hha*I fragment). Brain and other tissues of line 621-4 exhibit a higher degree of methylation at the restriction sites analysed than in the liver. This is analogous to the situation in line 621-2 (Figure 4b).

In order to prove that the differences between the lines were not due to incomplete digestions, all filters were reprobed with an unrelated endogenous mouse probe. An example of such an analysis where all DNAs show the same restriction pattern is shown in Figure 4c.

CHROMATIN OF THE EXPRESSING AND NON-EXPRESSING LINES IS HYPERSENSITIVE TO DNASE I DIGESTION

To investigate whether the differences in methylation in lines 621-2 and 621-4 were also reflected in the formation of different DNase I hypersensitive sites in chromatin, we performed a series of DNase I digestions of nuclei from liver with increasing amounts of enzyme. DNA was isolated and subsequently restricted with *Eco*RI or *Sal*I. In the following indirect end-labeling hybridization experiments the 1.1 SB fragment was used as a probe. In line 621-2 one intense band at 3.1 kb can be seen after restriction with *Eco*RI (Figure 5a). Several very faint bands are detectable at 12.3 kb, 6.1 kb, 3.6 kb, 3.2 kb, and 4 bands below 3.1 kb ranging from approximately 2.8 kb to 1.2 kb. These sites seem to be less sensitive to nuclease digestion. Even at higher DNase I concentrations (> 10 μg/ml) this pattern remains unaltered (data not shown). After restriction with *Sal*I (Figure 5b) a strong signal at 2.8 kb and a number of faint bands at 6 kb, 3.2 kb, 2.6 kb, 2.3 kb, and 1.2 kb can be seen. The upper bands arise from undigested methylated *Sal*I sites (see above).

The signals obtained after restriction with *Sal*I are nearly identical with those after digestion with *Eco*RI. Again no changes in the restriction patterns at increasing amounts of DNase I were observed. At least 10 bands of different intensities can be seen in liver DNA from line 621-4 after digestion with *Eco*RI at DNase I concentrations ranging from 1 to 7μg/ml (Figure 5c). One band is at approximately 2.6 kb and the others are grouped in triplets at 3.1kb (3.2 kb, 3.6 kb), 6.1 kb, and 9.3 kb (note that the 5′ *Eco*RI sites of the construct are lost during integration). At higher concentrations (> 7 μg/ml) additional faint bands appear which are smaller than 2.6 kb. After digestion with *Sal*I, three bands at 2.8 kb, 2.6 kb, and 1.2 kb are detectable (Figure 5d). The band at 6.1 kb results from undigested methylated *Sal*I sites in the transgenes (see above).

As depicted at the bottom of Figure 5 the triplets observed in line 621-4 correspond to fragments caused by nuclease hypersensitive sites within the metallothionein-I promoter that have already been described (Senear and Palmiter, 1983). In at least one-third of the transgene copies, the site centered at -30 relative to the transcriptional start site seems to be the most nuclease sensitive as estimated from the intensity of the band at 3.1 kb. Interestingly the hypersensitive region near the *Sac*I site in the promoter (-148), which is normally detected only in cell lines with an amplified MT-I gene (Senear and Palmiter, 1983), is present. The signals below 3.1 kb result from sensitive sites near the 5′ end of the constructs. In this respect the DNaseI/*Eco*RI restriction pattern in liver DNA of line 621-2 differs (Figure 5a) significantly from that of line 621-4. Although the site at -30 in the metallothionein-I promoter of the constructs is also predominantly hypersensitive in over 90% of all copies in line 621-2, chromatin is less sensitive at the other sites proximal the transcription start site. Compared to line 621-4, DNase I hypersensitivity seems to be exclusively concentrated at the -30 position, whereas other sites are more or less resistant. Only a minor fraction of the transgene copies are sensitive at the 5′ ends.

RANDOM NUCLEOSOME DISTRIBUTION IN THE NON-EXPRESSING LINE AND PHASED NUCLEOSOMES IN THE EXPRESSING LINE

In order to extend the studies on methylation and DNase I hypersensitivity to the nucleosomal organization at the insertion *loci* the nucleosomal distribution of the expressing and non-expressing lines was studied. Isolated nuclei from different tissues were digested with micrococcus nuclease, monosomes or disomes were prepared, and aliquots were digested with *Alu*I. Three MTGHRH-1 fragments (1.9 ES, 0.2 SP, and 1.1 SB, see Figure 2) were used to analyse the nucleosomal distribution along the entire constructs by comparing the hybridization patterns on sequencing gels after electrotransfer onto nylon membranes. Banding patterns of undigested and *Alu*I-digested monosomes from line 621-2 and 621-4 were compared with the pattern of deproteinized and *Alu*I digested DNA from control transgenic mice.

After hybridization with the 1.1 SB fragment 4 *Alu*I fragments at 357 bp (A9), 329 bp (A10), 180 bp (A12), and 155 bp (A11) are resolved in isolated control DNA (Figure 6a). Upon hybridization to the 0.2 SP fragment, which spans the transcription start site (see Figure 2), only one signal is detected at 199 bp, corresponding to the *Alu*I fragment A5 (Figure 6b). After hybridization with the 1.9

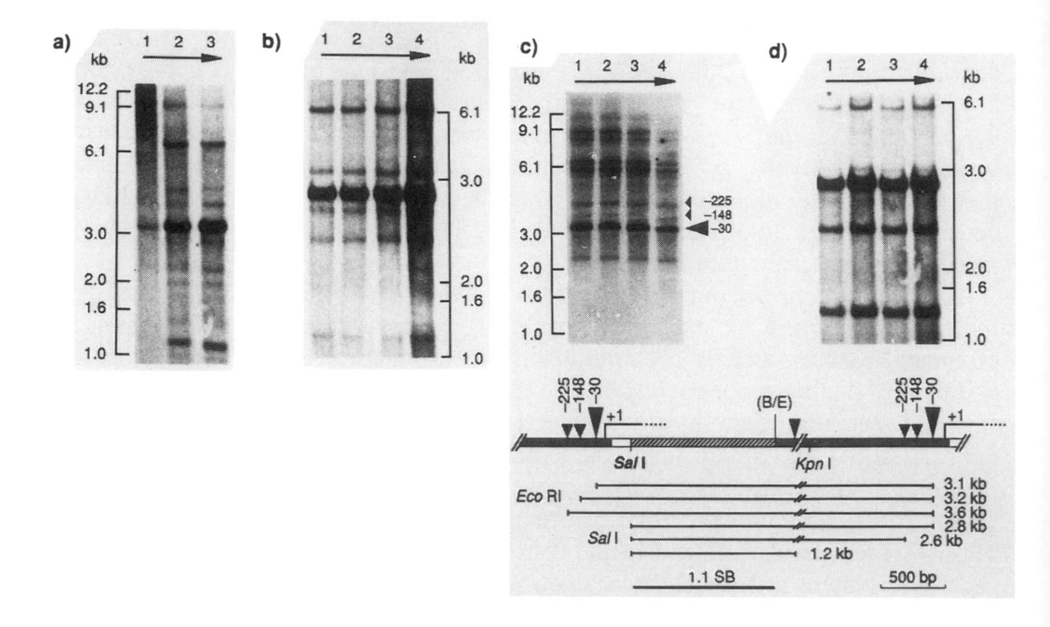

Figure 5 DNase I hypersensitivity of transgenes in lines 621-2 and 621-4.
Nuclei from liver tissues were prepared and digested with increasing amounts of DNase I (direction of increasing DNase I concentration indicated by arrows above). DNA was extracted, restricted with *Eco*RI (a: 621-2, c: 621-4) or *Sal*I (b: 621-2, d: 621-4), and electrophoresed on 0.8 % agarose gels. Sizes are indicated in kilobases (kb). DNase I concentrations, lanes 1: aliquots taken directly after addition of 1 μg/ml, lanes 2: 2 μg/ml, lanes 3: 3 μg/ml, lanes 4: 7 μg/ml. The diagram above shows the fragment junction region of two adjacent copies after digestion with *Eco*RI or *Sal*I (bold). Fragments hybridizing to the 1.1 SB probe after DNase I digestion are shown as discontinuous lines. DNase I hypersensitive sites are indicated by closed triangles. Shaded boxes: mMT-I promoter, white boxes: GHRH Leu27, hatched box: SV40, (B/E): fragment junction site.

ES probe three bands are resolved in isolated transgenic control DNA which originate from the promoter. These fragments (A2-4) have not been mapped precisely. The small *Alu*I fragments (A6-8) are not visible in this figure. The cross-hybridizations in Figure 5c are due to a highly reiterated region near the *Eco*RI site at the 5′ end of the metallothionein-I promoter (Senear and Palmiter, 1983).

Hybridization with the three MTGHRH-1 fragments to undigested monosomes from line 621-2 shows only a smear around the size of nucleosomal core DNA (130 bp-180 bp). Even after digestion with *Alu*I only faint bands were resolved (Figure 6a-c). A phasing of nucleosomes is observed with fragment 0.2 SP (Figure 6b). Very faint bands can be seen in Figure 6c in disomal DNA at the sizes of the signals in isolated control DNA.

Surprisingly in undigested monosomes of line 621-4 a band at approximately 800 bp hybridizes to the 1.1 SB fragment additional to the core DNA smear (Figure 5a). After digestion with *Alu*I two distinct bands arise differing from those in isolated chromatin, which presumably originate from the 800 bp band. The 329 bp fragment (A10) is the only fragment that is seen in both lines. The large band at 800 bp disappears when hybridizing to the 0.2 SP promoter fragment (Figure 6b). It should be noted however that the hybridization pattern still differs from that of controls (A5 fragment). At the 5′ end (1.9 ES probe) the situation is somewhat different.

The three signals at approximately 230 bp (Figure 6b) remain, but two strong bands at 580 bp and 450 bp, and 6 smaller bands between 50 bp and 110 bp appear in *Alu*I digested mono- and disomes of line 621-4 (Figure 6c). Our data demonstrate that the level of expression of transgenes is subject to a very complex organization of the integration *locus* reflected by a distinct pattern of DNA methylation, DNase I hypersensitivity and nucleosomal arrangement. We have chosen the murine metallothionein-I promoter because it contains constitutive DNase I hypersensitive sites and directs expression of coupled genes to a variety of tissues after induction with heavy metals or glucocorticoid hormones (Senear and Palmiter, 1983). The transgenic *loci* analysed here do not exhibit a uniform structure regarding methylation and nucleosomal distribution. Furthermore the organization of the *loci* differs between the tissues analysed. Therefore a position effect regulating expression via enhancers alone might be excluded. Although most of the transgenes in line 621-2 were methylated in the promoter region there were still a number of the transgenes in an "active", *ie* hypo- or unmethylated, conformation. The region at the transcription start site was shown to be particularly hypomethylated in at least 20% of the transgene copies. In both lines the degree of methylation was higher in tissues that are normally not active for metallothionein expression. Surprisingly, restriction with *Pst*I, *Sal*I, *Sac*I, and *Msp*I resulted in a number of partially digested fragments. These partial digestions may be attributed to methylation at C residues (5-mCpT, 5-mCpC) not present in the CpG dinucleotide which is normally the site of methylation in vertebrate genomes. It is known from other studies that G may not be an exclusive 3′ neighbour of 5-mC (Drahovsky and Pfeifer, 1988). In this respect it is also interesting to note that the maintenance and stability of 5-mC in dinucleotides other than 5′-CpG-3′ is still unclear (Drahovsky and Pfeifer, 1988). Methylation in line 621-2 seems to have only minor effects on the formation of DNase I hypersensitive sites. The 5′ ends of the constructs in line 621-2 are nearly as sensitive to DNase I digestions as in line 621-4. The promoter region upstream of the transcription start site remains hypersensitive in the majority of all copies. The hypersensitive sites detected in both lines reflect the situation present in cultured cell lines with amplified MT-I genes. Especially the site around -148 is normally only detected in mouse liver tumor cell line Cd^R (40-fold amplification), and a derivative of mouse thymoma cell line W7, Cd^R_{75} W7 (Senear and Palmiter, 1983). Interestingly in cell line W7 the MT-I gene is highly methylated and resistant to DNase I digestion. We detected an additional DNase I hypersensitive site at the 5′ end of the metallothionein-I promoter that has not been described before, probably due to the location in the highly reiterated region adjacent to the *Eco*RI site.

Major differences between both lines were observed in the chromatin structure at the integration *loci*. In line 621-2 most of the nucleosomes are arranged randomly, as the same cleavage sites are observed in chromatin and in deproteinized DNA. Nucleosome phasing was detectable only around the transcription start site and throughout the structural gene. Therefore in line 621-2 all sequence informations have a chance to be exposed in linkers or to be hidden in a nucleosome. While nucleosomes were distributed randomly throughout the integration *locus* in line 621-2, line 621-4 exhibits a very complex chromatin structure. As deduced from the indirect end-labeling experiments nucleosomes are phased at the transcription start site and the GHRH cDNA. Towards the 3′ ends and the adjacent 5′ promoter regions the constructs seem to form a higher order structure which is resistant to micrococcus nuclease digestion. Even at micrococcus nuclease concentrations >100 U/ml and prolonged incubations (30-40 min) the 800 bp band was not reduced when hybridizing to the 1.1 SB fragment. This fragment was repeatedly recovered in the monosomal fraction using preparative agarose gel electrophoresis as well as isokinetic sucrose gradients. The nucleosome organization of the transgenes is reiterated in every copy of the integration *locus* and remains the same in all tissues irrespective of the different degrees of methylation.

Manipulating genes by homologous recombination

Lack of transgene expression can be overcome by different approaches. It has been shown that the use of matrix (MAR) or scaffold attachment sites (SAR) within transgene constructs results in a position independent always on type of expression (Stief, Winter, Strätling and Sippel, 1989). These regions, approximately 0.6 to 1 kb in length, usually contain short (15 bp) sequence motifs homologous to topoisomerase II binding and cleavage sites. Binding of topoisomerase II renders closely linked genes prone to expression by interconverting different topological states of DNA through double-strand breaks and rejoining (Adachi, Käs and Laemmli, 1989). A similar situation is observed in the ß-globin gene *locus*, where a dominant control region (DCR) 6-18 kb upstream the ζ-globin gene containing four DNase I hypersensitive sites positively controls gene expression (Grosveld, Blom van Assendelft, Greaves and Kollias, 1987). When cloned upstream of heterologous genes, this region also stimulates expression at very high levels independent of orientation and chromosomal integration site. In the following we will focus on the use of homologous recombination to manipulate a specific gene within the complex eukaryotic genome (see Melton, 1990).

The advantage of homologous recombination over conventional gene transfer (microinjection) is that a gene can be targeted to a defined chromosomal region, hence the process of integration is no longer random. The incoming gene can either be used for disruption (insertion vector) or replacement (replacement vector) of its target. For the production of transgenic animals this technique is only applicable when pluripotent embryonal stem cells (ES cells) which can be cultured and treated *in vitro* have been established. In this respect it is also important to realize that a targeting event is usually observed less than 1 per 10^6 integrations. However, specific

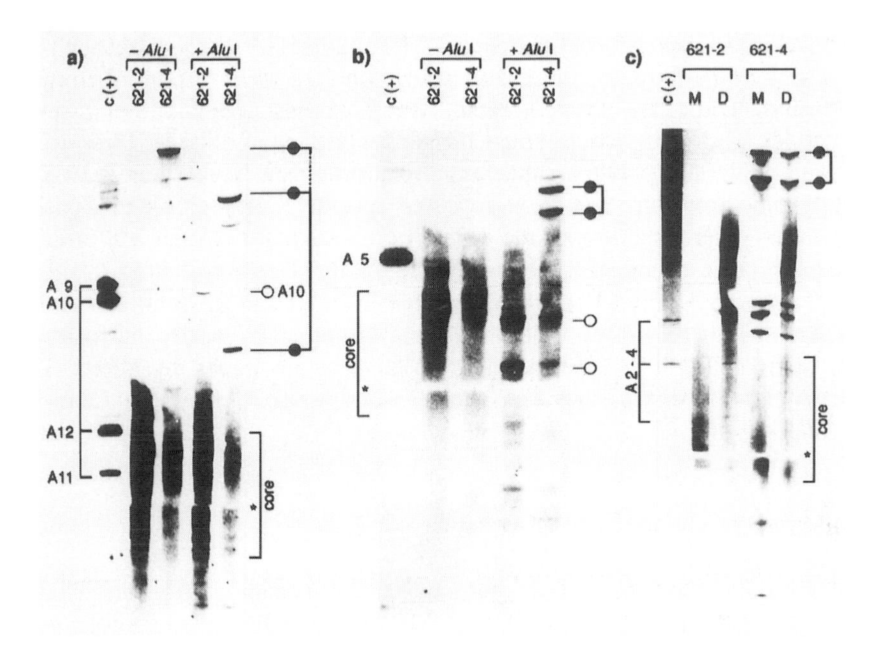

Figure 6 Nucleosome distribution at the transgene *loci*.
Core DNA (monosomes or disomes) was either restricted with *Alu*I (+ *Alu*I) or directly (-*Alu*I) separated on 7 M urea/5% polyacrylamide sequencing gels. After electrophoresis gels were cut into two (cuts are indicated by asterisks) and electroblotted onto nylon membranes. Membranes were hybridized to the (a) 1.1 SB fragment, (b) 0.2 SP fragment, and (c) 1.9 ES fragment. Lines with closed circles mark positions of supranucleosomal fragments only present in line 621-4. Lines with open circles indicate fragments present in both lines originating from phased nucleosomes or nucleosome-free regions (A10). Positions of core DNA are indicated. A2-12: positions of *Alu*I fragments; c (+): positive control (extracted genomic DNA from a transgenic mouse); M: monosomal DNA, D: disomal DNA.

in vitro selection procedures, *eg* the polymerase chain reaction (PCR) or selection for neomycin resistance, have facilitated rapid screening for targeting events. Once a homologous recombination has been achieved the targeted cells are injected into host embryos and then reimplanted into foster mothers, giving birth to chimeras (Brenig and Brem, 1991a;b). So far ES cells have only been established from mice and therefore homologous recombination is still limited to this species. But future progress in isolation and *in vitro* culture of pluripotent cells derived from porcine and ovine blastocysts will make homologous recombination also applicable to genetic manipulation of livestock. Candidates for homologous recombination are genetic disorders which result from mutations of a single gene. The Lesch-Nyhan syndrome in man is a severe neurological disorder which results from a hypoxanthine-guanine phosphoribosyltransferase (HPRT) deficiency. HPRT-deficient mice have been used as a model system to study this disease in man, although mice respond differently. HPRT-deficiency in mice has been corrected by homologous recombination in several ways (see Koller, Hageman, Doetschman, Hagaman, Huang, Williams, First, Maeda and Smithies, 1989). To create a subtle modification of the HPRT gene a two step "in-out" targeting procedure has been reported recently (Valancius and Smithies, 1991). However, other gene *loci*, *eg* the $ß_2$-microglobulin *locus* (Koller and Smithies, 1989), the *c-src* proto-oncogene (Soriano, Montgomery, Geske and Bradley, 1991), and the *En*-2 homeobox (Joyner, Herrup, Auerbach, Davis and Rossant, 1991), have also been targeted successfully. Although gene targeting is usually performed in embryonic stem cells, it is also possible to manipulate genes by direct DNA microinjection into fertilized oocytes (Brinster, Braun, Lo, Avarbock, Oram and Palmiter, 1989).

Acknowledgements

The authors are indebted to Ernst-L. Winnacker for providing excellent working conditions in his laboratory. Waldemar Wetekam is thanked for the generous gift of human GHRH clone LZPGRF 1/3 (Leu27). We are also grateful to Charles Weissmann for providing murine Mx cDNA. Appreciation is extended to Haralabos Zorbas, Horst Ibelgaufts, Brian Salmons, Walter H. Günzberg, and Mathias Müller for critical readings and helpful comments on the manuscript.

References

Adachi, Y., Käs, E. and Laemmli, U.K. (1989) *The EMBO Journal*, **13**, 3997-4006

Al-Shawi, R., Kinnaird, J., Burke, J. and Bishop, J.O. (1990) *Journal of Molecular and Cellular Biology*, **10**, 1192-1198

Brem, G., Brenig, B., Goodman, M., Selden, R.C., Graf, F., Kruff, B., Springmann, K., Hondele, J., Meyer, J., Winnacker, E.-L. and Kräußlich, H. (1985) *Zuchthygiene*, **20**, 251-252

Brem, G., Brenig, B., Hörstgen-Schwark, G. and Winnacker, E.-L. (1988) *Aquaculture*, **68**, 209-219

Brem, G., Wanke, R., Wolf, E., Buchmüller, Th., Müller, M., Brenig, B. and Hermanns, W. (1989) *Molecular Biology and Medicine*, **6**, 531-547

Brenig, B. and Brem, G. (1989) *Acta Endocrinologica suppl.*, **120**, 137-138

Brenig, B. and Brem, G. (1991a) *Reproduction of Domestic Animals*, **16**, 14-21

Brenig, B. and Brem, G. (1991b) In *Animal Biotechnology and the Quality of Meat Production*, (eds L.O. Fiems, B.G. Cottyn and D.I. Demeyer), Amsterdam, Elsevier Science Publishers B.V., pp 1-16

Brenig, B., Müller, M. and Brem, G. (1989) *Nucleic Acids Research*, **17**, 6422

Brinster, R.L., Braun, R.E., Lo, D., Avarbock, M.R., Oram, F. and Palmiter, R.D. (1989) *Proceedings of the National Academy of Sciences USA*, **86**, 7087-7091

Briskin, M.J., Hsu, R.-Y., Boggs, T., Schultz, J.A. and Rishell, W. and Bosselman, R.A. (1991) *Proceedings of the National Academy of Sciences USA*, **88**, 1736-1740

Cartwright, I.L. and Elgin, S.C.R. (1988) In *Architecture of eukaryotic genes*, (ed G. Kahl), Weinheim, VCH

Doerfler, W., Langner, K.-D., Knebel, D., Hoeveler, A., Müller, U., Lichtenberg, U., Weisshaar, B. and Renz, D (1988) In *Architecture of eukaryotic genes*, (ed G. Kahl), Weinheim, VCH

Doi, T., Striker, L.J., Quaife, C., Conti, F.G., Palmiter, R.D., Behringer, R., Brinster, R.L. and Striker, G.E. (1988) *American Journal of Pathology*, **131**, 398-403

Drahovsky, D. and Pfeifer, G.P. (1988) In *Architecture of eukaryotic genes*, (ed G. Kahl), Weinheim, VCH

Ebert, K.M., Low, M.J., Overstrom, E.W., Buonomo, F.C., Baile, C.A., Roberts, T.M., Lee, A., Mandel, G. and Goodman, R.H. (1988) *Molecular Endocrinology*, **2**, 277-283

Gordon, J.W., Scangos, G.A., Plotkin, D.J., Barbosa, A. and Ruddle, F.H. (1980) *Proceedings of the National Academy of Sciences USA*, **77**, 7380-7384

Gorman, C.M., Moffat, L.F. and Howard, B.H. (1982) *Journal of Molecular and Cellular Biology*, **2**, 1044-1051

Graessmann, M., Graessmann, A., Wagner, H., Werner, E. and Simon, D. (1983) *Proceedings of the National Academy of Sciences USA*, **80**, 6470-6474

Gross, D.S. and Garrard, W.T. (1988) *Annual Review of Biochemistry*, **57**, 159-197

Grosveld, F., Blom van Assendelft, G., Greaves, D. and Kollias, G. (1987) *Cell*, **51**, 975-985

Hamer, D.H. and Walling, M.J. (1982) *Journal of Molecular and Applied Genetics*, **1**, 273-288

Jove, R., Sperber, D.E. and Manley, J.L. (1984) *Nucleic Acids Research*, **12**, 4715-4730

Joyner, A.L., Herrup, K., Auerbach, B.A., Davis, C.A. and Rossant, J. (1991) *Science*, **251**, 1239-1243

Koller, B.H. and Smithies, O. (1989) *Proceedings of the National Academy of Sciences USA*, **86**, 8932-8935

Koller, B.H., Hageman, L.J., Doetschman, T., Hagaman, J.R., Huang, S., Williams, P.J., First, N.L., Madea, N. and Smithies, O. (1989) *Proceedings of the National Academy of Sciences USA*, **86**, 8927-8931

Kolsto, A.-B., Kollias, G., Giguere, V., Isobe, K.-I., Prydz, H. and Grosveld, F. (1986) *Nucleic Acids Research*, **14**, 9667-9678
Lois, R., Freeman, L., Villeponteau, B. and Martinson, H.G. (1990) *Journal of Molecular and Cellular Biology*, **10**, 16-27
Lorch, Y., LaPointe, J.W. and Kornberg, R.D. (1987) *Cell*, **49**, 203-210
McGowan, R., Campbell, R., Peterson, A. and Sapienza, C. (1989) *Genes and Development*, **3**, 1669-1676
Melton, D.W. (1990) *Biochemical Society Transactions*, **18**, 1035-1039
Michalska, A., Vize, P., Ashmann, R.J., Stone, B.A., Quinn, P., Wells, J.R.E. and Seamark, R.F. (1986) *Proceedings of the Australian Society of Reproduction Biology*, **18**, 13
Palmiter, R.D., Brinster, R.L., Hammer, R.E., Trumbauer, M.E., Rosenfeld, M.G., Birnberg, N.D. and Evans, R.M. (1982) *Nature*, **300**, 611-615
Palmiter, R.D., Norstedt, G., Gelinas, R.E., Hammer, R.E. and Brinster, R.L. (1983) *Science*, **222**, 809-814
Palmiter, R.D., Sandgren, E.P., Avarbock, M.R., Allen, D.D. and Brinster, R.L. (1991) *Proceedings of the National Academy of Sciences USA*, **88**, 478-482
Pursel, V.G., Pinkert, C.A., Miller, K.F., Bolt, D.J., Campbell, R.G., Palmiter, R.D., Brinster, R.L. and Hammer, R.E. (1989) *Science*, **244**, 1281-1287
Roupas, P., Chou, S.Y., Towns, R.J. and Kostyo, J.L. (1991) *Proceedings of the National Academy of Sciences USA*, **88**, 1691-1695
Senear, A.W. and Palmiter, R.D. (1983) *Cold Spring Harbor Symposia on Quantitative Biology*, **47**, 539-547
Soriano, P., Montgomery, C., Geske, R. and Bradley, A. (1991) *Cell*, **64**, 693-702
Staeheli, P., Haller, O., Boll, W., Lindenmann, J. and Weissmann, Ch. (1986) *Cell*, **44**, 147-158
Stief, A., Winter, D.M., Strätling, W.H. and Sippel, A.E. (1989) *Nature*, **341**, 343-345
Valancius, V. and Smithies, O. (1991) *Molecular and Cellular Biology*, **11**, 1402-1408

16
THE INFLUENCE OF THE MANIPULATION OF CARCASS COMPO ON MEAT QUALITY

J.D. WOOD and P.D. WARRISS
Department of Meat Animal Science, University of Bristol, Langford, Bristol, BS18 7DY

Introduction

It is now well accepted that consumers wish to buy meat which is leaner, *ie* less fat, than formerly. This is because they dislike the texture and taste of the obvious adipose deposits in meat, and also they believe that animal fats are harmful to health.

This change in consumer attitudes has had impacts all along the meat chain. Retailers demand leaner cuts from processors who in turn are introducing premiums and penalties into their contracts with farmers so as to discourage the production of overfat animals. This has led to a search for genetically leaner stock, diets which reduce fat deposition, techniques for objectively assessing fatness in live animals and carcasses and also new ways of modifying growth physiology so that a leaner animal results.

However, as with other agricultural products, improvements in yield (in this case of lean meat) sometimes lead to reductions in 'quality'. In the case of meat, reducing fat content especially of the muscle tissue itself can adversely affect eating quality. There is also evidence that certain treatments which reduce fatness have an even greater impact on quality than would be predicted from the reduction in fatness alone. For these reasons the consequences for meat quality of reducing fatness are of increasing interest in many countries.

Recent changes in carcass fat content

In the UK, other European countries, Australasia and Canada the most marked recent reductions in fatness have been in pigs rather than cattle or sheep. There are several reasons for this including: the financial benefits to be gained from lower feed costs are greater (lean deposition requires less feed energy than fat and concentrated pig diets are expensive); consumers probably perceive fatness as a bigger problem in pigmeat than in beef and lamb; and a simple measurement of fat thickness in the carcass is a reliable guide to carcass lean content on which a grading system can be based.

Table 1 ESTIMATES OF THE COMPOSITION OF THE AVERAGE BRITISH PIG CARCASS (WEIGHT 63 KG)

	1975	1989
P_2 fat thickness (mm)	18	12
Separable fat in carcass (%)	31	20
Lipid in defatted lean (%)	5.5	3.7
Lipid in subcutaneous fat (%)	85.4	77.2
Marbling fat (%)[a]	1.1	0.8
Fatty acids (%) - saturated	42.6	40.7
- polyunsaturated	9.7	13.2

[a] Ether-extractable lipid in *m. longissimus*. From Kempster, Dilworth, Evans and Fisher (1986); Lowe *et al* (1990) and Wood *et al* (1989).

The data in Table 1 are taken from recent UK studies. Grading schemes are based on the thickness of backfat measured at the P_2 position 65 mm from the dorsal mid-line at the level of the last rib. Farmers have responded to price incentives for low P_2 values by using genetically leaner stock, feeding more balanced diets and rearing a higher proportion of entire males. As a result all the indices of fat content have declined during the last fifteen years including the lipid content of muscle and fat tissues. Lean tissue as dissected from the carcass (defatted lean) includes some adhering subcutaneous and intermuscular fat and has a higher lipid content than cores of *m.longissimus.* This intramuscular lipid, located between muscle fibre bundles in the perimysial connective tissue and extracted using diethyl ether, is termed 'marbling fat' because of its characteristic appearance in steaks when present at high concentrations.

The changes in backfat lipid fatty acid composition shown in Table 1 are to be expected from the reduction in fat thickness. The concentration of polyunsaturated fatty acids, of which linoleic acid (C18:2) is the major constituent, always increases as the amount of fat is reduced (Wood, 1984). This is mainly because of a decline in the relative importance of *de novo* fatty acid synthesis, as opposed to dietary incorporation of the essential fatty acid linoleic acid, as fat deposition is reduced.

Recent estimates of the fat content of carcasses and the lipid content of defatted lean in UK cattle, sheep and pigs are given in Table 2. Whereas there has been a marked recent reduction in the fat content of the average pig carcass this has not occurred in beef and lamb.

Table 2 RECENT ESTIMATES OF THE CARCASS FATNESS AND THE LIPID CONTENT OF DEFATTED LEAN IN BRITISH CATTLE, SHEEP AND PIGS

	Cattle	Sheep	Pigs
Separable fat (%)	26.0	25.6	20.0
Lipid in defatted lean (%)	6.3	6.2	3.7

From Kempster, Cook and Grantley-Smith (1986); Meat and Livestock Commission (1989a) and Lowe *et al* (1990).

Effect of fat on eating quality

In processed meat products such as beefburgers in which a wide range of fat content can be found, the effect of fat inclusion on eating quality is easily seen. Typical results are shown in Table 3. In general juiciness and tenderness differences are easiest to detect.

For fresh meat, where samples are taken from animals produced in similar systems, the range in fat content is generally smaller than shown in Table 3 and lipid levels are certainly lower than 28%. However, especially when comparisons involve meat from very lean pig carcasses (P_2 values less than 10 mm), the effects seen in Table 3 can also be found. The first clear evidence of a detrimental effect on eating quality of reducing fat content too severely in the UK was given by Wood, Mottram and Brown (1981). They selected 16 pork carcasses of different fat content from those displayed at the Royal Smithfield Show. Both tenderness and juiciness in the leaner meat (8 mm P_2) were perceived by taste panellists to be lower than in the carcasses averaging 14 mm P_2 (Table 4) although only the effect on juiciness was significant.

Table 3 EFFECT OF FAT CONTENT OF HAMBURGERS[d] ON EATING QUALITY

	Extractable lipid (%)		
	9	20	28
Juiciness	7.2[a]	10.5[b]	11.3[b]
Tenderness	9.6[a]	11.8[b]	13.1[c]
Flavour	9.9[a]	9.3[a]	9.0[a]

Means with different superscripts differ significantly ($P<0.05$).
[d] Chuck from US Choice grade carcasses grilled to endpoint temperature of 71 or 77°. Taste panel scores 1 - 20 (higher scores indicate more juicy, more tender, stronger flavour). From Kregel *et al* (1986).

Table 4 EATING QUALITY IN 'LEAN' AND 'VERY LEAN' PORK CARCASSES (WEIGHT 56 KG)

	Lean	Very lean	
P_2 fat thickness (mm)	14	8	
M. longissimus depth (mm)	46	52	***
Lean in carcass (%)	57.2	62.9	***
Toughness (J)	0.14	0.16	NS
Tenderness (-7 to +7)	1.8	0.9	NS
Flavour (-7 to +7)	2.4	2.5	NS
Juiciness (0 to 3)	1.4	0.9	*

From Wood *et al* (1981).

In this and subsequent tables, * $P < 0.05$, ** $P < 0.01$, *** $p < 0.001$

Similar results have since been found in much larger tests conducted jointly by workers at Langford and the Meat and Livestock Commission (MLC) (Table 5). Although the scores of trained taste panellists were not greatly different in meat cuts prepared from carcasses with either 8 or 16 mm P_2 fat thickness, consumers detected important shifts in the acceptability of tenderness and juiciness when they prepared pork chops from the two groups in their own homes. In this work, a panel of butchers prepared the carcasses as for retail sale and gave scores for various aspects of carcass and meat quality. They strongly believed that the meat from the leanest carcasses would have poor eating quality whereas the differences actually noted by consumers were small. This discrepancy between the views of butchers and consumers has been observed before.

Table 5 EATING QUALITY IN CARCASSES OF DIFFERENT FAT THICKNESS

	P_2 fat thickness (mm)		
	8	16	
Trained taste panel[a]			
Tenderness	1.0	1.1	NS
Juiciness	1.1	1.3	**
Consumer panel[b]			
Tenderness	35	37	**
Juiciness	16	23	***

[a] Tenderness scored -7 to +7, juiciness 0 to 4. [b] Percentage of samples rated very tender or very juicy. From Wood *et al* (1986); Kempster, Dilworth, Evans and Fisher (1986).

Danish research has revealed highly significant differences in eating quality between pork loin steaks with different marbling fat (muscle lipid) concentrations. Results for tenderness assessed instrumentally and by taste panel are shown in

Table 6. In this work, extraction of muscle lipid was done using diethyl ether extraction and acid hydrolysis which gives values approximately 20% greater than for diethyl ether extraction alone (the normal method of analysis). In most European countries very few carcasses contain more than 2.0% ether extractable lipid in *m.longissimus*.

Table 6 THE EFFECT OF MARBLING FAT (MUSCLE LIPID) ON TENDERNESS IN GRILLED PORK STEAKS

Muscle lipid (%)	Tenderness (-5 to +5)	Shear force (units)
0.86	0.6	100
1.24	1.7	86
1.73	1.9	78
2.37	2.2	79
2.76	2.7	76
3.94	2.7	69

From Bejerholm and Barton-Gade (1986).

Factors controlling the tenderness/toughness of meat

PRODUCTION AND PROCESSING EFFECTS ON TENDERNESS

Research has shown that several factors in production and processing affect the tenderness of meat which is arguably the most important component of eating quality. A list of some of the factors affecting the tenderness of pigmeat is given in Table 7. In the case of cattle an additional factor would be sex, bulls having slightly tougher meat than steers and heifers. In all species, the effects of handling during the preslaughter period and the stunning method are important but poorly defined.

Although generalisations such as those contained in Table 7 can easily be made, the nature and level of each input required to elicit optimum eating quality are the subject of intensive research effort at present, co-ordinated in the UK by MLC.

Table 7 SOME FACTORS IN PIG PRODUCTION AND PROCESSING AFFECTING MEAT TENDERNESS

Factor	Effect
Breed	Tenderness slightly higher in Duroc breed, lower in Pietrain
Rearing method	Tenderness higher in pigs grown rapidly, lower in those given beta-adrenergic agonists
Carcass chilling rate	Muscles develop 'cold shortening' toughness if chilled too rapidly
Suspension method	Suspending the carcass from the pelvis rather than the achilles tendon produces more tender meat, especially in rapid chilling systems
Conditioning time	Extending the period between slaughter and retail sale increases tenderness

From Wood (1990).

The factors outlined in Table 7 operate through biochemical and physiological mechanisms only some of which are understood. In the case of breed effects, the important characteristics appear to be the concentration of marbling fat in relation to overall fatness (Duroc pigs have higher concentrations) and the genetic propensity to develop pale soft exudative (PSE) muscle. Marbling fat is thought to improve tenderness by providing a soft substitute for fibrous muscle proteins or by interrupting the bonding between fibre bundles allowing breakdown in the mouth to occur more easily (Wood 1990). In stress susceptible pigs (*eg* the Pietrain breed) toughness probably arises as the result of a reduction in the water-holding capacity of PSE muscle. Changes in tenderness associated with stunning method are also possibly connected with variation in the incidence of PSE muscle.

The effect of chilling rate on tenderness is caused by the action of temperature on the muscle fibres. In pre-rigor muscle (approximated by pH values above 6.0), temperatures below about 10°C cause the fibres to contract severely, causing 'bunching' and consequent toughening. Placing vulnerable muscles such as the *m.longissimus* under tension during chilling, which happens when the carcass is suspended from the pelvis, prevents this shortening. Electrical stimulation of the carcass also protects against cold shortening by depleting energy stores so that muscle contractions cannot occur (Dransfield, Ledwith and Taylor, 1991).

Changes in the activities of proteolytic enzymes could possibly explain the effects of growth rate and beta-adrenergic agonist drugs on tenderness. These enzymes are responsible for the increased tenderness which occurs when the conditioning period is extended from 2 to about 12 days in pigs. When growth rate is slowed through underfeeding, or beta-adrenergic agonist drugs are administered, there is a reduced rate of protein breakdown in muscle which may also occur in the postmortem period leading to reduced tenderness. Also, there may be a higher proportion of older, *ie* more cross-linked connective tissue proteins in muscle tissue which has turned over less rapidly and this would also reduce tenderness.

The evidence for some of the causes of the effects described in Table 7 is given in the next sections.

GROWTH RATE

The association between growth rate and eating quality in pigmeat has recently been demonstrated in trials conducted by MLC (Table 8). Growth rate was varied in 'white-type' and 'meat-type' pigs from four commercial breeding companies by changing feed intake between 20 and 80 kg live weight from *ad libitum* to about 80% of *ad libitum*. This reduced growth rate by about 20%. Restricted pigs were leaner and had a lower concentration of marbling fat in muscle but this did not entirely explain their lower tenderness and juiciness as determined by taste panellists. It was suggested that growth rate itself was a major factor (see also Warkup and Kempster, 1991).

We have recently re-examined early data collected at Langford by Wood and Riley (1982) and Mottram, Wood and Patterson (1982). Entire and castrated male pigs were given different levels of feed intake in order to produce different concentrations of fat in the carcass. The aim was to determine whether differences in sex or fatness were more important as causes of variation in quality traits. The results (Table 9) showed that slower growth in both sexes produced tougher bacon. The lower tenderness scores in castrates compared with entires could also be the result of slower growth. A plot of treatment mean values for tenderness score against estimated lean tissue growth rate (Figure 1) suggests these characteristics are closely related.

Table 8 EFFECTS OF *AD LIBITUM* VERSUS RESTRICTED FEEDING IN PIGS TO 80 KG LIVE WEIGHT ON CARCASS AND MEAT QUALITY

	Ad libitum	Restricted	
P_2 fat thickness (mm)	12.8	11.1	*
Lean in carcass (%)	55.5	57.5	*
Marbling fat (%)	0.85	0.75	*
Fat firmness (units)	615	590	*
C18:2 in backfat (%)	14.6	15.5	*
Tenderness[a]	5.20	4.73	*
Juiciness[a]	4.44	4.25	*
Pork flavour intensity[a]	4.52	4.57	NS

* Indicates differences significant at 5% level or greater. [a] Taste panel scores (1-8) in roast loin muscle. From Meat and Livestock Commission (1989b).

Table 9 GROWTH AND MEAT QUALITY IN LITTERMATE ENTIRE AND CASTRATED MALE PIGS GIVEN DIFFERENT AMOUNTS OF FEED AND THEREFORE GROWING AT A HIGH (H) OR LOW (L) RATE BETWEEN 27 AND 87 KG LIVE WEIGHT

	Entire males		Castrated males	
	H	L	H	L
Average daily gain (g)	920	748	601	439
P_2 fat thickness (mm)	14.0	11.8	16.5	14.4
Lean in carcass (%)	57.8	62.5	55.8	58.6
Lean tissue growth rate(g/d)	320	300	214	160
Eating quality of bacon[a]				
Tenderness	2.5	2.4	1.7	1.5
Flavour	3.2	2.8	3.5	2.9

[a] Trained taste panel scores -7 to +7. Tenderness significantly different between sexes ($P<0.05$) but not growth rates. From Wood and Riley (1982); Mottram *et al* (1982).

In beef cattle trials in the US, a positive association between tenderness and growth rate, particularly lean tissue growth rate, has been identified (reviewed by Wood, 1990). The view that US beef is generally tender in comparison with that from other countries where mainly extensive feeding systems are used could be explained by the common practice of feedlot finishing on high energy diets. Aberle, Reeves, Judge, Hunsley and Perry (1981) suggested that faster growth could result in a lower proportion of older, cross-linked collagen in muscle as the result of a faster turnover of muscle proteins, which would lead to increased tenderness. Other evidence suggests that conditioning rate is higher in beef animals which have grown rapidly (Lochner, Kauffmann and Marsh, 1980). However the higher concentration of marbling fat in fast growing beef cattle which generally have fatter carcasses is also a factor in their greater tenderness.

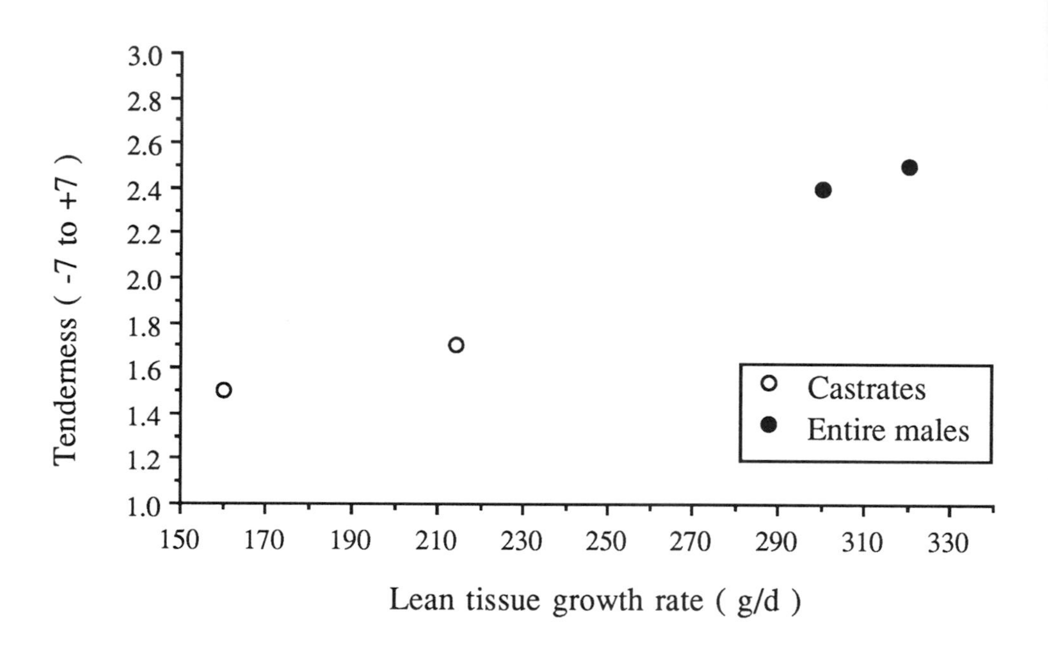

Figure 1 The relationship between the tenderness score for bacon given by trained taste panellists and lean tissue growth rate in the carcass joints between 27 and 87 kg live weight. Each point is the mean for 16 pigs. (See Wood and Riley, 1982 and Mottram *et al*, 1982).

ADMINISTRATION OF BETA-ADRENERGIC AGONISTS

These materials, which are structurally similar to the stress hormones adrenaline and noradrenaline, are effective 'repartitioning agents', causing dietary energy to be directed towards protein rather than fat deposition (Mersmann, 1990). The most widely studied compounds are clenbuterol and cimaterol but the effects of others, *eg* salbutamol and ractopaminc, have also been described. Although the mode of action of each of these differs slightly, they all appear to exert their repartitioning effect through stimulation of lipolysis and inhibition of muscle protein degradation (Thornton and Tume, 1988). The latter observation is particularly significant in view of the tendency of these materials to produce slightly tougher meat.

The results in Table 10 (Warriss, Kestin and Brown, 1989) illustrate some of the many effects of beta-adrenergic agonists and explain their attractiveness to meat producers wishing to increase the yield of lean meat. Lambs were given a concentrate diet during a 49-day growth period, containing either no beta-adrenergic agonists or either cimaterol or clenbuterol. Both treatment groups converted feed into weight more efficiently, had a higher yield of carcass relative to live weight, a more 'blocky', well-conformed carcass (this indicates thicker muscles) and contained a lower amount of fat.

Table 10 EFFECTS OF THE BETA-ADRENERGIC AGONISTS CIMATEROL (CIM, 10 PPM) AND CLENBUTEROL (CLEN, 2 PPM) ON GROWTH, FATNESS AND MEAT QUALITY IN MALE SHEEP FED A PELLETED DIET (49 DAY GROWTH PERIOD)

	0	Cim	Clen
Live weight at slaughter (kg)	32.8	33.5	33.0
Kg feed/kg gain	6.6[a]	5.2[b]	5.3[b]
Killing-out proportion	0.51[a]	0.55[b]	0.55[b]
Conformation score[c]	7.5[a]	11.4[b]	10.8[b]
Dissected fat in loin (%)	27.7[a]	19.7[b]	21.3[b]
Ultimate pH in LD[d]	5.94[a]	6.38[b]	6.46[b]
ST	6.20[a]	6.58[b]	6.43[b]
Lightness LD	43.3[a]	41.3[ab]	40.6[b]
Saturation LD	11.1[a]	6.7[b]	6.7[b]
Marbling fat LD (%)	2.7[a]	1.3[b]	1.4[b]
Haem pigments LD (mg/g)	3.85[a]	2.62[b]	2.19[b]

Means with different superscripts are significantly different ($P<0.05$). [c] 1-15 point scale. [d] LD *m.longissimus dorsi* ST *m.semitendinosus*. From Warriss *et al* (1989).

There were some marked effects on characteristics related to meat quality. Thus, the pH of 2 major muscles remained high 24 hours after slaughter which was interpreted as indicating a greater utilization of glycogen *in vivo*, resulting in reduced glycogenolysis after slaughter. The reduced haem pigment concentration would have tended to produce lighter coloured muscles but this effect was counterbalanced by the higher ultimate pH values. Since the haem pigments reach higher concentrations in red oxidative rather than white glycolytic muscle fibres it was suggested that a shift in the proportion of muscle fibre types had occurred, towards a higher proportion of white fibres. This in turn would explain the greater *in vivo* utilization of glycogen.

Other work has demonstrated the effects of beta-adrenergic agonists on muscle pH and the incidence of 'dark cutting' meat. The results in Table 11 show that as the concentration of cimaterol in the diet given to lambs increased, the incidence of unacceptably dark meat also increased in association with an increase in ultimate pH.

Table 11 EFFECT OF THE BETA-ADRENERGIC AGONIST CIMATEROL ON ULTIMATE pH AND INCIDENCE OF DARK CUTTING IN *M.LONGISSIMUS* (LD) MUSCLES OF 60 SUFFOLK X GALWAY WETHER LAMBS

	Cimaterol (ppm in diet)			
	0	0.6	2.3	11.4
Ultimate pH in LD	5.67	5.74	5.87	5.90
% dark cutting[a]	8	7	20	27

[a] pH > 6.0 From Allen *et al* (1985).

Early work showing the effects of beta-adrenergic agonists on muscle tenderness was conducted in Ireland (Table 12). In studies on steers, the juiciness, and particularly tenderness scores given by taste panellists were reduced as the concentration of cimaterol in the diet increased. The objective measurement of toughness was also significantly increased with cimaterol dose. A similar effect was reported for lambs (Table 13).

Table 12 EFFECTS OF THE BETA-ADRENERGIC AGONIST CIMATEROL ON EATING QUALITY IN *M.LONGISSIMUS* STEAKS OF FRIESIAN STEERS. SCORES OF TRAINED TASTE PANELLISTS AND OBJECTIVE MEASUREMENT OF TOUGHNESS

	Cimaterol (mg/day)			
	0	35.0	49.5	66.0
Juiciness[a]	2.5	1.8	1.2	1.7
Flavour[b]	3.8	2.7	2.8	3.2
Tenderness[c]	4.3	2.1	1.9	2.1
Overall acceptability[d]	4.8	3.0	2.7	3.1
Shear force (N)	45.8	95.4	105.8	110.1

[a] Score 0-4 [b] Score 0-5 [c] and [d] Score 0-7 From Allen *et al* (1986).

Table 13 EFFECTS OF THE BETA-ADRENERGIC AGONIST CIMATEROL (CIM, 2 PPM IN DIET) ON GROWTH AND TOUGHNESS OF *M.LONGISSIMUS DORSI*, (LD) IN SCOTTISH BLACKFACE LAMBS. TRIAL PERIOD 49 DAYS

	Control	Cim
Average daily gain (g)	241	274
Subcutaneous fat (%)	10.5	9.0
Shear force LD (N)	52.3	74.5

From Hanrahan *et al* (1988).

Explanations for the marked reductions in tenderness have centred mainly on the connective tissue proteins and the conditioning enzymes. The lower concentration of soluble (as opposed to insoluble or cross-linked) collagen following treatment with cimaterol in one study (Table 14) was suggested to have resulted from an inhibition of protein degradation in muscle *in vivo* causing a higher proportion of older collagen to remain in the tissue (also Bailey, 1988). In another study, the higher concentration of troponin-T in cimaterol-treated bull muscle remaining after 11 days conditioning, as identified by electrophoresis of myofibrillar proteins, was interpretated by Belgian workers as indicating reduced protease activity post-mortem (Table 15).

Table 14 EFFECTS OF THE BETA-ADRENERGIC AGONIST CIMATEROL (CIM, 0.06 MG/KG LIVE WEIGHT/DAY) ON COLLAGEN CONTENT AND SOLUBILITY IN 3 MUSCLES[a]. FRIESIAN STEERS TREATED FOR 70 DAYS (12-22 WEEKS OF AGE)

	Control	Cim
Collagen content (mg/g)	15.2	12.5
Soluble collagen (g/100g collagen)	18.8	13.0

[a] Mean values for *m.vastus lateralis*, *m.semitendinosus* and *m.longissimus dorsi*. From Dawson *et al* (1990).

Table 15 EFFECTS OF CIMATEROL (CIM, 4 PPM IN FEED FOR 246 DAYS) ON TOUGHNESS AND MYOFIBRILLAR PROTEINS (μG/MG)[a] IN *M. LONGISSIMUS DORSI* OF BELGIAN BLUE BULLS

	Control	Cim	
Toughness (N)	48.2	70.2	*
Troponin-T	2.4	7.5	*
30k dalton	12.8	6.2	*

[a] Following storage at 2°C for 11 days. From Fiems *et al* (1990).

A more detailed investigation of the activities of protease enzymes following slaughter of lambs treated with beta-adrenergic agonist drugs was conducted by Kretchmar, Hathaway, Epley and Dayton (1990). Wether lambs treated with the compound L-644, 969 were heavier than controls following 6 weeks treatment and had significantly tougher *m.longissimus* muscles (Table 16). The activity of the calcium-dependant protease, calpain I, which is active in the neutral pH range, seemed to be reduced in the treated animals and the activity of the enzyme inhibitor calpastatin was increased. Cathepsin-B, active at lower pH, also seemed to be reduced. The myofibrillar fragmentation index which is an index of muscle fibre breakdown during conditioning was lower in treated lambs especially 6 days after slaughter (however the large increase in fragmentation index between 1 and 6 days post-mortem in controls was not matched by a proportionate decrease in shear force).

Table 16 EFFECTS OF THE BETA-ADRENERGIC AGONIST L-644, 969 (4 PPM IN DIET, 6 WEEK TREATMENT PERIOD FROM 32 KG LIVE WEIGHT) ON FAT THICKNESS, MUSCLE TENDERNESS (*M.LONGISSIMUS DORSI*, LD) AND PROTEASE ENZYME ACTIVITIES IN CROSSBRED WETHER LAMBS

	Control	L-644, 969	
Final live weight (kg)	42.7	46.2	*
Shear force LD (kg) - 3 days	2.8	6.0	***
6 days	2.4	4.9	***
Enzyme activity[a]			
Calpain I	1.5	1.3	
Calpain II	1.8	2.5	
Calpastatin	3.6	5.7	
Cathepsin B	31	21	
Myofibril fragmentation index[a]			
(units) - 1 day	37	31	
6 days	74	33	

[a] Nominal units, taken from graphs in text, enzymes measured on day of slaughter. From Kretchmar *et al* (1990).

Recent studies of the effects of the beta-adrenergic agonist salbutamol in pigs are given in Tables 17 to 19. In the first study (Table 17) the same general effects on muscle quality as were observed in lambs (Table 10) were found. Thus, the ultimate pH of treated muscles remained high and the saturation of the *m.longissimus* was reduced in association with the decrease in haem pigment concentration. These results were again interpreted as indicating a shift in the population of muscle fibre types towards a higher concentration of white glycolytic fibres. The increase in toughness of treated pigs could be a consequence of this change as other studies have found (*eg* Solomon, Campbell, Steele, Caperna and McMurty, 1988).

Table 17 EFFECTS OF THE BETA-ADRENERGIC AGONIST SALBUTAMOL (SAL, 3 PPM IN THE FEED BETWEEN 32 AND 85 KG LIVE WEIGHT) ON MEAT QUALITY IN PIGS

	Control	Sal	
Final live weight (kg)	84.7	84.3	NS
P_2 fat thickness (mm)	10.5	8.7	***
Liver glycogen (mg/g)[a]	12.6	6.4	***
Ultimate pH in LD	5.52	5.56	**
Ultimate pH in AD	5.76	5.93	***
Lightness[b]	56.1	56.9	NS
Saturation[b]	7.5	5.5	***
Toughness (kg)[b]	4.11	5.04	***
Collagen[b]	3.8	3.3	***
Marbling fat[b]	8.1	8.6	NS
Glycogen[b]	5.4	4.6	NS
Haem pigments[b]	0.82	0.69	***
Fat firmness (0-1000)	612	540	**

LD *m. longissimus dorsi*. AD *m.adductor*.

[a] 6-hour food withdrawal, differences not significant after 24 hour food withdrawal. Measurements made 15 min after slaughter.

[b] Measurements made in LD. All concentrations in mg/g. Glycogen measured 15 min after slaughter, all others at 24 hours. From Warriss, Kestin, Rolph and Brown (1990).

Further studies conducted at Langford examined the effects of salbutamol in different pig muscles. Treated pigs had increased toughness in *m.longissimus, m.semimembranosus* but not the *m.supraspinatus* (Table 18). The latter muscle has a higher concentration of red oxidative fibres as the results for haem pigment concentration show. In all five muscles, salbutamol reduced the concentration of haem pigments indicating a shift in fibre types towards a higher proportion of white glycolytic fibres. As inferred from the toughness differences between muscles, such a change would tend to reduce the tenderness of the muscle when eaten.

Table 18 EFFECTS OF THE BETA-ADRENERGIC AGONIST SALBUTAMOL (SAL, 2.7 PPM IN THE FEED BETWEEN 28 AND 92 KG LIVE WEIGHT) ON TOUGHNESS AND THE CONCENTRATION OF HAEM PIGMENTS IN *M.LONGISSIMUS* (LD), *M.SEMIMEMBRANOSUS* (SM) AND *M.SUPRASPINATUS* (SS) MUSCLES IN 80 GILTS

		Control	Sal	
Toughness (kg)	LD	5.06	5.83	*
	SM	5.08	5.51	*
	SS	4.69	4.31	NS
Haem pigments (mg/g)	LD	0.62	0.50	***
	SM	0.78	0.63	***
	SS	1.90	1.71	***

From Warriss, Brown, Rolph and Kestin (1990).

A detailed study of the effects of salbutamol on the eating quality of pigmeat was undertaken by Warriss, Nute, Rolph, Brown and Kestin (1991). The results (Table 19) show that only in the case of toughness measured instrumentally was there a significant difference between treated pigs and controls. Although the texture (tenderness) score was slightly lower in pigs given salbutamol the difference was not significant. The distributions of tenderness scores showed considerable overlap between the two groups.

Table 19 EFFECTS OF THE BETA-ADRENERGIC AGONIST SALBUTAMOL (SAL, EITHER 2.7 OR 3.0 PPM IN THE DIET) ON EATING QUALITY OF ROAST PORK AS DETERMINED BY TRAINED TASTE PANEL

	Control	Sal	
Texture (1-8)	4.92	4.71	NS
Juiciness (0-3)	1.02	1.11	NS
Flavour (0-3)	1.28	1.26	NS
Overall acceptability (1-8)	4.67	4.67	NS
Toughness (kg)	5.01	5.86	***

From Warriss *et al* (1991).

In conclusion, it appears that the reduction in tenderness seen in some studies in which beta-adrenergic agonists have been given is related to changes in muscle protein metabolism and particularly the inhibition of protein degradation. An additional factor may be an induced change in the proportions of muscle fibre types.

SOMATOTROPIN

Bovine and porcine somatotropin have also been used to modify growth rate and carcass composition in meat animals often with dramatic results (Table 20). In the study illustrated, P_2 fat thickness was reduced more by porcine somatotropin (PST) than by fifteen years of selection and management changes in Britain (Table 1).

In the studies described in Tables 20 and 21 a slight increase in toughness was noted as PST dose increased. A larger increase in toughness has recently been reported by Solomon, Campbell, Steele and Caperna (1991). At 90 kg live weight following the administration of 100 μg PST/kg live weight/day between 30 and 60 kg, shear force in *m.longissimus* was 6.7 kg compared with 5.6 kg in controls. At 60 kg live weight, the point at which PST administration ceased, the difference was larger, 6.8 kg in treated pigs compared with 5.3 kg in controls. Since PST greatly increases protein deposition and turnover rates these results are not consistent with the positive association between turnover rate and tenderness noted with other treatments.

Table 20 EFFECTS OF PORCINE SOMATOTROPIN (PST) ON GROWTH, FATNESS AND MEAT QUALITY IN 120 CROSSBRED PIGS, BETWEEN 57 AND 103 KG LIVE WEIGHT

	PST (mg/day)				
	0	1.5	3.0	6.0	9.0
Average daily gain (g)	774	828	865	886	838
Feed intake (kg/day)	2.93	2.69	2.50	2.28	2.26
P_2 fat thickness (mm)	25.8	17.8	14.1	13.3	10.7
Toughness (kg)[a]	3.2	3.3	3.4	3.6	3.8
Tenderness[b]	9.3	8.9	8.6	8.7	9.6
Off-flavour[b]	14.0	13.8	13.7	13.8	14.2

[a] Using Warner-Bratzler shear press. [b] Taste panel scores 0 (low tenderness or high off-flavour intensity) to 15 (high tenderness or low off-flavour intensity). From Bechtel *et al* (1988).

Table 21 EFFECTS OF PORCINE SOMATOTROPIN (PST) ON *M.SEMITENDINOSUS* (ST) MUSCLE WEIGHT AND FIBRE TYPES AND TOUGHNESS OF *M.LONGISSIMUS* (LD)

	PST (μg/kg body weight/day)				
	0	30	60	120	200
LD toughness (kg)	3.2	2.9	3.6	3.0	3.8
ST weight (g)[a]	385	413	423	464	466
ST red fibres (%)[b]					
Deep medial portion	43.1			46.8	
Superficial portion	3.8			4.5	

[a] Corrected to same carcass weight [b] Mean of castrated males and females. From Beermann *et al* (1990).

BREED

As already mentioned, the effects of breed on meat tenderness have been ascribed in earlier work to the effects of marbling fat and, in pigs, the genetic susceptibility to stress, causing variation in the incidence of PSE muscle. There is however some evidence that differences in lean tissue growth rate, associated with the activities of conditioning enzymes, are also involved (Table 22). Bos indicus cattle had tougher *m.longissimus* muscles than Bos taurus, a lower myofibrillar fragmentation index and a tendency towards an increased activity of the inhibitor enzyme calpastatin. The Sahiwal cattle also grew more slowly than the Herefords.

Table 22 BREED EFFECTS ON TENDERNESS OF *M.LONGISSIMUS DORSI* (LD) IN CATTLE

		Hereford x Angus	$^3/_8$ Sahiwal	$^5/_8$ Sahiwal
Shear force LD (kg) -	day 1	7.0	9.3	9.6
	day 14	4.7	6.4	7.7
Myofibril fragmentation index[a] (units) -	day 1	52	40	38
	day 14	80	74	60
Enzyme activity (units)[a] day 1				
Cathepsin B			Similar in all groups	
Calpains I and II			Similar in all groups	
Calpastatin			Slightly higher in Sahiwal	

[a] Estimated from graphs presented in text. From Whipple *et al* (1990).

Conclusions

As carcass fatness is reduced and leanness is increased through a variety of means there is a tendency for the eating quality and particularly the tenderness of meat to decline. The effect is particularly marked in pigs at or around P_2 levels of 10 mm. Some of this reduction can be ascribed to the fall in intramuscular (marbling) fat which also occurs as carcass fatness is reduced. However the low correlations between marbling fat concentration and tenderness (typically around 0.2) suggest the involvement of other factors. Some of these have been identified, for example the tendency of muscles in lean carcasses to cold shorten if chilling rates are rapid. Even so, significant variation in tenderness still remains after the effects of all known factors are removed and a search for the cause of this is becoming more important as the demands for high quality meat increase.

It has been suggested that variation in tenderness is associated with variation in protein (lean tissue) growth rate, the rate of protein turnover and specifically the activities of the enzymes controlling protein degradation *in vivo* and after slaughter. Thus, animals in *ad libitum* or high energy feeding systems and certain fast growing breeds produce tender meat and animals given beta-adrenergic agonists produce tough meat, the link between these being the balance between the synthesis and degradation rates of muscle proteins. However this cannot be the whole explanation for variation in tenderness because administration of somatotropin, although increasing protein deposition with no inhibition of degradation, does not increase it, probably the reverse.

Whatever the cause of variation in tenderness it is clear that it can be improved by modifications to practices in production and processing. This is the 'blueprint' approach to quality control, improving quality through the proper implementation of key practices.

References

Aberle, E.D., Reeves, E.S., Judge, M.D., Hunsley, R.E. and Perry, T.W. (1981) *Journal of Animal Science*, **52**, 757-763

Allen, P., Tarrant, P.V., Hanrahan, J.P. and Fitzsimson, T. (1985) *Food Science and Technology Research Report*, Dublin, An Foras Taluntais, pp 6-7

Allen, P., Tarrant, P.V., Joseph, R.L. and Quirke, J.F. (1986) *Food Science and Technology Research Report*, Dublin, An Foras Taluntais, pp 34-35

Bailey, A.J. (1988) *Proceedings of the 34th International Congress of Meat Science and Technology*, pp 152-160

Bechtel, P.J., Easter, R.A., McKeith, F.K., Novakovski, J., McLaren, D.G. and Grebner, G.L. (1988) *Proceedings of the 34th International Congress of Meat Science and Technology*, pp 603-604

Beermann, D.H., Fishell, V.K., Roneker, K., Boyd, R.D., Armbruster, G. and Souza, L. (1990) *Journal of Animal Science*, **68**, 2690-2697

Bejerholm, C. and Barton-Gade, P.A. (1986) *Proceedings of the 32nd European Meeting of Meat Research Workers*, pp 389-391

Dawson, J.M., Buttery, P.J., Gill, M. and Beever, D.E. (1990) *Meat Science*, **28**, 289-297

Dransfield, E., Ledwith, M.J. and Taylor, A.A. (1991) *Meat Science*, **29**, 129-139

Fiems, L.O., Buts, B., Boucque, Ch. V., Demeyer, D.I. and Cottyn, B.G. (1990) *Meat Science*, **27**, 29-39

Hanrahan, J.P., Allen, P. and Sommer, M. (1988) In *Control and Regulation of Animal Growth*, (eds J.F. Quirke and H. Schmid), Wageningen, Pudoc, pp 149-160

Kempster, A.J., Cook, G.L. and Grantley-Smith, M. (1986) *Meat Science*, **17**, 107-138

Kempster, A.J., Dilworth, A.W., Evans, D.G. and Fisher, K.D. (1986) *Animal Production*, **43**, 517-533

Kregel, K.K., Prusa, K.J. and Hughes, K.V. (1986) *Journal of Food Science*, **51**, 1162-1165

Kretchmar, D.H., Hathaway, M.R., Epley, R.J. and Dayton, W.R. (1990) *Journal of Animal Science*, **68**, 1760-1772

Lochner, J.W., Kauffmann, R.G. and Marsh, B.B. (1980) *Meat Science*, **4**, 227-241

Lowe, D.B., Kempster, A.J., Fogden, M.W. and White, C.F. (1990) *Animal Production*, **50**, 560 (abstract)

Meat and Livestock Commission (1989a) *Sheep Yearbook*, Milton Keynes, Meat and Livestock Commission

Meat and Livestock Commission (1989b) *Stotfold Pig Development Unit, First Trial*, Milton Keynes, Meat and Livestock Commission

Mersmann, H.J. (1990) In *Reducing Fat in Meat Animals*, (eds J.D. Wood and A.V. Fisher), London, Elsevier Applied Science, pp 101-144

Mottram, D.S., Wood, J.D. and Patterson, R.L.S. (1982) *Animal Production*, **35**, 75-80

Solomon, M.B., Campbell, R.G., Steele, N.C., Caperna, T.J. and McMurty, J.P. (1988) *Journal of Animal Science*, **66**, 3279-3284

Solomon, M.B., Campbell, R.G., Steele, N.C. and Caperna, T.J. (1991) *Journal of Animal Science*, **69**, 641-645

Thornton, R.F. and Tume, R.K. (1988) *Proceedings of the 34th International Congress of Meat Science and Technology*, pp 6-14

Warkup, C.C. and Kempster, A.J. (1991) *Animal Production*, **52**, 559

Warriss, P.D., Kestin, S.C. and Brown, S.N. (1989) *Animal Production*, **48**, 385-392

Warriss, P.D., Kestin, S.C., Rolph, T.P. and Brown, S.N. (1990) *Journal of Animal Science*, **68**, 128-136

Warriss, P.D., Brown, S.N., Rolph, T.P. and Kestin, S.C. (1990) *Journal of Animal Science*, **68**, 3669-3676

Warriss, P.D., Nute, G.R., Rolph, T.P., Brown, S.N. and Kestin, S.C. (1991) *Meat Science*, **30**, 75-80

Whipple, G., Koohmaraie, M., Dikeman, M.E., Crouse, J.D., Hunt, M.C. and Klemm, R.D. (1990) *Journal of Animal Science*, **68**, 2716-2728

Wood, J.D. (1984) In *Fats in Animal Nutrition*, (ed J. Wiseman), London, Butterworths, pp 407-453

Wood, J.D. (1990) In *Reducing Fat in Meat Animals*, (eds J.D. Wood and A.V. Fisher), London, Elsevier Applied Science, pp 344-397

Wood, J.D. and Riley, J.E. (1982) *Animal Production*, **35**, 55-63

Wood, J.D., Mottram, D.S. and Brown, A.J. (1981) *Animal Production*, **32**, 117-120

Wood, J.D., Jones, R.C.D., Francombe, M.A. and Whelehan, O.P. (1986) *Animal Production*, **43**, 535-544

Wood, J.D., Enser, M.B., Whittington, F.M. and Moncrieff, C.B. (1989) *Livestock Production Science*, **22**, 351-362

LIST OF POSTER PRESENTATIONS

GLUCOCORTICOID SENSITIVITIES OF DIFFERENT BOVINE SKELETAL MUSCLES

H. Sauerwein, I. Dürsch and H.H.D. Meyer
Institut für Physiologie der Süddeutschen Versuchs- und Forschungsanstalt für Milchwirstschaft, TU München, D-8050 Freising-Weihenstephan, Vöttingerstr. 45. Germany

THE EFFECT OF INSULIN STATUS UPON THE ACTIONS OF GROWTH HORMONE IN SHEEP

A.S. Bowman and J.M. Bassett
The University of Oxford, Growth and Development Unit, University Field Laboratory, Wytham, Oxford OX2 8QJ

ATTENUATION OF RESPONSIVENESS TO ADRENALINE DURING PROLONGED ADMINISTRATION OF BETA AGONIST DRUGS TO FETAL AND GROWING SHEEP

J.M. Bassett
The University of Oxford, Growth and Development Unit, University Field Laboratory, Wytham, Oxford OX2 8QJ

INTRA AND INTERSPECIFIC DIFFERENCES IN THE ATYPICAL BETA-ADRENERGIC RESPONSE OF ADIPOSE TISSUE

M.P. Portillo[+], D. Langin[*], M. Berlan[*] and M. Lafontan[*]
[+]Dept Nutrition and Food Science, University of País Vasco, Vitoria, Spain
[*]INSERM U-317, Institut de Physiologie, Toulouse, France

BONE FORMATION IN RATS SUBCUTANEOUSLY ADMINSTERED WITH BONE MATRIX DERIVED PROTEINS

M. Elorriaga, F. Lecanda, M. Marquínez, J. Alfredo Martínez and J. Larralde
Dept Physiology and Nutrition, University of Navarra, Pamplona, Spain

BONE GROWTH IN INTACT FEMALE RATS TREATED WITH SOMATOTROPIN (rGH)

A.S. Del Barrio, J. Alfredo Martínez and J. Larralde*
Dept Nutrition and Food Science, University of País Vasco, Vitoria, Spain
Dept Physiology and Nutrition, University of Navarra, Pamplona, Spain

EFFECTS OF GLUCAGON-LIKE PEPTIDE 1 (7-36)AMIDE ON SERUM INSULIN CONCENTRATIONS IN SHEEP

A. Faulkner
Hannah Research Institute, Ayr KA6 5HL, Scotland

RESPONSES IN GASTRIC INHIBITORY POLYPEPTIDE TO NUTRIENT ABSORPTION IN PRE-RUMINANT AND RUMINANT ANIMALS

P.A. Martin, A. Faulkner, J.P. McCarthy and D.J. Flint
Hannah Research Institute, Ayr KA6 5HL, Scotland

EFFECT OF THE LACK OF MET + CYS AND MET + CYS + 50% ENERGY-RESTRICTION ON DNA METABOLISM OF GASTROCNEMIUS MUSCLE IN RATS

J.L. Rey de Viñas, M.J. Sánchez-Barco, R. Hinojosa Melero, C. Osuna, B. De Arribas
Instituto de Nutrición y Bromatología, Facultad de Farmacia, Ciudad Universitaria, 28040 Madrid, Spain

EFFECT OF THE LACK OF MET + CYS AND MET + CYS + 50% ENERGY-RESTRICTION ON RNA METABOLISM AND ENZYME ACTIVITIES OF GASTROCNEMIUS MUSCLE IN RATS

J.L. Rey de Viñas, M.J. Sánchez Barco, R. Hinojosa Melero, C. Osuna and B. De Arribas
Instituto de Nutrición y Bromatología, Facultad de Farmacia, Ciudad Universitaria, 28040 Madrid, Spain

GUT HORMONES AND THE CONTROL OF ADIPOSE TISSUE METABOLISM IN PIGS

J. Oben, R. Elliott, L. Morgan, J. Fletcher* and V. Marks
School of Biological Sciences, University of Surrey, Guildford, Surrey GU2 5XH
*Unilever Research, Colworth Laboratory, Bedford MK44 1LQ

MODIFICATION OF GIP RESPONSE TO LIPOPROTEIN LIPASE ACTIVITY IN RATS INTUBATED WITH TRIOLEIN

J. Oben, L. Morgan, J. Fletcher* and V. Marks
School of Biological Sciences, University of Surrey, Guildford, Surrey GU2 5XH
*Unilever Research, Colworth House, Sharnbrook, Bedford MK44 1LQ

THE ENERGY COST OF CARRYING FAT: A CONTROL OF BODY WEIGHT

C.D.R. Jones
MRC Dunn Nutrition Unit, Keneba, PO Box 273, The Gambia

VOLUNTARY MILK INTAKE IS THE BASIS OF GENETIC SELECTION FOR WEANING WEIGHT OF MERINO LAMBS

V.H. Oddy
NSW Agriculture & Fisheries, Elizabeth Macarthur Agricultural Institute, PMB 8, Camden, NSW, 2570, Australia

LACK OF A SPECIFIC GROWTH HORMONE BINDING PROTEIN IN PERIPHERAL PLASMA OF RUMINANTS

D.A. Shutt, P.A. Speck and V.H. Oddy
NSW Agricultural and Fisheries, Elizabeth Macarthur Agricultural Institute, PMB 8, Camden, NSW, 2570, Australia

EFFECT OF AGE AND WEIGHT ON MUSCLE COMPOSITION AND CARCASS CHARACTERISTICS IN YOUNG FATTENING BULLS

A. Clinquart, C. Van Eenaeme, L. Istasse, I. Dufrasne, A. Mayombo and J.M. Bienfait
Department of Nutrition, Faculty of Veterinary Medicine, University of Liège, Rue des Vétérinaires, 45, 1070 - Brussels, Belgium

IN VIVO AND IN VITRO MUSCLE PROTEIN TURNOVER DURING THE GROWING FATTENING PERIOD IN TWO BREEDS OF YOUNG BULLS OF DIFFERENT AGES

C. Van Eenaeme, A. Clinquart, L. Istasse, P.Baldwin, V. Hollo, J.M. Bienfait
Dept Animal Nutrition, Veterinary Faculty, University of Liège, 45, Rue des Vétérinaires, 1070 - Brussels, Belgium

EFFECTS OF PORCINE SOMATOTROPIN ADMINISTRATION ON LIPOGENESIS IN GROWING PIGS

J. Mourot, L. Lefaucheur and M. Bonneau
INRA, Station de Recherches Porcines, 35590 L'Hermitage, France

EFFECTS OF PORCINE SOMATOTROPIN ADMINISTRATION ON GROWTH PERFORMANCE, MUSCLE HISTOCHEMICAL AND BIOCHEMICAL CHARACTERISTICS AND PORK MEAT QUALITY

L. Lefaucheur, A. Missohou and M. Bonneau
INRA, Station de Recherches Porcines, 35590 L'Hermitage, France

STIMULATION OF ADIPOCYTE PRECURSOR PROLIFERATION BY GROWTH FACTORS IN COMBINATION WITH VERY LOW DENSITY LIPOPROTEIN: A LINK BETWEEN ADIPOCYTE HYPERPLASIA AND SERUM LIPID CONCENTRATION

S.C. Butterwith and D. Peddie
Department of Cellular and Molecular Biology, AFRC Institute of Animal Physiology and Genetics Research, Edinburgh Research Station, Roslin, Midlothian EH25 9PS

REGULATION OF ADIPOCYTE PRECURSOR CELL PROLIFERATION - SYNERGISTIC EFFECT OF TRANSFORMING GROWTH FACTOR ß1 AND THE INSULIN-LIKE GROWTH FACTORS

S.C. Butterwith and C. Goddard
Department of Cellular and Molecular Biology, AFRC Institute of Animal Physiology and Genetics Research, Edinburgh Research Station, Roslin, Midlothian EH25 9PS

THE EFFECT OF BIPIPERIDYL MUSTARD ON THE CONTROL OF FAT DEPOSITION IN T/O MICE

S.A. Jagot* and G.P. Webb
Division of Physiology and Pharmacology, Polytechnic of East London, Romford Road, Stratford, London, E15 4LZ
*Current address: Academic Surgery Unit, School of Postgraduate Medicine and Biological Sciences, University of Keele, Thornburrow Drive, Hartshill, Stoke-on-Trent, ST4 7QB

SEX AND FAT DISTRIBUTION IN C57B1/6 MICE

S.A. Jagot* and G.P. Webb
Division of Physiology and Pharmacology, Polytechnic of East London, Romford Road, Stratford, E15 4LZ
*Current address: Academic Surgery Unit, School of Postgraduate Medicine and Biological Sciences, University of Keele, Thornburrow Drive, Hartshill, Stoke on Trent, ST4 7QB

EFFECT OF SOMATOTROPIN ADMINISTRATION, *IN OVO*, ON MYOSIN ISOFORM EXPRESSION IN TURKEYS

N. Kanemaki and K. Maruyama
Avian Physiology Laboratory, Agricultural Research Service, US Department of Agriculture, Beltsville, Maryland, USA

AN APPRAISAL OF ACID ETHANOL EXTRACTION FOR THE DETERMINATION OF INSULIN LIKE GROWTH FACTOR 1 IN SHEEP SERUM

A.R.G. Wylie
Department of Agriculture for Northern Ireland, Food and Agricultural Chemistry Research Division, Newforge Lane, Belfast BT9 5PX, Northern Ireland

BETA-CELL TROPIN: A TRUE OBESITY HORMONE?

J.L. Morton and M. Davenport
Clore Laboratory for the Biologicial Sciences, University of Buckingham, Buckingham MK18 1EG

STRUCTURAL AND METABOLIC DIFFERENCES BETWEEN NINE IDENTIFIED ADIPOSE DEPOTS IN RODENTS

C.M. Pond and C.A. Mattacks
Department of Biology, The Open University, Milton Keynes MK7 6AA

PHYSIOLOGICAL RESPONSES TO SELECTION FOR CARCASS LEAN CONTENT IN A TERMINAL SIRE BREED OF SHEEP

N.D. Cameron
Institute of Animal Physiology and Genetics Research, Edinburgh Research Station, Roslin EH25 9PS, Scotland

RECOMBINANTLY-DERIVED IGF-1 HAS NO EFFECT ON THE GROWTH, FATNESS OR MUSCULARITY OF LAMBS

R.W. Purchas, Y.H. Cottam, S.N. McCutcheon, H.T. Blair, B.H. Breier, P.D. Gluckman and A.Y. Abdullah
Department of Animal Science, Massey University, Palmerston North
and Paediatrics Department, University of Auckland Medical School, Auckland, New Zealand

THE EFFECT OF THE α_2 AGONIST GUANFACIN ON GROWTH AND METABOLISM OF CATTLE

R.A. Hunter and D.B. Lindsay
CSIRO Division of Tropical Animal Production, Box 5545 Rockhampton Mail Centre, Queensland 4702, Australia

LIPOGENESIS AS A PROPORTION OF TOTAL GLUCOSE UTILISATION IN ADIPOSE TISSUE

J. Islwyn Davies and B.M. Grail
School of Biological Sciences, University of Wales: Bangor, Bangor, Gwynedd LL57 2UW

A THERMOGENIC EFFECT OF A NEUTRAL NON-STARCH POLYSACCHARIDE ASSESSED IN THE RAT BY FAT AND LEAN ACCUMULATION: EFFECT OF AMBIENT TEMPERATURE AND DIETARY FAT INTAKE

J.C. Brown, G. Livesey and E.K. Lund
AFRC Institute of Food Research, Norwich Laboratory, Norwich Science Park, Colney, Norwich NR4 7UA

GROWTH AND CARCASS COMPOSTION OF FRIESIAN, CANADIAN HEREFORD X FRIESIAN AND SIMMENTAL X FRIESIAN STEERS

M.G. Keane
Teagasc, Grange Research Centre, Dunsany, Co. Meath, Ireland

RESIDUES AFTER ANABOLIC DOSAGE OF CLENBUTEROL

H.H.D. Meyer and L. Rinke
Institut für Physiologie der Süddeutschen Versuchs- und Forschungsanstalt für Milchwirtschaft, TU München, D-8050 Freising-Weihenstephan, Vöttingerstr. 45, Germany

INFLUENCE OF PEPTIDES ASSOCIATED WITH THE GASTROINTESTINAL TRACT ON OVINE PERIRENAL FAT LIPOGENESIS

A.S. Haji Baba and P.J. Buttery
Department of Applied Biochemistry and Food Science, University of Nottingham, Sutton Bonington, Loughborough LE12 5RD, Leicestershire

LIST OF PARTICIPANTS

Mr S Allcock	University of Nottingham School of Agriculture, Sutton Bonington, Loughborough LE12 5RD
Miss S Austin	Seale-Hayne Faculty of Agriculture, Food and Land Use,Polytechnic South West, Newton Abbot, Devon, TQ12 6NQ
Dr R Bardsley	University of Nottingham School of Agriculture, Sutton Bonington, Loughborough LE12 5RD
Dr J Bass	Ruakura Agricultural Centre, New Zealand
Dr J M Bassett	University Field Laboratory, Wytham, Oxford, OX2 8QJ
Dr P C Bates	IAPGR, Cambridge Research Station, Babraham, Cambs, CB2 4AT
Prof D H Beermann	53 Morrison Hall, Cornell University, Ithaca, NY 14853, USA
Dr D Beever	IGER, Hurley, Maidenhead, Berks
Dr A W Bell	Dept of Animal Science, 262 Morrison Hall, Cornell University, Ithaca, NY 14853 - 4801, USA
Mr J Berg	Department of Animal Science, Norway
Ir P Bikker	Dept of Animal Nutrition, Wageningen Agricultural University, 6708 PM Wageningen, The Netherlands
Prof G Bono	Istituto Produzione Animale, via S.Mauro 2 - 33010 Pagnacco, Udine, Italy
Dr K N Boorman	University of Nottingham School of Agriculture, Sutton Bonington, Loughborough LE12 5RD
Dr Alan S Bowman	University Field Laboratory, Oxford
Mr J Brameld	University of Nottingham School of Agriculture, Sutton Bonington, Loughborough LE12 5RD
Dr B Brenig	Institute of Biochemistry, Karlstrasse 23, D-8000 Munich 2, Germany
Miss J C Brown	Institute of Food Research, Colney, Norwich, NR4 7UA
Dr F Buonomo	Monsanto Agricultural Company, St Louis, USA

Dr S C Butterwith	IAPGR, Roslin, Midlothian EH25 9PS
Prof P J Buttery	University of Nottingham School of Agriculture, Sutton Bonington, Loughborough LE12 5RD
Mr R M V H Caldeira	Faculda de Medicina Veterinaria, Ru Gomes Freire, 1199 Lisboa Cedex, Portugal
Dr N Cameron	IAPGR, Edinburgh Research Station, Roslin, EH25 9PS
Mr L A Cardoso	Centro de Veterinaria E, Zootecnia Faculdade de Medicina Veterinaria, Rua Gomes Freire, 1199 Lisbon Cedex, Portugal
Mr M Castejon	Wye College, Wye, Ashford, Kent
Dr J C Caygill	MAFF, Room G9, Nobel House, 17 Smith Square, London SW1 3JR
Dr R Clarke	University of Nottingham School of Agriculture, Sutton Bonington, Loughborough LE12 5RD
Mr A Clinquart	Nutrition Fac Med Vet, 45 Rue des Veterinaires, 1070 Bruxelles, Belgium
Dr D Cole	University of Nottingham School of Agriculture, Sutton Bonington, Loughborough LE12 5RD
Mrs K M Collingwood	IAPGR, Cambridge Research Station, Babraham, Cambs, CB2 4AT
Miss S E Cooper	Butterworth Heinemann, Lineacre House, Jordan Hill, Oxford, OX2 8DP
Ms M L Cruz	Metabolic Research Laboratory, Radcliffe Infirmary, Woodstock Road, Oxford OX2 6HE
Dr J I Davies	School of Biological Sciences, University of Wales, Bangor, Gwynedd LL57 2UW
Dr J Dawson	University of Nottingham School of Agriculture, Sutton Bonington, Loughborough LE12 5RD
Ir K H De Greef	Dept of Animal Nutrition, Wageningen Agricultural University, 6708 PM Wageningen, The Netherlands
Mr A S Del Barrio	Dpto Nutricion y Bromatologia, University of Pais Vasco, Portal de Lasarte s/n, Vitoria, Spain
Dr G N DeMartino	University of Texas South Western Medical Center, 5323 Harry Hines Blvd, Dallas TX 75235, USA
Prof D I Demeyer	RUG-CVI, Proefhoevestraat 10, 9090-Melle, Belgium
Dr N Dumelow	University of Nottingham School of Agriculture, Sutton Bonington, Loughborough LE12 5RD
Mr M Elorriaga	Dpt Fisiologia y Nutricion, Univ de Navarra, Irunlarrea s/n, Pamplona, Spain
Dr W J Enright	TEAGASC, Grange Research Centre, Dunsany, Co.Meath, Ireland
Dr M B Enser	University of Bristol Department of Meat Science, Langford, Bristol BS18 7DY

Ms C Essex	University of Nottingham School of Agriculture, Sutton Bonington, Loughborough LE12 5RD
Mrs F Evans	OECD/Agriculture, 2 rue Anore-Pascal, 75775 Paris Cedex 16, France
Dr N A Evans	Pfizer Central Research, Sandwich, Kent CT13 9NJ
Dr F Fairhurst	Rowett Research Institute, Bucksburn, Aberdeen, AB2 9SB
Dr A Faulkner	Hannah Research Institute, Ayr KA6 5HL
Mrs S Fiddaman	University of Nottingham School of Agriculture, Sutton Bonington, Loughborough LE12 5RD
Dr L Fiems	National Institute for Animal Nutrition, Scheldeweg 68, 9090 Melle, Belgium
Dr J M Fletcher	Unilever Research, Colworth House, Bedford MK44 1LQ
Dr D J Flint	Hannah Research Institute, Ayr, KA6 5HL
Dr J Z Foot	Dept Ag Pastoral Res Institute, Hamilton, Vic 3300, Australia
Dr P Garnsworthy	University of Nottingham School of Agriculture, Sutton Bonington, Loughborough LE12 5RD
Dr G J Garssen	Research Institute for Animal Production "Schoonord", PO Box 501, 3400 AM Zeist, The Netherlands
Mr G Geesink	VVDO, Postbus, Utrecht, The Netherlands
Revd M J Gibb	IGER, Hurley, Maidenhead, Berks SL6 5LR
Dr S Gilmour	IAPGR, Cambridge Research Station, Babraham, Cambs, CB2 4AT
Dr C Goddard	AFRC IAPGR, Edinburgh Research Station, Roslin, Midlothian EH25 9PS
Miss G R Goldberg	Dunn Clinical Nutrition Centre, 100 Tennis Court Road, Cambridge CB2 1QL
Dr D E Goll	Dept of Nutrition and Food Science and Biochemistry, University of Arizona, Tucson, Arizona 85721, USA
Mr H Greathead	University of Nottingham School of Agriculture, Sutton Bonington, Loughborough LE12 5RD
Dr H Griffin	IAPGR, Roslin, Midlothian, EH25 9PS
Mr A Haji Baba	University of Nottingham School of Agriculture, Sutton Bonington, Loughborough LE12 5RD
Miss E Hajj	ENSA INRA, 34060 Montpellier Cedex, France
Dr S W Hardy	Unilever Research, Colworth House, Sharnbrook, Beds MK44 1LQ
Dr W Haresign	University of Nottingham School of Agriculture, Sutton Bonington, Loughborough LE12 5RD
Dr J M M Harper	University of Nottingham School of Agriculture, Sutton Bonington, Loughborough LE12 5RD
Dr I C Hart	American Cyanamid Co, PO Box 400, Princeton, NJ 08543, USA

Mr R Hawkey	University of Nottingham School of Agriculture, Sutton Bonington, Loughborough LE12 5RD
Dr C E Hinks	University of Edinburgh Institute of Ecology and Resource Management, School of Agriculture, West Mains Rd, Edinburgh EH9 3JG
Mr J-F Hocquette	INRA, UR Metabolismes Energetique et Lipidique, Theix 63122 Ceyrat, France
Dr D F Houlihan	Dept of Zoology, University of Aberdeen, Tillydrone Ave, Aberdeen AB9 2TN
Miss C Huggett	University of Nottingham School of Agriculture, Sutton Bonington, Loughborough LE12 5RD
Dr R A Hunter	CSIRO, Division of Tropical Animal Production, Box 5545, Rockhampton Mail Centre, Queensland 4702, Australia
Dr K Incze	Hungarian Meat Research Inst, 17 Bp Gubacsi ut 6/b, Hungary
Mr K L Ingvartsen	Ministry of Agriculture National Institute of Animal Science, Dept of Research in Cattle & Sheep, Foulum PO Box 39, DK-8830, Tjele, Denmark
Dr L Istasse	Nutrition Fac Med Vet, 45 Rue de Veterinaires, 1070 Bruxelles, Belgium
Miss W Ivings	IGER, Hurley, Maidenhead, Berks SL6 5LR
Dr S-A Jagot	Dept Postgraduate Medicine, University of Keele, Thornburrow Drive, Hartshill, Stoke on Trent, ST4 9QB
Dr S James	Pitman-Moore Ltd, Harefield, Middx UB9 6LS
Dr D Jewell	1401 S Hanley, St Louis, MO 63144, USA
Dr N Kanemaki	Avian Physiology Lab, USDA, ARS, Beltsville, MD 20705-2350, USA
Dr M G Keane	TEAGASC, Grange Research Centre, Dunsany, Co. Meath, Ireland
Dr J A Kirk	Seale-Hayne Faculty of Agriculture, Food and Land Use, Polytechnic South West, Newton Abbot, Devon, TQ12 6NQ
Dr M Koohmaraie	USDA, ARS, Roman L Hruska, US Meat Anim Res Center, PO Box 166, Clay Center, NE 68933, USA
Dr I Kyriazakis	SAC-Edinburgh, Animal Sciences Division, Bush Estate, Penicuik, Midlothian EH26 0QE
Dr P Lapeyronie	ENSA INRA, 34060 Montpellier Cedex, France
Dr L Lefaucheur	Station de Recherches Porcines, INRA de St Gilles, 35590, L'Hermitage, France
Mr J P Lemos	Faculda de Medicina Veterinaria, Ru Gomes Freire, 1199 Lisboa Cedex, Portugal
Dr J E Lindberg	Swedish Univ of Agric Sci, Dept of Animal Nutrition and Management, Box 7024, S-75007 Uppsala, Sweden

Prof D B Lindsay	CSIRO Division of Tropical Animal Production, Rockhampton Q 4702, Australia
Dr T Lindsey	SmithKline Beecham Animal Health, 1600 Paoli Pike, West Chester, PA 19380, USA
Dr G E Lobley	Rowett Research Institute, Bucksburn, Aberdeen
Dr A Long	14 Woodland Rise, Greenford, Middx UB6 0RD
Miss D Longley	University of Nottingham School of Agriculture, Sutton Bonington, Loughborough LE12 5RD
Dr M Luck	University of Nottingham School of Agriculture, Sutton Bonington, Loughborough LE12 5RD
Ms E K Lund	Institute of Food Research, Colney, Norwich, NR4 7UA
Mr N Lynn	MAFF, ADAS, Block 2 Government Buildings, Otley Road, Lawnswood, Leeds LS16 5PY
Mrs T H Macedo Da Costa	Metabolic Res Laboratory, Nuffield Dept of Clinical Medicine, Radcliffe Infirmary, Woodstock Road, Oxford OX2 6HE
Dr J C Macrae	Rowett Institute, Bucksburn, Aberdeen
Miss S Maltby	IGER, Hurley, Maidenhead, SL6 5LR
Dr C Maltin	Rowett Research Institute, Greenburn Road, Bucksburn, Aberdeen
Dr P A Martin	Hannah Research Institute, Ayr, KA6 5HL
Prof J A Martinez-Hernandez	Dept Nutrition and Food Science, University of Pais Vasco, 01007, Vitoria, Spain
Dr R Mawson	Unilever Research Ltd, Colworth House, Sharnbrook, Beds MK44 1LQ
Mr J P McCarthy	Hannah Research Institute, Ayr, KA6 5HL
Dr D N McMillan	Rowett Research Institute, Greenburn Road, Bucksburn, Aberdeen
Mr P J McTiffin	SPA Ltd, Avenue 3, Station Lane, Witney, Oxon.
Dr E Melloni	Instituto polycatreda di Chimica Biologica, Universita degli studi di Genova, Via Balbi 5, Viali Benedetto XV No 1, 16130 Genova, Italy
Dr T B Mepham	University of Nottingham School of Agriculture, Sutton Bonington, Loughborough LE12 5RD
Dr H H D Meyer	Institut fur Physiologie, Techn. Universitat Munchen, Vottinger Strasse 45, 8050 Freising-Weihenstephan, Germany
Dr L M Morgan	School of Biological Sciences, University of Surrey, Guildford, Surrey, GU2 5XH
Dr J L Morton	Dept of Biological Sciences, University of Buckingham, Hunter St, Buckingham MK18 1EG
Mrs H S Munday	Waltham Centre for Pet Nutrition, Freeby Lane, Waltham on the Wolds, Melton Mowbray, Leicestershire LE14 4RT

Dr R J Neale	University of Nottingham School of Agriculture, Sutton Bonington, Loughborough LE12 5RD
Dr B S Noble	Rowett Research Institute, Greenburn Road, Bucksburn, Aberdeen
Dr J Oben	School of Biological Sciences, University of Surrey, Guildford, GU2 5XH
Dr A Oberbauer	Dept of Animal Science, University of California, Davis, California, 95616-0175, USA
Dr H Oddy	NSW Agriculture & Fisheries, Elizabeth Macarthur Agricultural Institute, PMB 8, Camden, NSW 2570, Australia
Dr A Ouali	Meat Research Station, INRA de Theix, 63122 Ceyrat, France
Dr R M Palmer	Rowett Research Institute, Bucksburn, Aberdeen, AB2 9SB
Miss R Parrinder	University of Nottingham School of Agriculture, Sutton Bonington, Loughborough LE12 5RD
Dr J M Pell	IAPGR, Babraham, Cambridge CB2 4AT
Dr J Perez-Lanzac	CEE DG VI BII.1, Loi 84, 1/25, B-1049, Brussels, Belgium
Dr C M Pond	The Open University, Milton Keynes, MK7 6AA
Ms M del Puy Portillo	Dept de Nutricion y Bramatologia, Facultad Farmacia, University of Pais Vasco, Portal de Lasarte s/n 01007, Vitoria, Spain
Dr R W Purchas	Department of Animal Science, Massey University, Palmerston North, New Zealand
Dr P T Quinlan	Unilever Research, Colworth House, Sharnbrook, Bedford, MK44 1LQ
Mr C Raichon	OECD/Agriculture, 2 rue Anore-Pascal, 75775 Paris Cedex 16, France
Dr P Reeds	Children's Nutrition Research Center at Baylor College, 1100 Bates, Houston, Texas, 77030 USA
Dr J L Rey de Vinas	Inst de Nutricion, Universidad Complutense de Madrid, 28040 Madrid, Spain
Prof I Romboli	Fac Med Veterinaria, Via Nizza 52, 10126 Torino, Italy
Dr A M Salter	University of Nottingham School of Agriculture, Sutton Bonington, Loughborough LE12 5RD
Mr J Santos Silva	Estacao Zootecnica Nacional, Vale Santarem - 2000 Santarem, Portugal
Dr H Sauerwein	Institut fur Physiologie, Techn. Universitat Munchen, Vottinger Strasse 45, 8050 Freising-Weihenstephan, Germany
Mr A W J Savage	University of Bristol, Dept of Meat Animal Science, Langford, Bristol, BS18 7DY

Mr L Sinclair	University of Nottingham School of Agriculture, Sutton Bonington, Loughborough LE12 5RD
Ir J Smulders	Mengroedir UT Delfia bv, Postbox 8, 3600 AA Maarssen, The Netherlands
Mr J Soar	University of Nottingham School of Agriculture, Sutton Bonington, Loughborough LE12 5RD
Mr P Speck	University of Nottingham School of Agriculture, Sutton Bonington, Loughborough LE12 5RD
Miss C Stewart	IAPGR, Cambridge Research Station, Babraham, Cambridge, CB2 4AT
Dr K Takahashi	Faculty of Agriculture, Hokkaido University, Sapporo, Japan
Dr F Thompson,	Rumenco, Derby Road, Burton on Trent, Staffs DE13 0DW
Mr A P Thompson	Pfizer Central Research, Sandwich, Kent CT13 9NJ
Dr B M Thomson	Rowett Research Institute, Bucksburn, Aberdeen, AB2 9SU
Dr F Toldra	Meat Products Dept, Instituto Agroquimica y Tecnologia de Alimentos, Jaime Rois 11, 46010 Valencia, Spain
Ir L Uytterhaegen	RUG-CVI, Proefhoevestraat 10, 9090 Melle, Belgium
Dr R G Vernon	Hannah Research Institute, Ayr, KA6 5HL
Mr C Warkup	MLC, PO Box 44, Snowden Drive, Winterhill, Milton Keynes, MK6 1AX
Miss S Wastie	Hannah Research Institute, Ayr, KA6 5HL and University of Nottingham School of Agriculture
Dr P Weller	IAPGR, Cambridge Research Station, Babraham, Cambs, CB2 4AT
Dr J Wiseman	University of Nottingham School of Agriculture, Sutton Bonington, Loughborough LE12 5RD
Dr M J Witty	Pfizer Central Research, Ramsgate Rd, Sandwich, Kent CT13 9NJ
Dr J D Wood	University of Bristol, Dept of Meat Animal Science, Langford, Bristol BS18 7DY
Miss D Wray-Cahen	IGER, Hurley, Maidenhead
Miss S Wright	University of Nottingham School of Agriculture, Sutton Bonington, Loughborough LE12 5RD
Dr A R G Wylie	Dept of Agriculture for N Ireland, Food and Agricultural Chemistry Research Division, Newforge Lane, Belfast BT9 5X
Miss L E Young	Dept of Zoology, University of Aberdeen, Tillydrone Ave, Aberdeen, AB9 2TN

INDEX

RELATED TITLES

BIOTECHNOLOGY IN GROWTH REGULATION Edited by R G Heap, C G Prosser and G E Lamming (1989)

NEW TECHNIQUES IN CATTLE PRODUCTION Edited by C J C Phillips (1989)

PIG PRODUCTION IN AUSTRALIA Edited by J A A Gardener, A C Dunkin and L C Lloyd (1990)

TRANSGENIC ANIMALS Edited by N First and F P Heseltine (1991)

BIOTECHNOLOGICAL INNOVATIONS IN ANIMAL PRODUCTIVITY Edited by BIOTOL (1991)

LEANNESS IN DOMESTIC BIRDS Edited by B Leclerq and C C Whitehead (1988)

All of these books may be purchased from your local bookseller, or in case of difficulty from: